WOOD
MAGAZINE®

HOW TO BUILD
A GREAT
HOME
WORKSHOP

Sterling Publishing Co., Inc. New York

Library of Congress Cataloging-in-Publication Data

WOOD® magazine: how to build a great home workshop / editors of WOOD® magazine.

p. cm.

Includes index.

ISBN-13: 978-1-4027-1177-0

ISBN-10: 1-4027-1177-8

1. Workshops—Design and construction.

2. Workshops—Equipment and supplies.

3. Woodworking tools. I. Wood magazine.

TT152.W65 2006

684'.08—dc22

2006005072

Edited by Peter J. Stephano

Designer: Chris Swirnoff

10 9 8 7 6 5 4 3 2 1

Published by Sterling Publishing Co., Inc.

387 Park Avenue South, New York, NY 10016

This edition is based on material found in WOOD® magazine articles

© 2007 by WOOD® magazine editors

Distributed in Canada by Sterling Publishing

℅ Canadian Manda Group, 165 Dufferin Street

Toronto, Ontario, Canada M6K 3H6

Distributed in the United Kingdom by GMC Distribution Services

Castle Place, 166 High Street, Lewes, East Sussex, England BN7 1XU

Distributed in Australia by Capricorn Link (Australia) Pty. Ltd.

P.O. Box 704, Windsor, NSW 2756, Australia

Printed in China

Sterling ISBN-13: 978-1-4027-1177-0

ISBN-10: 1-4027-1177-8

For information about custom editions, special sales, and premium and corporate purchases, please contact Sterling Special Sales Department at 800-805-5489 or specialsales@sterlingpub.com.

CONTENTS

HOW TO BUILD A GREAT HOME WORKSHOP

1-1.

1-2.

CHAPTER 1

A Woodworker's Dream Shop

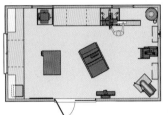

WHAT CONSTITUTES THE *perfect woodworking shop? Frankly, there probably isn't just one because woodworking tends to become personalized. A woodworker who specializes in turning, for instance, will set up a shop somewhat differently than a furnituremaker. Yet, there are benchmarks for woodworking shops that provide you with ideal standards for work flow, machine space, electric power, lighting, ventilation/dust control, and other factors (1–1 and 1–2). Many of these guidelines are discussed in this chapter, as well as elsewhere in this book. (For instance, see Chapter 3 for information on dust control.) Keep these guidelines in mind when you set out to plan your new workshop or upgrade your present one.*

(Opposite page). Viewed through the 72"-wide opening of the French doors, this 240-square-foot hobby woodworking shop looks spacious and inviting. Skylights and windows add lots of natural light.

HOW TO FIND SPACE FOR THE WOODWORKING SHOP

Having a space specifically dedicated to your woodworking makes a lot of sense because you'll feel more comfortable sharpening your skills in a place you've designed just for that purpose. To begin with, you'll ideally need an area that's convenient, well-organized, and with sufficient space for work and storage. Depending on your experience level, the number and size of the tools you have (or expect to have), and the size of most of the projects you build or plan to build, that space could be large or small, basic or elaborate.

Of course, available space, your needs, and no doubt your budget, will end up dictating where you situate your shop. For most woodworking, the ideal shop is a dedicated building created just for woodworking. But barring that, the most likely candidates are basements and garages—even attics and seldom-used rooms can accommodate a workshop. Also, you don't have to create your shop all at once: You can add amenities and personal touches to a basic shop as you expand your skills and woodworking interests. The following information shows how such a progression might happen.

How Shop Needs Grow

Many, if not most, woodworkers graduate to the craft from the do-it-yourself level of home repair and improvement. Already owning a basic complement of portable power and hand tools, they've made do with working space that occupied either a garage end wall or a small basement room. Then, at entry-level woodworking, they see the need for some basic stationary tools and a larger space.

A shop to meet such needs might easily be in a 14 x 20' basement room or a single-car garage. That's enough space (280 square feet) for a centered assembly bench with a benchtop tablesaw and an end-wall bench with a grinder, belt sander, vise, and benchtop drill press. A side-wall bench becomes a power mitersaw workstation, and there's still room for material storage on the remaining wall. Such a shop could keep a beginning woodworker content for quite a while.

Eventually, though, more tools need to be added to build bigger and better projects, and space is needed to accommodate them. Let's say that at this intermediate woodworking stage, he or she sees the need for a stationary 10" tablesaw, a 6" jointer, an air compressor, and maybe a dust collector (besides the shop vacuum). This level of woodworking requires a larger area, approximately 22 x 24' (528 square feet), roughly the size of a two-car garage. Many hobby woodworkers might be permanently satisfied with this space.

As often happens, however, the intermediate woodworker gains skills and confidence and now requires more space for new, as well as improved, stationary tools that permit the building of larger and more complicated projects. At this point, the experienced woodworker may also think about building things for profit. A shop to house the added equipment, such as a table-mounted router, a stationary bandsaw, a thickness planer, a wood lathe, and an improved sanding station needs to occupy about 768 square feet—the space in a three-car garage or similar-sized separate building.

The next step up in size is the professional workshop, which can be larger than 1,000 square feet. That level, however, is beyond the scope of this book. But in Chapter 2, you'll see several examples of shops built to meet varying needs and budgets, and one of them may turn out to be just the right plan for you. Meanwhile, the standards presented on the following pages will help you consider all the aspects that must be factored in before you begin planning your shop.

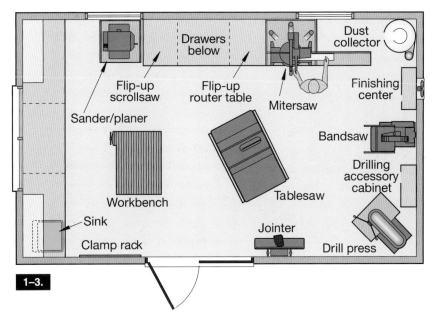

Flip-up scrollsaw

Sander/planer

Drawers below

Flip-up router table

Mitersaw

Dust collector

Finishing center

Bandsaw

Drilling accessory cabinet

Tablesaw

Workbench

Sink

Clamp rack

Jointer

Drill press

1–3.

How Will Your Work Flow?

Just as a kitchen's layout determines how efficiently meals are prepared, building projects comfortably and efficiently depends largely on shop layout (**1–3** and **1–4**). Your power tools should be situated according to your working patterns and the size and shape of your space. And there are certain minimums for passageways and clearances. For guidelines, plan on at least 24" walkways between benches and machinery or between stationary tools. To work between benches and equipment rather than just walk, you need 30" to 36".

There's also a prevailing strategy for shop layouts that's based on the workstation concept.

The stationary machines at one end of this long basement shop are meant to break down and machine lumber before it goes into subassemblies.

You'll see a great example of this with the Idea Shop 1 detailed in Chapter 2. The concept is based on work flow, and it stems from application in a small shop space where eliminating clutter is highly important. With the workstation concept, work is organized into

specialized machining areas, such as sawing, drilling, sanding, and so on. Cleanup is done after each session, so there's enough room to get each machining step completed. Here's how it works in a small shop, but you'll soon realize how it applies to larger spaces:

Step 1. Rip, joint, plane, and cut off stock into manageable parts, so you won't have to handle large pieces in the middle of the piece's construction.

Step 2. Complete cuts, joinery, drilling, and assembly necessary for all subassemblies.

Step 3. Use your assembly bench to join the subassemblies.

Step 4. Apply finish in the assembly area, but use only those finishes in the shop that don't require a dust-free atmosphere. Spray outside or in a separate area.

1–4.

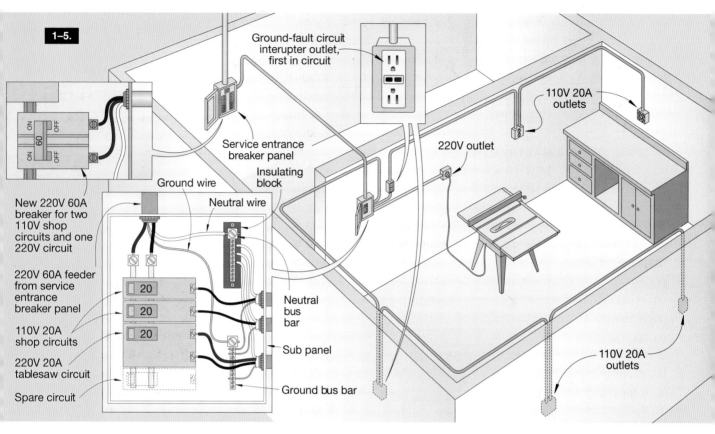

1–5.

Ground-fault circuit interupter outlet, first in circuit

110V 20A outlets

Service entrance breaker panel

Insulating block

220V outlet

Ground wire

Neutral wire

New 220V 60A breaker for two 110V shop circuits and one 220V circuit

220V 60A feeder from service entrance breaker panel

110V 20A shop circuits

220V 20A tablesaw circuit

Spare circuit

Neutral bus bar

Sub panel

Ground bus bar

110V 20A outlets

The floor plan of the 12 x 20' shop shown in **1–3** was laid out for effective workflow. It makes the most of every square foot with logically located worksta- tions and tuck-away tool storage. Machines are positioned with infeed and outfeed room, and can be moved to accommodate working with long boards because they're mounted on mobile bases. Mobility also allows you to quickly reorganize your work space by relocating equipment while working on a big project. In contrast to the moderately sized shop just discussed, take a look at the one shown in **1–4.** It's a 1,750-square-foot basement shop consisting of one large room, interrupted only by posts

supporting a lengthwise beam. Tools are squeezed into the space in logical workstations. Large stationary machines are grouped for breaking down stock. A central workbench area is flanked by portable tools, bits, and clamps. Storage and other tools occupy every remaining square inch.

In the following pages, you'll discover that there are indeed countless ways to set up a woodworking shop. But in sorting through the examples to find out or adapt what might work best for you, keep in mind what you read here regarding basic planning.

WIRING GUIDELINES FOR YOUR SHOP

Whether your woodworking shop is a dedicated space in your home, part of a garage, or a room in the basement, adequate and safe wiring is a necessity (**1–5**). So if you are considering a new shop or reworking the old, here are some helpful guidelines.

Is Your Present Electrical Service Adequate?

The electricity your local energy company supplies is distributed throughout your residence from a service panel located in your basement or other utility area.

If you live in a house 30 or more years old with the original wiring, you may have only a 60-amp service panel. With such little capacity, you probably won't be able to run more wires to power your shop.

Newer homes—and electrically updated remodeled ones—have 100- or even 200-amp service. With such large capacity, there may be unused circuits available for your shop. Or, if all the circuits appear to be in use, it's often possible to split an existing circuit or two to satisfy minimum requirements. Consult local codes.

Depending on the number of large stationary power tools in your shop and the shop's distance from the main service panel, you may want to add a subpanel to handle the load. For this kind of a job, you should seek an electrician to install it.

With 15-amp circuits, you may want to build in some overload insurance with 20-amp circuits. The cost difference is negligible, and the higher capacity lowers your chances of popping a circuit breaker. Two such circuits (plus one for lighting) are adequate for most home shops because all the tools won't be running at the same time.

If you suspect you'll need more than that, read the owner's manuals that came with your tools, or check the tools' labels to see what amperage they draw. You'll need this information, called load, to plan the size and number of circuits in your shop.

For example, if one of your tools requires 20 amps, a 15-amp circuit won't do. (**Table 1–1** lists approximate amperage ratings for tools.) Also, if two large tools normally run at the same time, such as a tablesaw and jointer, they shouldn't be on the same branch circuit. And, if one of your tools requires 220 volts, it needs a separate circuit.

Always put lighting on a separate circuit. To avoid black-outs, don't combine circuits that serve power tools with those that provide electricity for lighting. One 15-amp circuit probably will handle the entire lighting job.

But just in case, here's how to calculate your lighting needs. Each square foot of shop space should get three watts of light. So if your shop measures 20 x 20', you have 400 square feet times 3, or 1,200 watts. Divide 1,200 watts by 115 volts (actual performance), and you get 10.43 amps. If you feel that you need more light, add up all the wattages (bulbs and fluorescents) you need or want. Then, divide that number by 115 volts to arrive at the number of amps drawn if all fixtures were on. Don't plan on exceeding 80 percent capacity of the circuit's rating, though. That is, for a 15-amp circuit, you wouldn't want to exceed a 12-amp demand for your lighting needs.

For 220-volt motors, calculate the load using half the 110-volt amperage rating.

Table 1–1: Power Tool Amperage		
HP	VOLTS	AMPS
¼	110	4.4 to 6.3
⅓	110	5 to 7.2
½	110	7 to 9.8
¾	110	11.5
1	110	13 to 15
1½	110	16 to 19
2	110	19 to 23

Use the Proper Wires

Always use wires large enough to do the job. And always use grounded wires, with one black insulated wire (the "hot" wire), one white insulated wire (the "neutral" wire), and a bare wire (the "ground"). For dry areas, run type NM wire. In damp locations, use type NMC.

The wire gauge (the wire's diameter—the smaller the number, the larger the wire) you use will be determined by the distance from the service panel to the outlet, as well as by the circuit amperage. For example, a 15-amp circuit with 14-gauge wire can safely reach 30 feet. With 12-gauge wire, you could go to an outlet 47 feet away from the panel.

In many cities, all wiring must be enclosed in metal, thin-wall conduit. Check what your local building code requires. Usually, though, the rule of thumb is if your wiring isn't enclosed in a wall, it must be run in conduit.

Generally, you should install conduit anywhere wiring can be seen, felt, touched, inadvertently reached, or could come in contact with moisture.

The conduit itself must be grounded. Metal outlet boxes and connectors make the grounding simpler. Don't use plastic boxes because they don't conduct electricity. Also, you'll need to reevaluate wire size with conduit because it reduces amperage capacity as much as 20 percent.

Grounded Outlets

Grounded outlets protect you and your tools. A GFCI (ground-fault circuit interrupter) is a device installed either as a breaker in the service panel or as a receptacle in the outlet box that senses electrical leakage in the ground circuit. If there's a leak, GFCIs shut off voltage instantly. They prevent electrocutions that can occur when a person working with a portable power tool becomes the electricity's path to the ground. GFCIs also protect electrical motors from power surges.

SHOP LIGHTING STRATEGIES

Nothing makes you feel as comfortable and safe working in your shop as good lighting (**1–6** and **1–7**). So just what makes good lighting?

Many lighting experts say that high-quality lighting provides

a visually comfortable environment, with little glare, that makes performing tasks easier. But there's no easy way to arrive at a universal formula for workshop lighting because it depends very much on the type of equipment you're using, the shape of your workshop, and where you perform tasks. Lighting needs also vary tremendously with a person's age, as examined in Older Woodworkers Need More Light on *page 13*. Lighting researchers do agree on some basic guidelines, though. They are described below.

Use the Proper Type of Light

Generally, you want to install fluorescent fixtures overhead in your shop because they produce better contrast, practically no shadows, and less glare for superior visibility. They're also three to four times more efficient than incandescents and last ten times longer.

However, incandescents have a place in the shop, too. As task lighting, the shadows their bulbs produce help more clearly define detail work like carving and operations such as drilling and scrollsawing. But for general lighting, they produce too much glare. (Incandescent reflector lamps make better use of light than standard A-type, round bulbs because they direct the light at your work. They also render color better.)

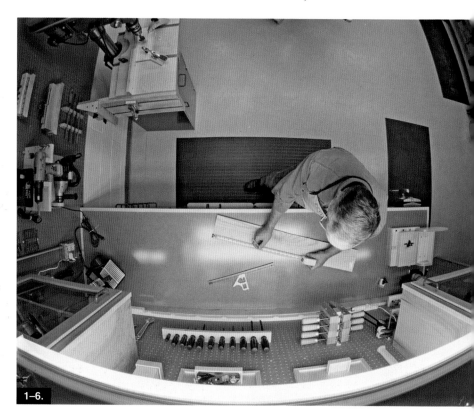

1–6.

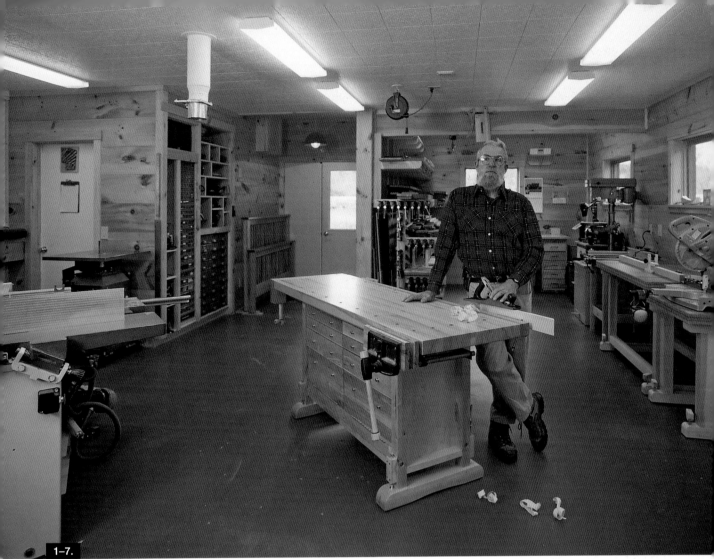

1–7.

Fred Collins' well-lighted Vermont workshop has tongue-and-groove pine paneling ("car siding") that adds warmth, sturdiness, and durability. The poured concrete floor was coated with a two-part epoxy for protection and ease of cleaning.

Determine the Amount of Light Needed

According to the Illuminating Engineering Society of North America (IESNA), the amount of light, measured in foot-candles (see Glossary of Lighting Terms on *page 14*) required for a visual task increases with its difficulty. For instance, IESNA suggests

SHOP TIP

Older Woodworkers Need More Light

Failing vision and other eye deterioration problems aside, studies by gerontologists indicate that your need for light drastically increases as you grow older. Research shows that maintaining the same reading speed you had at 20 years of age takes 50 percent more light when you reach 50. At age 60, it requires twice as much, and at 80, three times more light. Glare also bothers older people a lot more. And a gradual yellowing of the lenses in the eyes changes color perception somewhat, making it more difficult to discern closely related colors in the blue-green end of the color spectrum.

This all means that if you're 50 or older, you need more lighting help. Here are some tips:

● Increase overhead illumination. You can do this by replacing standard two-lamp fluorescent fixtures with three- or four-lamp ones to beef up lighting without much extra rewiring.

● Reduce glare by replacing (or retrofitting) fixtures that have bare fluorescents with ones featuring diffusing covers.

● Install warmer Spec 30 (SP30 or Designer 30) fluorescent tubes in overhead fixtures to correct for color.

Table 1–2:
Lighting Requirements of Different-Sized Shops

SHOP DIMENSIONS IN FEET*	NUMBER OF FIXTURES NEEDED**	TOTAL WATTS
10 x 10	3	240
10 x 15	4.5	360
10 x 20	6	480
15 x 15	7	560
15 x 20	9	720
20 x 20	12	960
20 x 30	18	1,440
30 x 30	27	2,160
30 x 40	36	2,880
40 x 40	48	3,840

* Dimensions assume an 8' ceiling height.
** Standard 4', two 40-watt lamp, fluorescent fixtures or equivalent in larger fixtures to get 75 foot-candles.

SHOP TIP

Glossary of Lighting Terms

Ballast. An electrical device used in fluorescent and high-intensity discharge fixtures to furnish the necessary starting voltage and maintenance current to the lamp for proper lighting performance. Electronic ballasts transform current at high frequency to operate discharge lamps, such as fluorescents, without flickering.

Fluorescent lamp. A bulb that produces light by passing electric current through a metallic gas to excite special chemicals called phosphors, causing them to glow or "fluoresce." The glowing phosphors coat the inside of the tube, converting about 80 percent of the electricity into light. A standard 40-watt, 4' fluorescent bulb puts out about 2,700 lumens.

Foot-candle. The quantity of light reaching a surface; equal to the number of lumens reaching a surface divided by the square footage of the surface, e.g., one lumen over one square foot equals one foot-candle, and a 100-foot-candle level equals 10,000 lumens for every 100 square feet of shop space.

Incandescent lamp. A bulb that uses a tiny wire called a filament that glows white hot when electricity passes through it. Ten percent of the electricity is converted into light; the rest is lost as heat. A 100-watt incandescent lamp produces about 1,750 lumens.

Lumen. The measurement of light output from a lamp.

Watt. A unit of electrical power produced by a current of one ampere across a potential difference of one volt.

that 30 to 50 foot-candles are adequate overall illumination for working around the home. Scrollsawing for extended periods of time, though, requires 100 to 200 foot-candles.

Based on this, some experts suggest that you shoot for an overall (ambient) lighting level of 75 foot-candles with fluorescents. Increase that as needed with 100-watt incandescent task lighting. (It's important to note that your shop lighting should be on a separate circuit just in case one of your machines overloads and trips a breaker.)

How does this lighting level translate into fixtures needed? Not all the light produced by a lamp reaches the work level. Some is absorbed by walls and ceilings, and some is lost in the inefficiency of the fixture. To compensate, use three 2-lamp, 40-watt, 4' fluorescent fixtures per 100 square feet (10 x 10') of workspace, locating them above work centers. (See **Table 1–2** for general lighting requirements of different-sized shops.)

But before you visit your electrical supplier, it's a good idea to use electronic ballasts with your fluorescent lamps because they don't flicker and cause a strobe effect with running woodworking machines. Then, too, it's great to have a light fixture that throws some illumination onto the ceiling and walls. This gives you balanced light that makes the shop more visually comfortable. You should also paint walls and ceilings white to maximize energy efficiency.

❖

CLIMATE CONTROL CONSIDERATIONS

Very hot and extremely cold weather can limit wood-working for those who situate their shops in a separate building or garage. (With basement shops, temperatures aren't usually as much of a problem as dampness.) But that's simply because the owner never quite gets around to installing heating and/or cooling units or systems. However, to be a productive woodworker, you must include climate control in your planning (**1–8**).

Your comfort, of course, is of high priority, but also think about the material you work with: wood. Unlike steel, plastic, and other man-made materials, wood is naturally hydroscopic. That means it has the ability to take on moisture as well as release it. So in damp or humid weather, it absorbs moisture from the atmosphere and swells. Conversely, when the air surrounding wood is very dry, it drains itself and shrinks. One result of this propensity is doors and drawers that stick when wood swells; another is failing joinery, even cracks and checks, when it shrinks.

To counter this tendency for wood to "move," woodworkers long ago began kiln-drying furniture-quality hardwoods and softwoods. This means that they slowly—through air-drying,

The two sets of double doors in Larry Evans' Dallas, Texas, shop provide super air flow to beat the summer heat. The doors also allow the wood-working owner to easily move and work with long stock.

1–8.

then heat—reduce the natural moisture content of wood to match the relative indoor humidity of modern homes and building with climate-control systems. Such moisture reduction lowers wood's moisture content to from 6 to 9 percent. However, even wood with that low of a moisture content will still draw some moisture from the atmosphere if not stored properly in a climate-controlled setting. Knowledgeable woodworkers understand this, and therefore employ joints in furniture and cabinets that allow for this swelling and shrinking of wood, such as frame-and-panel construction. So that's the major reason other than comfort to install heating and cooling units or systems in a workshop. Chapter Four has some contemporary shop-heating solutions for various-sized spaces. Meanwhile, here are some options to ponder:

Extend an existing system.

If your shop space will be attached to your home, it's easy enough to extend ducts of a forced-air system or the piping for a water-heating system, if the central unit has the capability to carry the added load.

Install a second system.

Putting a small forced-air system to work in a shop might make sense if space allows. Of course, cold-air return ducts will require filters to keep dust away from the furnace's pilot light and burner. Ideally, you should wall-off the new system.

Wire-in electric baseboard heaters.

For this, the heating cost is low if you use the shop only occasionally.

Portables and Woodstoves.

Portable, temporary heaters are available, too. These range from radiant electric to propane- and kerosene-fueled. There are drawbacks: They occupy floor space; usually don't distribute heat well; and, in the case of fueled types, produce noxious fumes and present a fire danger. In rural areas, woodstoves can be an option. However, their dry heat can be disadvantageous, and they also present a fire danger as well as insurance problems.

Ventilation also plays a role in climate control. You can easily provide clean, fresh air to your workshop with dilution ventilation. That simply means you bring in a large volume of air to reduce the concentration of dust and fumes, and then remove the air. The easiest form of dilution ventilation is a wide open window at one end of the room and another open window or door opposite. Fans, of course, increase air flow.

STORAGE SPACE

Storage—how much you'll need and what shape it takes—is one of the most important aspects of shop planning. That's because without sufficient storage, your spanking new work-

(Right). The mobile cabinet at left above houses a small planer and belt/disc sander. Next to it, there's a lift-up scrollsaw stand and cabinet. The wall cabinets, featuring a hanging system of mating beveled strips, easily relocate if needed, providing moveable storage. And there's plenty of Pegboard storage for hand tools.

(Inset). A flip-up router table neatly stores in its own cabinet for space savings. Above the bench top, tools hang on screw-mounted Pegboard hooks and holders that won't pop loose and drop.

shop can quickly turn inefficient as well as ugly!

Plan storage to accommodate hand tools; supplies, such as sandpaper; hardware of all types; lumber and sheet goods; and finishing materials. Otherwise, your shop will quickly become cluttered. Chapters 7 through 10 offer plenty of plans for shop storage units and organizers for tools and supplies.

Wall-hung, perforated hardboard (Pegboard) panels provide a perfect, within-reach storage solution for hand tools, as in **1–9**. And the available tool holders and other accessories are nearly limitless.

Drawers beneath a workbench can hold sanding supplies, and small accessories, such as drill and router bits. Inexpensive plastic organizing bins, either fitted in drawers or hung on the wall, keep fasteners in their place.

Acrylic-fronted cabinets hung in strategic locations, such as those shown in **1–10**, can house portable power tools. Standard

1-9.

1-10.

SHOP TIP

Have All Your Lumber on Hand When You Begin

Go through the plan's materials list, or make one up for projects that you design, and add everything up to determine your lumber needs. Then, get all of your lumber, plus at least 20 percent extra, at one time. Doing this allows you to match grain and color.

Don't think of the extra that you buy as waste, because you shouldn't discard any decent-size cut-offs until all the parts of a project are cut and assembled. This wood is perfect for testing setups and techniques, and trying out finishes. Plus, having identical stock on hand can save a project if you need to recreate a spoiled piece or make an inconspicuous patch.

1–11.

Sheet goods can stand on edge if they are situated so they won't bend and protected from concrete by a pad or pallet. See how to build this rack in Chapter 9.

Because unless you buy only just enough boards or sheet goods for the project at hand (and who ever does that?), they take up a lot of space. They also must be kept dry so as not to warp or otherwise degrade. That's why all wood should never be in prolonged contact with a concrete floor. Construct flat racks for your woods if you have the available space. Otherwise, sheet goods can stand on edge, file-card fashion, if propped so they won't bend or bow and protected from concrete by a pad or pallet (**1–11**). If extra floor space will be scarce, look above for constructing wall- or ceiling-mounted racks. In Chapter 9 you'll find several other solutions for storing lumber and sheet goods.

Now, you're ready to move on and see how the guidelines and principles you've just read about were put to work in several shops designed by *WOOD®* magazine.

kitchen cabinets, both base and wall, are available at home centers. Or build your own, such as the custom router-table cabinet in **1–9**, as you'll learn how to do step-by-step in Chapter 7.

Flammable and toxic liquids, like thinner, glues, and paints, should be stored in a lockable cabinet, preferably a fire-proof, metal one. If not that, at least keep it out of reach from your curious youngsters.

Material storage may present the most serious challenge to workshop planning. Why?

CHAPTER 2

Customize Your Space

Now THAT YOU HAVE AN IDEA OF *all the elements that need consideration when you plan a woodworking shop—and you've seen some examples of what others have done—we're going to introduce you to some very special shops. Each was designed by the editors of* WOOD® *magazine to meet certain parameters, from small space to work flow, and convenience to double-duty. You'll discover a downright delightful basement shop; an ultra-efficient shop sited in a single-car garage; a shop that shares space in a two-car garage; a dedicated-space shop in a separate building; and, for comparison, a professional's work place on an office building's third floor! One of them just might meet your needs, but even if not, you'll see and understand how the benchmarks presented in Chapter 1 come into play. Then, we'll give you the how-tos and tools for creating your own space.*

1 A BOUNTIFUL BASEMENT

How many woodworkers do their woodworking in a basement shop? Plenty. Even if that's not where your shop is or will be, you'll find all kinds of small-space solutions in the 12 × 16' basement shop shown in **2–1** to **2–13**. You'll want to adapt many of them to your home workshop.

The space that became this shop represents a room of the size that you'll find in many basements, about 12 x 16'. Because of its location, an effective, yet quiet, dust-collection system was of prime importance. Then, of course, there was the limited space to deal with. But as you'll see, our design team came up with doable solutions. We'll share many of them with you here and in other chapters of this book, with plans and step-by-step instructions so that you can adapt them to your workshop.

(Below). The main goal in this basement shop was to squeeze in all the tools and machines needed for a complete workshop and still provide room to move around. Another was to reduce dust. That's why, for instance, the tablesaw's extension table doubles as one for sanding. Holes in it lead to an air-tight collection box beneath with a 4" dust-collection port. The basement window also was outfitted with a fan designed for dusty conditions. Yet it doesn't block light, and the window still opens.

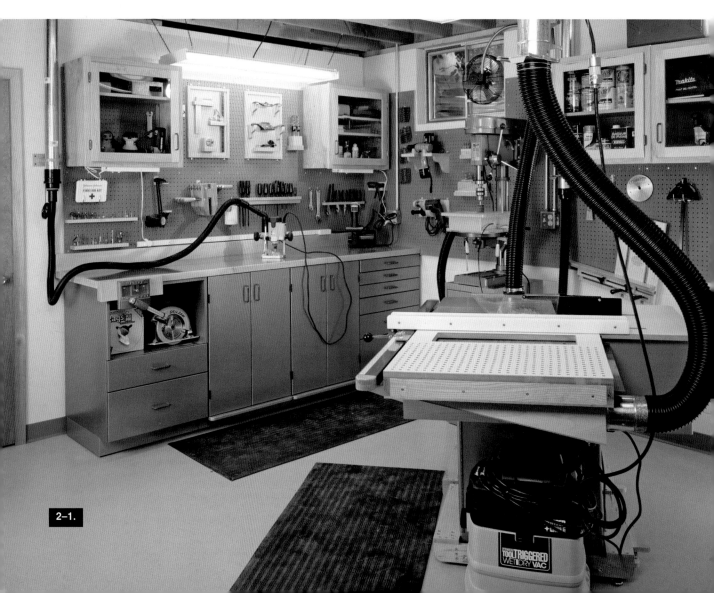

2–1.

A Bountiful Basement Floor Plan

1	10" tablesaw
2	Sanding table
3	6" jointer
4	Shop vacuum
5	Workbench
6	Router table
7	Scrollsaw
8	Exhaust fan
9	Drill press
10	Modular wall cabinets
11	Perforated hardboard
12	14" bandsaw
13	Cyclone dust collector
14	Planer/sander stand
15	36" lathe and cabinet
16	Stock storage on wall
17	Mitersaw
18	Floorsweep

2–2.

2–4.

This custom-made maple tool holder cradles a cordless drill, charger, and an assortment of driving bits. The design adapts to many portable power tools. Look for it in Chapter 10.

Out-of-the-Way Storage

Storage received more than a passing nod. And here, it consists of more than wall-mounted perforated hardboard (although there's plenty of that, with custom-made tool holders as well).

The Plexiglas-fronted cabinets hanging about let you see their contents at a glance. But, you're not stuck with their placement once they're up, as shown in **2–9**. Keeping the cabinets toward the top of the room also gives you more usable work space below, and that really counts in a workshop in a small room.

A host of full-extension drawers in the workbench (**2–1**) provide storage for hand tools, hardware, measuring devices, and all those items you need to keep track of. The tablesaw's extension table provides storage, too, with a recessed tray to keep pushsticks, tape measures, safety goggles, and whatever else you need within easy reach yet out of the way of your work (**2–6**).

We used joist space to accommodate the lighting fixtures and to run the dust-collection piping, too. If your shop headroom is limited to 8' or less, you'll want to consider open joists, also.

A WOOD® staff member demonstrates the space-saving, flip-up storage of the router table beneath the workbench. See how to make it in Chapter 7.

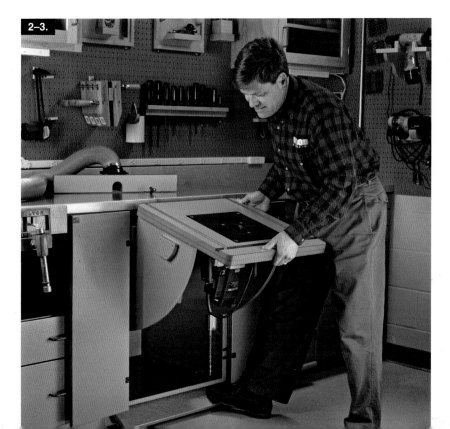

2–3.

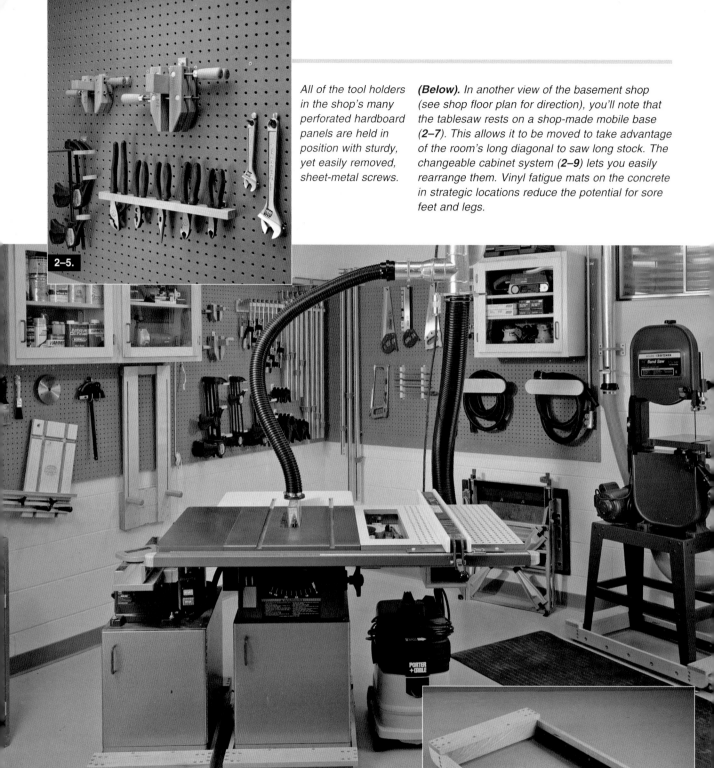

All of the tool holders in the shop's many perforated hardboard panels are held in position with sturdy, yet easily removed, sheet-metal screws.

2–5.

(Below). In another view of the basement shop (see shop floor plan for direction), you'll note that the tablesaw rests on a shop-made mobile base *(2–7)*. This allows it to be moved to take advantage of the room's long diagonal to saw long stock. The changeable cabinet system *(2–9)* lets you easily rearrange them. Vinyl fatigue mats on the concrete in strategic locations reduce the potential for sore feet and legs.

2–6.

(Right). Built of maple and heavy-duty fasteners for industrial strength, the mobile bases for each of the shop's woodworking machines have foot-operated locking mechanisms. The wooden lever uses a cam to rotate the stop against the casters. You'll find complete how-to instructions for building them on page 250.

2–7.

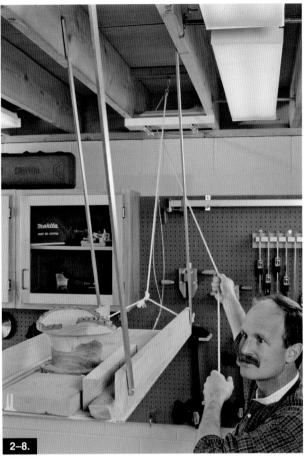

2–8.

One section of joist space was borrowed to house a handy pull-down storage unit. Its balanced pulley system won't dump the contents.

A bevel-ripped support strip on the back of each cabinet allows them to be easily moved. A mating strip rides atop the wall panels. In Chapter 7, we'll show you how to build this system.

2–9.

Dust-Collector Systems

Dust is a problem in every woodworking shop. But in a basement shop, collection becomes top priority because wood dust can invade the house. Therefore, our criteria for selecting power tools was that each had to have a dust-collection port and run as quietly as possible.

Traditional central dust-collection systems don't operate quietly, though. And the larger the horse-power, the louder the noise level. When your basement shop might be located under your family room or living room, noise becomes a real concern.

Long used in industrial and commercial applications, cyclone dust collectors operate more quietly than others (**2–11,** on *page 24*). In fact, our 1½-hp. one operates below the 90 dBA danger level. And it clears the air at a rate of 760 cubic feet per minute (CFM).

Wherever possible, machines and workstations in the shop have dust-collection ports attached to the cyclone system. The motor of the contractor's style tablesaw, for instance, was boxed in to enclose a garbage bag that catches falling sawdust, and the lid acts as an outfeed extension. Then, there's a dust port attached to the saw guard that corrals sawdust off the top of the blade. And the saw's right-hand extension table doubles as a sanding station.

Mitersaws pose a problem in connecting to a central dust-collection system. So we built a cabinet for ours to stand on that features a dust-collecting hood. Sawdust falls through an opening directly into a garbage can that rests below.

Lathes also are a dust-collection headache. Our cabinet catches the dust and shavings behind the machine, where they're quickly gathered with a flexible hose connected to the central system. (Quick-disconnect fittings—the type used for recreational vehicle drain hoses—work well for the couplings of extension hoses.) Finally, there's a floorsweep connected to the dust-collection system. With a broom, you sweep floor debris right into the floor-level take-up!

(Far left). To use the thickness planer, you release the catches and rotate the top. After machining stock, you simply rotate the planer back into position.

(Near left). The 1½-hp. cyclone dust collector operates quietly (62 to 82 dBA). Fitted with a cartridge-type air filter, it's 99.9 percent efficient at removing even the smallest dust particles from the air. They're available from major woodworking suppliers.

2–10.

2–11.

2–12.

2–13.

A lathe takes up a lot off wall space. This solution not only helps housekeeping, but makes cutting to length an easier job in a tight space. **(Inset).** Space to work the lathe was made by fitting the machine with a flip-down wooden lid. The maple cover then performs as a cutoff table for the mitersaw at the right.

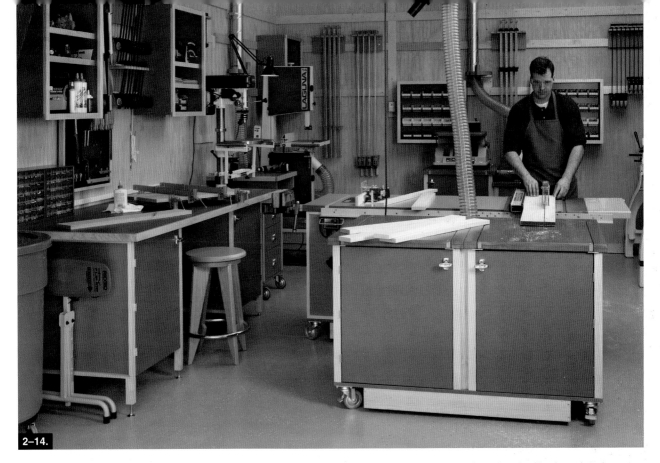

2   EFFICIENCY IN ONE STALL

Affordable, commonly available materials—medium-density fiberboard and medium-density overlay plywood—were used for the workbench and all the cabinetry. Find out how to build the basics in Chapter 7.

When the shop you see in **2–14** and **2–15** was designed, the key guide words were adaptable, affordable, mobile, modular, multi-use, and space-saving. In a nutshell, we wanted to create a full-featured shop in a compact space. So we began with a third stall of a three-car garage, a feature found in many newer homes, as shown in **2–16** and **2–17**. Then we went to work.

Of course, you may not have this type of garage at your house. That's okay. Wherever you place

*Rearranging the wall components is just about effortless. We changed from the original configuration, shown in **2–14**, to this one in less than 20 minutes.*

2–16.

2–17.

We walled off one 15 x 22' bay from the rest of the garage to create shop space, and then covered the drywall with a more durable, attractive surface: beadboard pine plywood.

your shop, we know you'll find a whole world of great ideas in this one that you can easily put to use.

As the floor plan in **2–18** shows, our design incorporates all of the major tools needed for a complete woodworking shop into this 15 × 22' space. Plus, the shop offers ample storage space, work surfaces that double as bases for benchtop tools, and a serious dust-collection system. Beyond that, almost everything in the shop is mobile.

Let's take a closer look at what makes this shop tick, and at some of the projects that'll be described in later chapters.

Keep It Simple, Make It Affordable

You may have noticed that all of our shop fixtures share a similar look. Sure, that consistency makes for a great appearance, but the real reason for their resemblance lies in our goal to make them as easy and affordable as possible to build. The workbench bases, flip-top cabinets, and router-table base, for example, are essentially identical in construction. Master building one, and you quickly can create them all. The same holds true for the wall cabinets— three sizes, one basic design.

We built almost everything from three materials: medium-density fiberboard (MDF),

Efficiency in One Stall Floor Plan

PROJECTS

1 Wall cleat system
2 Workbench
3 Wall cabinet
4 Tool-storage board
5 Clamp rack
6 Mobile drawer cabinet
7 Sanding center
8 Lamp holder
9 Mitersaw station
10 Flip-top cabinets
11 Tablesaw/routing center
12 Mobile base

TOOLS

A Air compressor
B Cyclone dust collector
C Drill press
D Bandsaw
E Oscillating sander
F Drum sander
G Lathe
H Floor sweep
I Lumber rack
J Mitersaw
K Jointer
L Dust collection duct
M D.C. muffler/filter
N Air filter (on ceiling)
O Planer
P Mortise
Q Tablesaw
R Router

2–18.

medium-density overlay (MDO) plywood, and soft maple. You'll find these durable, inexpensive, materials in home centers. Add simple hardware and an easy-to-renew clear finish, and you get high function on a low budget.

Go Mobile to Get Versatile

These fixtures aren't just easy to build, they're a cinch to move around. Equipped with heavy-duty casters, shop-built mobile bases allow you to place the tools you need at center stage, push others out of the way against a wall, and lock everything securely in place. This lets the workshop function like a much larger space.

We discovered, too, that shop fixtures don't need wheels to be mobile. All of the wall cabinets, clamp racks, and the perforated-hardboard tool-storage board quickly slip onto and off of a simple but secure cleat system mounted to the walls. Why? For one, it makes mounting these items a snap, even for one person. Also, the system allows you to easily reconfigure the entire shop as your needs change. Compare **2–14** with **2–15**.

Finishing Touches

In addition to being functional and affordable, we wanted this woodworking shop to be comfortable and attractive.

As **2–16** and **2–17** show, we added ⅜" beadboard plywood panels over the existing drywall. This attractive material completely transforms the shop's atmosphere, and makes a sturdy anchorage for wall-mounted accessories. In addition, the extra layer helps deaden noise.

We painted the floor using a water-based epoxy that's very tough and easy to sweep clean.

Ceiling-mounted fluorescent lights cost little to install, run economically, and provide great light. Task lights brighten areas in need of more-intense illumination.

Also mounted out of the way on the ceiling, a natural-gas heater makes the shop cozy during cold months. It pulls combustion air in from outside, so there's never a worry about fumes or dust causing a spark hazard.

3 TWO-CAR GARAGE WORKSHOP

2–19.

Security can be a problem in an unattached building, where its contents are on view when the door is open. This plan features lockable cabinets that conceal tools and supplies.

Many woodworkers create a shop in a garage (usually occupied by a vehicle or two). To determine the best way to handle this scenario, we took a standard 24 x 24' double-car garage and organized it from scratch (**2–19** to **2–22**). For year-round comfort, we insulated the walls and ceiling, sheathed the walls with beaded plywood panels, and covered the ceiling with drywall. Both were painted with a reflective white semigloss enamel. The floor received a tough coating of epoxy paint. For heating, we installed two small, wall-mounted natural-gas furnaces (20,000 and 30,000 BTU) with sealed combustion chambers. Their positioning on one wall also conserves precious floor space. An open garage door and an exhaust fan combine for summer ventilation. In winter, air recirculates through a ceiling-mounted air-filtration unit

For storage units, we chose reasonably priced yet sturdy wall and base cabinets available at most home centers. To make the space more cheery, we sprayed the cabinet carcasses and doors of the shop with a bright red enamel and the face frames with clear urethane.

Even with a 24 x 24' area, you need to save space where you can. So we used a portable workbench with fold-down wings that tucks neatly away between base cabinets. This frees up another parking spot just for the tablesaw and its mobile base cabinet.

Commercially made mobile bases keep the jointer and the bandsaw moving easily. The tablesaw, planer, sander, and scrollsaw feature shop-built mobile bases. We also built a special two-sided tool chest on casters to hold an assortment of clamps on one side and hand and power tools on the other; it gets parked along one wall. For dust collection, we installed a shop-vacuum-based system with a fixed duct run between base and wall cabinets along the back wall. The system utilizes two countertop connections and

As if by magic, a complete woodworking shop appears in the garage after the vehicles move out. Fluorescent fixtures provide plenty of light, and the bright paint on the ceiling and walls enhances it. Two-part epoxy floor paint proves durable and resists grease. Electrical drop cords with twist-lock plugs supply power to machines. For air tools and vehicle care, there's a compressor hose on a ceiling reel.

2–20.

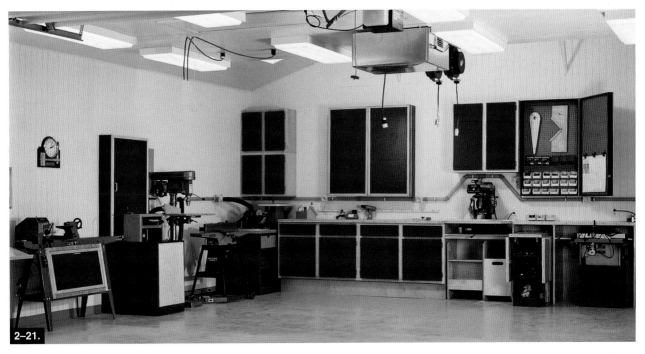

2-21.

Two-car Garage Workshop Plan

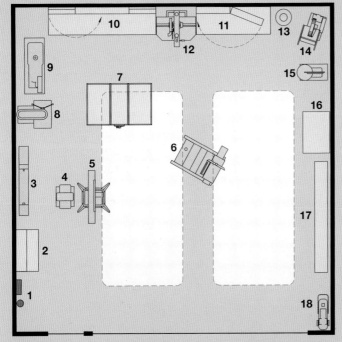

2-22.

FLOOR PLAN

1 Phone/intercom
2 Design and reference center
3 Lathe
4 Planer
5 Jointer
6 Tablesaw
7 Workbench
8 Drill press
9 Sander
10-11 Base cabinets with upper swing open cabinets
12 Radial-arm saw
13 Shop vacuum
14 Bandsaw
15 Scrollsaw
16 Mobile toolchest
17 Lumber storage
18 Portable air compressor

Equipment stores conveniently out of the way so vehicles can be parked with ease.

directly serves the sander, radial-arm saw, drill press, and bandsaw. For dust collection to other machines, just roll out the vacuum and attach it.

Although there isn't a clearly defined spray-finishing area, we allowed for one. In the far right corner of the shop, above one of the heaters, we mounted a through-the-wall power ventilator (covered when not in use). Right next to the vent is an oil-less connection for compressed air, for when you need to spray.

Along the wall, a cabinet provides storage for books, magazines, plans, and drawing utensils. Below the cabinet, a wall-hung drafting table lifts up to provide a work surface. Here you also find the phone, with a first-aid kit and an ABC-rated fire extinguisher placed conveniently nearby.

❖

2–23.

NO-NONSENSE DEDICATED SPACE

Paneled in solid hemlock over drywall, this dedicated shop includes such features as a concealed dust-collection system with drops, plenty of natural and fluorescent lighting, and adaptable shop-made storage projects.

Here's a great plan for maximizing space while offering flexibility, safety, and comfort in a complete, self-contained shop (**2–23** to **2–27**).

When *WOOD®* magazine's staff designers wanted a dedicated shop—one that doesn't share space with other activities—they began with an empty, bare-bones 14 × 28' room and one important concept in mind: Woodworkers tend to break a project down into specialized tasks, such as cutting, sanding, and assembly. So, the workstation concept takes precedence in this setup.

Accessories and jigs needed for a project are found within arms' reach of every major stationary woodworking machine.

Fitting a full complement of stationary power tools into the room was challenging. (See **2–24** for tool placement.) To conserve

No-Nonsense Dedicated Space Plan

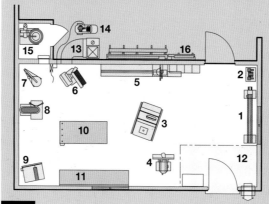

FLOOR PLAN

1 Lathe
2 Grinder
3 Tablesaw
4 Jointer/Planer
5 Mitersaw
6 Stationary belt sander
7 Scrollsaw
8 Drill press
9 Bandsaw
10 Assembly workbench
11 Wall workbench
12 Finishing area
13 Heating/cooling unit
14 Air compressor
15 Dust collector
16 Lumber storage

2–24.

2–25.

The lumber storage (see Chapter 9), heating, air compressor, and dust collection closet are situated to minimize shop noise.

2–26.

A mitersaw stand on one side for cut-off purposes is very handy when it's near the tablesaw area.

space, we located the air-handling equipment, dust-collection unit, air compressor, and lumber-storage bin just outside the room in an adjacent garage (**2–25**).

The sliding compound mitersaw station (**2–26**) sees a lot of action, so we built blade and jig storage overhead. We also added a cabinet-length table

extension and fence, and made a handy pull-out scrap bin beneath the saw. Other compartments hold portable tools, such as routers and sanders.

Cabinets near the lathe (**2–27**) hold turning tools and protective equipment, such as a face shield and respirator.

Air-tool storage also follows the workstation concept. Pneumatic tools are kept in a cabinet, then attached as needed to a 25' length of hose that lies flat on the floor, making it easy to move around.

We set the 10" tablesaw at an angle in the center of the shop to handle longer stock, but it needn't stay there. Like the lathe, bandsaw, jointer/planer, and belt/disc sander, it rests on a mobile base, so you can use it anywhere.

For comfort and safety, plenty of natural and artificial

2–27.

The wood lathe is placed beneath a window to provide good lighting while turning. Mobility is always a nice feature!

lighting pours through two skylights and two windows, as well as fluorescent ceiling fixtures and incandescent task lights. Leg strain, which could be caused by the concrete floor, is minimized by rubber anti-fatigue mats placed in strategic spots around the room.

For winter heating, a 50,000-BTU forced-air gas furnace was installed in the garage, a few feet behind the wood-storage rack. For summer cooling, a companion air-conditioning compressor sits outside.

Locating the furnace in an adjoining space creates a protective wall between the pilot light and any dust and volatile fumes. A 4"-thick pleated filter inside the furnace's cold-air return provides added precaution.

To control dust, a 2-hp collector fits into adjacent available garage space. But, for additional dust and noise protection, we housed the collector in its own insulated closet. The system features metal ducting because, unlike PVC pipe, it doesn't need a ground wire to prevent buildup of static charges.

At the spray-finishing area (**2–24**), a filter-shielded, explosion-proof exhaust fan mounted in the wall removes toxic fumes via a 16 × 6" vent. A wraparound vinyl curtain contains the area and protects the rest of the shop from overspray.

A PROFESSIONAL'S PLACE

Prior to *WOOD*® magazine's premier, the decision was made to build every project slated for its pages in an on-site shop adjacent to the editorial offices. This was no easy task because staff members were ensconced on the top floor of a three-story building! Yet, it came to be, and since Day One, full-time project builders have been at the very heart of the magazine's operation.

Here, you'll see the space (**2–28** and **2–29**) where hundreds of projects were created—many of which appear in this book. It's a model of organization and storage, and is well-suited for two woodworkers creating at once, as frequently happens. Although *WOOD*® magazine shop's floor plan (**2–30**) may not be ideal for you, there are plenty of ideas to incorporate into your space.

Shop Layout

The workshop occupies a 47 × 22½' space in a downtown Des Moines, Iowa, office building. Its location provides quite a view of the river and the park beyond, but it also meant planning around the necessary freight elevator that cuts into the room (**2–29**).

2–28.

Chuck Hedlund, WOOD® magazine's project builder, clamps an assembly together at one of the workbenches. This view looks north from the double entry doors. At right you can see the ducts and blast gates of the central dust-collection system.

A Professional's Shop Plan

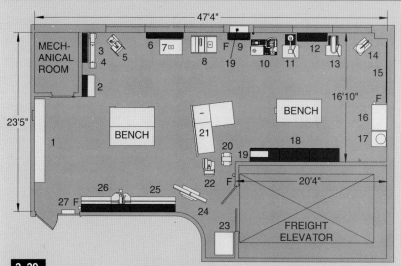

2–29.

SHOP LAYOUT

1. Lumber storage
2. Drawing table
3. Lathe-tool cabinet
4. Lathe
5. Bandsaw
6. Router-bit cabinet
7. Router table
8. Tablesaw
9. Sandpaper cabinet
10. Sanding center
11. Drill press
12. Drill-bit cabinet
13. Benchtop
   drill press
14. Scrollsaw
15. Clamp-storage wall
16. Paint storage
17. Paint booth
18. Tool cabinets
19. Air-filtration unit (2)
20. Planer
21. Tablesaw
22. Belt/disc sander
23. Dust collector
24. Jointer
25. Hand-tool cabinets
26. Radial-arm saw
27. First-aid center
28. Fire extinguisher

of those that readers have or can afford.

We long ago learned the value of mobility. That's why we mobilized the stationary power tools and work centers along the window wall (at top in **2–29**). The bandsaw, tablesaw, and one large drill press sit on mobile bases. The shop-built router table, sanding center, and small drill-press cabinet feature casters. Depending on the space requirements of the task at hand, each of these machines and work centers easily relocate.

We placed the tablesaw with the 52" extension table, the thickness planer, the 8" jointer, and the belt/disc sander in the center of the shop (although each machine also has a mobile base).

Because of the possible number of concurrent activities—such as producing a shop-tested technique story on routing dadoes while projects are being built—going on at any one time, the shop required tool and layout considerations different from those of the average home woodworker. Yet, it's a long way from being a commercial production shop because we use tools typical

A fold-down tabletop and a cork bulletin board with a storage cabinet above serve as the shop's planning center. Note that the adjacent lathe and the bandsaw— like most machines in the shop— have mobile bases.

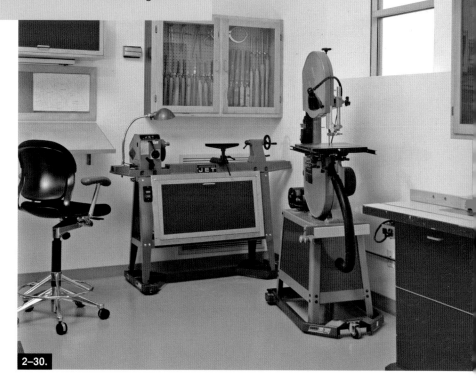

2–30.

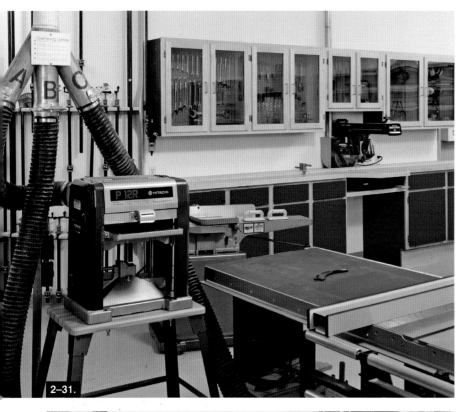

2-31.

2-32.

The central dust-collection system serves the planer, tablesaw, and stationary belt/disc sander. For the radial-arm saw on the shop's east wall, we constructed a 12'-long cutoff table with base cabinet. Cabinets above hold accessories and safety equipment.

This central position provides plenty of room for ripping sheet goods and machining stock. A workbench at each end of the shop makes it possible for two project builders to work on different projects at the same time.

The radial-arm sawing center occupies the shop's short wall (between the elevator and the entrance doors). It's surrounded by cabinetry that houses pertinent accessories and hand tools as well as the shop's safety equipment.

At the opposite end wall from the clamp-storage rack, there's a space for lumber and sheet goods storage. That end of the shop also accommodates a planning area and turning center, as shown in **2–30**.

The shop also features a dust-collection system that serves the central machines, the tablesaw and router table along the wall, and the radial-arm saw (**2–31**). The bandsaw and the sanding center rely on shop vacuums in their cabinet bases (**2–32**). Two particulate filters at different locations catch fine, air-borne dust. Most finishing is done in a separate, sealed room.

The sanding center features an oscillating spindle sander and a benchtop belt/disc sander. The storage cabinet below contains a portable shop vacuum to power the work station's self-contained dust-collection system.

2–33.

A 12'-long bank of 8'-high cabinets provide storage for portable-power tools, saws, and other equipment in the top tier. The lower units have bins for hardware and fasteners. All shelving is adjustable.

The 8 x 8' clamp rack on the north wall handily groups similar style clamps with the help of shop-built, maple hanging fixtures.

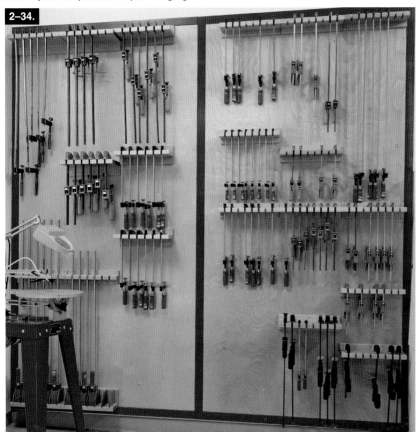

2–34.

HOW TO LAY OUT YOUR SHOP

As we mentioned in the very beginning of this book, when it comes to workshops, the description "one size fits all" just doesn't fit! Even if we own similar tools, each of our needs, methods of working, and available space may be completely different.

Regardless of your special situation, we'll show you how to lay out a good workshop without ever moving around your heavy tools. To begin, gather up a pencil and paper, and inventory what tools you own or want to own for your shop (**2–35**). Then grab the super-handy grid sheet (¼" = 1') and scaled icons of 18 common shop tools and fixtures that you'll find on *pages 38* and *39*. We used this system to create the 24 x 24' and 12 x 20' shops shown in **2–36** and **2–37**. Here's how you put it to work.

Pick a Place

If you're setting up a new shop, as opposed to redoing an existing one, figure out where it should be located. Basements hold great appeal because they have electrical service, heat, and at least some lighting. On the downside, challenges exist getting big machines in and projects out. Plus, noise and mess can invade the house. A stand-alone building solves those problems, but requires space and a sizable investment.

2–35.

Make an inventory of what tools you own or want to own for your shop.

A garage represents a reasonable compromise. To turn one into a shop, you'll have to beef up the electrical service and lighting, and you may need to add heat and insulation. But a garage offers many advantages of a dedi-cated building, and can still accommodate cars on occasion, as you saw on *pages 28* and *29*.

Prioritize Your Tools

Now take time to think about the projects you build. Many wood-workers set up with a standard "furnituremaker's" layout—

2–36. **ORGANIZE TO MAXIMIZE YOUR SHOP'S POTENTIAL**

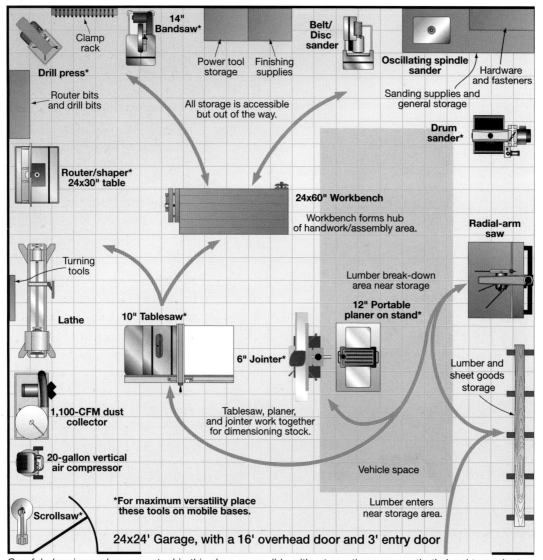

Clamp rack

14" Bandsaw*

Belt/ Disc sander

Drill press*

Power tool storage

Finishing supplies

Oscillating spindle sander

Hardware and fasteners

Router bits and drill bits

All storage is accessible but out of the way.

Sanding supplies and general storage

Drum sander*

Router/shaper* 24x30" table

24x60" Workbench

Workbench forms hub of handwork/assembly area.

Radial-arm saw

Turning tools

Lumber break-down area near storage

Lathe

10" Tablesaw*

6" Jointer*

12" Portable planer on stand*

Lumber and sheet goods storage

1,100-CFM dust collector

Tablesaw, planer, and jointer work together for dimensioning stock.

20-gallon vertical air compressor

Vehicle space

Scrollsaw*

***For maximum versatility place these tools on mobile bases.**

Lumber enters near storage area.

24x24' Garage, with a 16' overhead door and 3' entry door

Arrows indicate workflow through the shop

Careful planning makes every tool in this shop accessible without creating a space that's hard to navigate. Enclosed storage areas keep the items within secure and away from dust, and they're conveniently located near the work centers they serve.

tablesaw and workbench in the middle surrounded by other tools—as we did in **2–36**. But if you spend your time creating boxes at the router table or turning bowls on the lathe, consider making that your center-piece tool. You decide which deserves prime space.

Surround that most-used tool with its supporting players. The less important its role, the farther a tool can sit from the hub. At this point, just get the tools on the page, not worrying about their exact final location. In a smaller shop, you may discover that you lack enough space for all the tools to be set up at one time. Don't fret, we'll bring you solutions shortly.

For now, find homes for the largest and most important tools.

Stock Up on Storage

With your rough positioning completed, turn your attention to storage. Make a list of all the items you need to keep put away. In addition to such obvious items as power tools and lumber, don't forget fasteners and hardware, clamps, and finishing supplies. These little items can eat up big space. Now figure out which should be stored together. For example, accessories, make good companions drill bits and router bits, plus drilling and routing. Paying attention to storage needs also helps you refine tool layout. In the shop in 2–36, the drill press and router table are closely located, with a single cabinet in between that serves both.

Note: The drill press resides in a corner for a reason. It takes only about 2' of space, but can handle a 4'-long board.

Consider the Work flow

Before moving tools into final position on your drawing, think about work flow in the shop, as shown by the orange arrows in **2–36**. You don't want to move lumber or project parts around any more than necessary as you work.

In both layouts shown here, raw lumber and supplies enter conveniently through the garage door opening. As machining breaks the lumber into parts, they

A complete shop will fit into a one-car garage—and still have room for the car—if you carefully plan where each tool resides during use and while in storage.

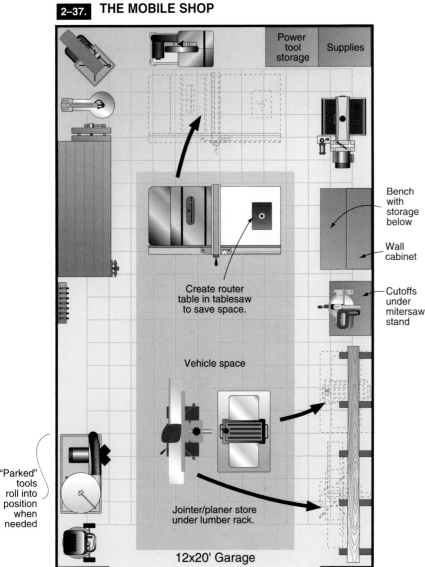

2–37. **THE MOBILE SHOP**

Power tool storage

Supplies

Create router table in tablesaw to save space.

Vehicle space

Bench with storage below

Wall cabinet

Cutoffs under mitersaw stand

"Parked" tools roll into position when needed

Jointer/planer store under lumber rack.

12x20' Garage

2–38. **WORKSHOP LAYOUT**

¼" = 1'

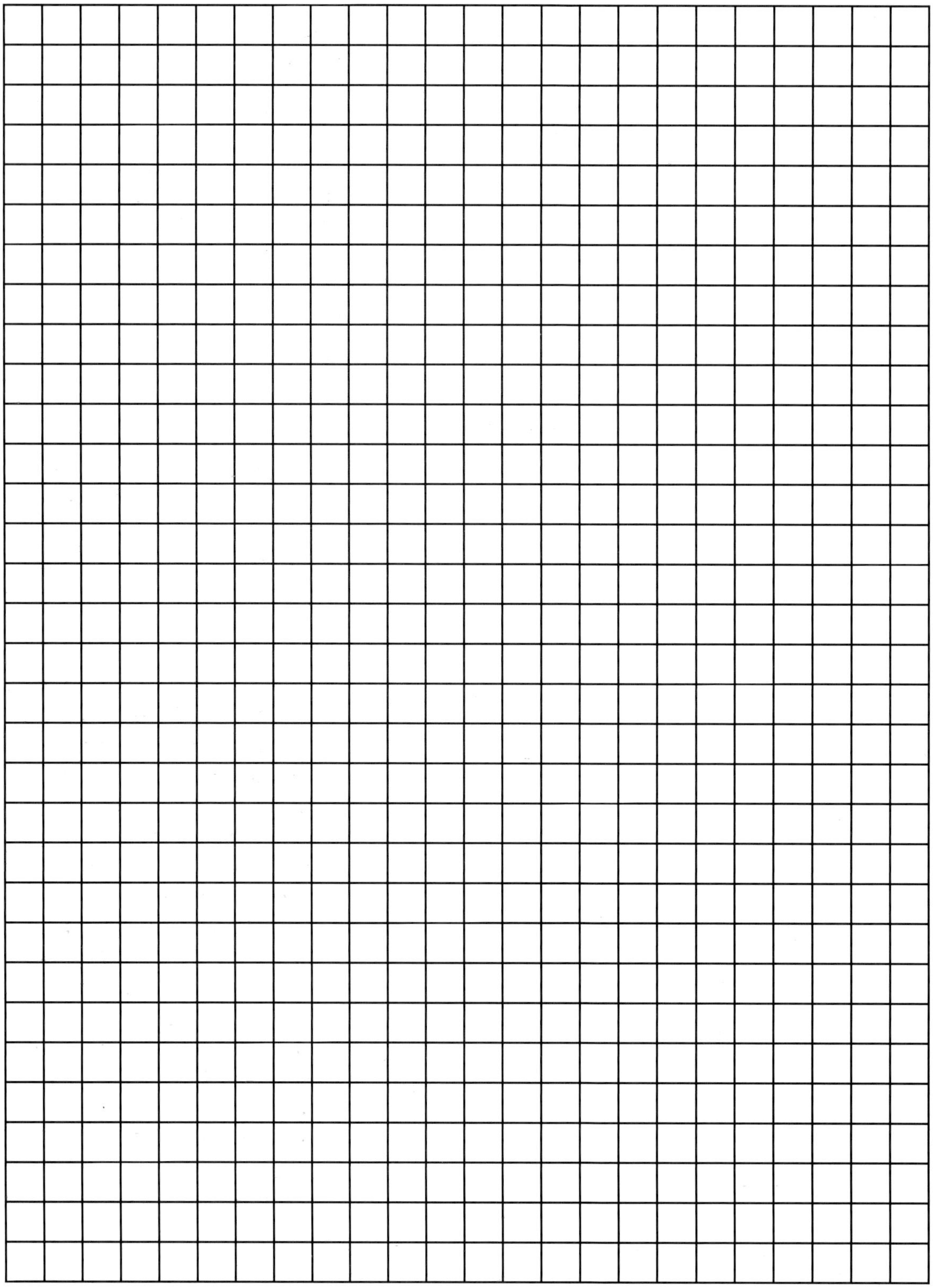

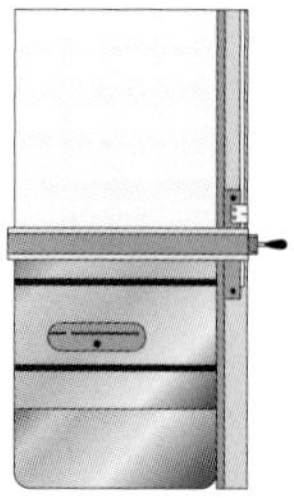

10" Tablesaw
with 30" fence

6x96"
Board

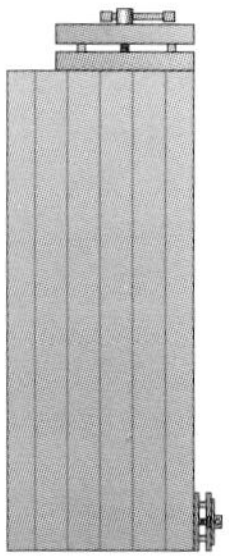

24x60"
Workbench

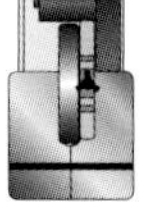

14" Bandsaw

To use this shop layout system, photocopy the icons and the grid. Adhere the photocopied icons to stiffer paper and cut out each to be used with the grid. (You may need several grids.) When you arrive at agreeable icon locations, trace around them with a pencil.

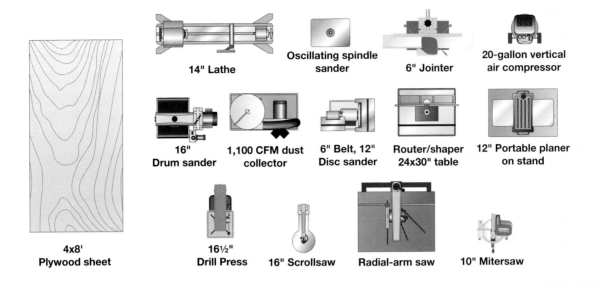

4x8'
Plywood sheet

14" Lathe

Oscillating spindle
sander

6" Jointer

20-gallon vertical
air compressor

16"
Drum sander

1,100 CFM dust
collector

6" Belt, 12"
Disc sander

Router/shaper
24x30" table

12" Portable planer
on stand

16½"
Drill Press

16" Scrollsaw

Radial-arm saw

10" Mitersaw

move toward the back of the shop. Think of three "zones" in the shop. In the first, you store and break down raw stock. Then, you dimension that stock to create project parts.

Finally, you machine and assemble the parts, plus sand and finish the project. Work centers, such as the sanding area at the upper right of 2–36, exist within each zone. These concentrate similar activities within a smaller area.

So, using your list of tools and storage needs, begin arranging your drawing to best accommodate work flow. You'll have to make a few compromises, especially if your shop is long and narrow or oddly shaped. As stated earlier, in the small shop, concentrate on positioning your main machines. Place secondary ones off to the side.

Draw in storage racks and cabinets based on what tools they serve, what they hold, and where

they'll fit. Rather than include icons for cabinets, we'll let you sketch your own to create custom sizes best suited to your layout.

Don't forget, too, that long boards, and especially sheet goods, require lots of infeed and outfeed space. That's why we've even included templates for a sheet of plywood and an 8' board that you can use to test the position of your tablesaw.

In a small shop, as shown in **2–37**, creating zones can prove challenging.

Concentrate most on establishing effective work flow patterns; locating ample storage space; and locating tools, such as a workbench, that are difficult to move. Then put the rest on wheels.

Get Mobile, Be Flexible

Any shop, from small to large, benefits from mobility. Why? Shops constantly evolve as you add tools or take on new

woodworking challenges. Or, you may simply need extra assembly space for a large project. And face it: In a garage shop, you might have to accommodate cars. Flexibility, then, is critical.

In **2–36**, all of the tools marked with an asterisk would ideally rest on mobile bases. In this shop, the tablesaw, planer, and jointer would get bases first so they can be moved to allow parking cars. With bases, such tools as the drum sander, router table, bandsaw, and drill press could roll into open areas when they need to accommodate large stock.

In the shop in **2–37**, mobile bases are just about mandatory. Placing tools on wheels allows them to be "parked" out of the way when not needed. Plus, getting a car in here requires pushing everything out of the way.

CHAPTER 3

Clean Air In Your Shop

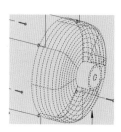

L ET'S FACE IT, IF YOU'RE AT *all active in your shop, you're either going to generate wood dust or stir it up. And dust comes in three forms: coarse sawdust, fine wood dust, and powder. They're all—at the least—irritating, but the powder form—the type generated by sanding—will endanger your health.*

Why is that? Powdery dust particles actually work their way into your lungs. Continued exposure coats them, and that's when health problems, such as emphysema, occur. Woodworkers also face the risk of developing nose cancer at a rate 1,000 times greater than nonwoodworkers. That's why controlling wood dust is so important, and the reason we've devoted this chapter to informing you on how to do it effectively.

In the following pages, you'll find out the best ways to control wood dust with modern collection and air-filtration systems. You'll also learn how to calculate your needs and the basics of dust-collection installation. Meanwhile, keep the following dust-control guidelines in mind.

BASIC DUST DEFENSE

Dust, the powdery kind, remains suspended in shop air for hours. So although you may like to have plenty of ventilation, keep in mind that free movement of air into and around the inside of your shop only raises and spreads dust. That's reason enough to keep doors and windows closed, unless you've installed an exhaust fan, such as the one shown in **3–1** and described on *page 56,* to pull air out of the shop.

A dust-collection system captures and channels dust from stationary power tools to a central collection point (**3–2**). And having one is the best step you can take toward controlling dust. However, don't forget that portable power tools create dust, too. Plan on eliminating as much as possible by replacing or upgrading portable power tools, such as sanders, with models equipped with dust-collection bags. While they won't round up all the fine dust, they'll help.

You also should wear safety goggles and a respirator with a filter designed to trap 99 percent or more of breathable dust. (Read more about these in Chapter 5.) Remember, though, that no matter what type of dust protection you choose to wear, make sure that you keep it on during the dust-producing operation and for a while afterward.

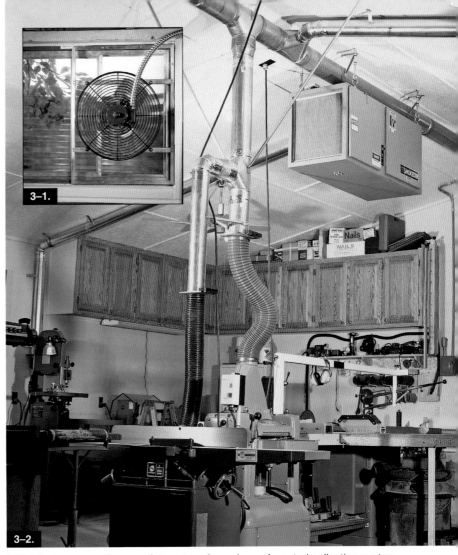

*Little dust escapes the complete setup shown here. A central collection system runs from all major tools, and a ceiling-hung unit collects and filters fine air-borne particles. **(Inset).** An exhaust-fan unit like this, with a dust-proof motor housing, will draw dust, as well as fumes, from your shop. See how to install one later in this chapter.*

Another sure way to get an edge in your battle with dust involves making less of it, or at least the finer kind. That means cutting down on sanding. Hand planing, then scraping, will smooth the wood without producing fine dust. Woodturners should hone their tools sharp, so they shear wood rather than scrape it. Sharp saw and jointer blades also mean smoother cuts, with less finish-sanding.

Then, too, there's a proper way to clean your shop to control dust.

At the end of each day or period of shop work, pick up the dust that has settled. Don't, however, do it by dry-sweeping or blowing with a compressor hose. That only stirs up the dust again. Instead, use wet mops and damp cloths, followed by a good vacuuming. And be sure to store your collected dust outside the shop in a large bag or sealable drum.

A CASE STUDY IN DUST COLLECTION

John McCausland, of Jamestown, Pennsylvania, taught woodworking, metal-working, and drafting until his retirement. So he knew a thing or two about what he needed when he designed and built his home workshop in a 2½-car garage. The 26 × 28' space, shown in **3–3**, provided John a place to pursue his hobby.

John built his dust-collection system with off-the-shelf components, and was happy with how it worked. But as our expert (an engineer with a leading maker of dust-collection equipment) found out, it could have been better. Read on to find out how it was improved. And don't worry, we've tried to keep it simple.

Let's Start with the Collector

A 3-hp dust collector with four bags powers John's system. To save space and reduce noise, John installed the collector in a shed attached to the shop, shown in **3–4**. He ran the main duct through a hole in the wall, just below the peak of the vaulted ceiling.

Placing the dust collector outside the shop brings another benefit. The machine's stock bags filter only particles 20 microns and larger. But it's the fine particles 20 microns and smaller that are most hazardous to breathe. To capture them before the air returns to the shop, John cut holes in the wall that separates the shed from the shop, and added frames that hold four high-efficiency furnace filters.

An expert's opinion: The dust collector does the job, but isn't

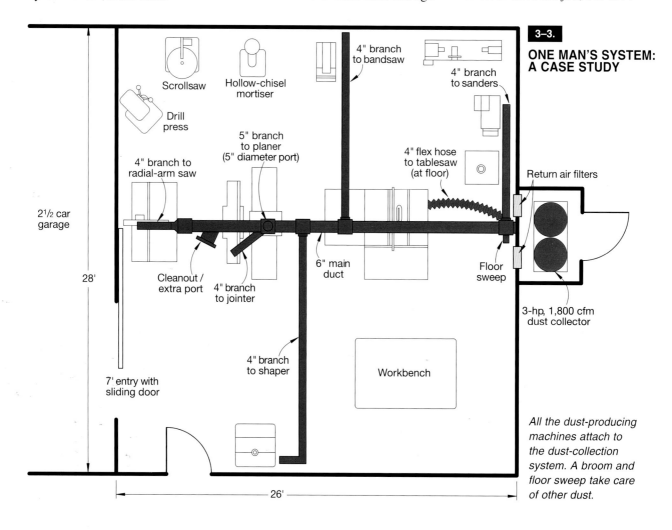

3–3.

ONE MAN'S SYSTEM: A CASE STUDY

All the dust-producing machines attach to the dust-collection system. A broom and floor sweep take care of other dust.

(Figure labels: Scrollsaw; Hollow-chisel mortiser; Drill press; 4" branch to radial-arm saw; 5" branch to planer (5" diameter port); 4" branch to bandsaw; 4" branch to sanders; 4" flex hose to tablesaw (at floor); Return air filters; 2½ car garage; Cleanout / extra port; 4" branch to jointer; 6" main duct; Floor sweep; 3-hp, 1,800 cfm dust collector; 28'; 4" branch to shaper; Workbench; 7' entry with sliding door; 26')

3–4.

A shed houses the dust collector outside the shop. The large bags on the left are oversize, 1-micron units; at right, are the collector's stock 20-micron bags.

SHOP TIP

Performance By the Numbers

To get an idea of how well John's dust-collection system worked, we started by drilling a small hole in the main duct, just beyond where it comes through the wall. That allowed our expert to insert the pilot tubes for two different gauges: one that measured airflow in cubic feet per minute (cfm), and one to check static pressure (sp). Using an ammeter, we also measured the motor's amperage draw. See the results in Table 3–1.

We took readings under three conditions: First, with all the blast gates open to check the system's maximum achievable flow. Next, we closed off all gates except the radial-arm saw to determine maximum flow farthest from the collector. Finally, we tested with just the bandsaw's gate open to determine flow at the most-restrictive port.

The first sets of numbers show performance with the collector's stock 20-micron filter bags. Performance was adequate everywhere but at the bandsaw. The second set reflects the impressive improvements made by installing 1-micron polyester felt bags.

Table 3–1: **Amperage Draw**

GATES OPEN	PORT DIAMETER	20-MICRON FILTER BAGS		1-MICRON FILTER BAGS	
		CFM	MOTOR AMPS	CFM	MOTOR AMPS
All	Various	745	8.5	1,054	10.5
Radial-arm saw	4"	484	7.4	736	9.0
Bandsaw	2"	235	6.2	323	7.5

performing to its full potential. With stock filter bags, it musters only 736 cfm. The collector is rated at 1,800 cfm, which it probably can't achieve, but it's still short of the mark. The motor generates only about 60 percent of its maximum amperage (14 amps at 220 volts) due to ductwork inefficiencies.

Our experts agreed that mounting the collector outside the shop saves space and reduces noise. But, he adds, "Why filter the air twice? Replacing the 20-micron bags with 1-micron bags would eliminate the small dust, and make the filters unnecessary."

John also could install a cyclone system in place of the current collector. A cyclone captures the big chips and most of

To reduce the number of long duct runs, a 5" section does double duty, serving the 15" planer and a 4" branch to the jointer.

3–5.

3–6.

Compared to the tee (installed), a wye and adjustable elbow provide smoother transition from a branch to the main duct.

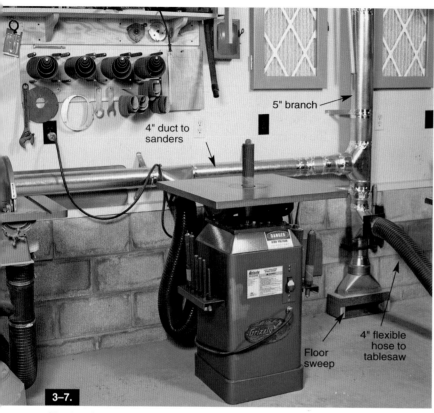

4" duct to sanders

5" branch

4" flexible hose to tablesaw

Floor sweep

3–7.

Sharing duties reduces duct length. Here, one 5" branch serves the tablesaw, floor sweep, and a 4" pipe to two sanders. Blast gates direct airflow to just the ports in use.

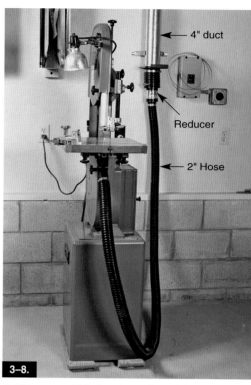

4" duct

Reducer

2" Hose

3–8.

A 2" hose restricts bandsaw duct pickup. Options for improvement include a 4" hose with a reducer or making a 4" port. For another bandsaw-collection option, see page 53.

the fine dust in an easy-to-empty tub. Therefore, the filter bags don't clog as quickly, and John will only face the mess of cleaning them every few months or so, rather than every few weeks.

We couldn't install a cyclone system then, but, following our expert's advice, we did install four 1-micron bags ($45 each) on John's collector. As the chart in Performance By the Numbers shows, airflow increased by more than 300 cfm due to the bags' greater airflow capability. That means the motor can pull more air and produce more of its potential power, as demonstrated by the increased amp draw.

Duct Dos and Don'ts

Because the price was right, John used HVAC (furnace-style) pipes to create the ducts in his dust-collection system. All the pieces consist of 26- and 28-gauge galvanized steel.

Our expert says: Snap-lock HVAC pipe works fine, and is economical. But preferred are 24- and 26-gauge galvanized steel for any ducts larger than 4" because it's strong enough to resist being sucked flat by a powerful collector. Never use lightweight 30-gauge or dryer-vent pipe. Even a modest collector can do them in.

Where ducts intersect and turn,

John used short-radius elbows and tee fittings designed for HVAC use. These increase resistance in the system, preventing it from moving all the air it could.

An expert's opinion: These fittings really hinder the performance of John's system. Fittings designed for dust collection would get the most from his collector, or possibly allow using a smaller one.

For best airflow, elbows should bend at a gentle radius—at least one and a half times (1.5×) the diameter of the pipe. The radius of HVAC elbows usually equals pipe diameter (1×). Where pipes join, wye fittings, seen in **3–6**, allow air and chips to flow through the transition with little turbulence.

Duct sizing in John's system is straightforward, starting with the 6"-diameter main line. A 5" branch, shown in **3–7**, serves the floor sweep, and has two additional 4" branches that serve the tablesaw and spindle- and belt/disc-sanders. At another point on the main, a 5" branch feeds in from the planer (with a 4" off-shoot to the jointer). All other branch lines are 4" diameter.

Machine Hook-Ups

John connected all of his machines to the ductwork using flexible hoses. Even machines he doesn't move around, such as the thickness planer and jointer, connect this way to simplify hookups and prevent machine vibration from rattling the duct-work. In most cases, the flexible hoses are less than 3' long.

John's bandsaw represents one exception to this rule. It's con-nected to the 4" branch line with approximately 8' of 2"-diameter flexible hose, as shown in **3–8**. This small hose fits the bandsaw's stock, under-table port.

Our expert's opinion: The long, small-diameter hose greatly reduces the cfm of airflow at the bandsaw, yielding marginal dust-pickup. Always run the largest appropriate hose as far as possible, and then reduce it to a smaller port size only at the end. Better yet, modify the machine, if you can, to accept a larger port.

Final Analysis

John's system serves him well, removing chips effectively, even if lacking in efficiency. There's room for improvement, as we pointed out. But this case proves that if you understand the basics, you can control workshop dust. Now, let's check out some duct-work and fitting options, as shown in **3–9** on *page 46*, and then go on to plan a dust-collection system for your shop.

GUIDELINES FOR SETTING UP YOUR OWN DUST-COLLECTION SYSTEM

To design your own system, you'll have to consider many factors, such as what machines you'll connect, where to place them, what size dust collector to buy, and what type and size duct system you'll need. Thankfully, you can build a great system following a few rules of thumb:

Note: The following rules may yield a system with more power than you actually need; but too much power is better than too little. Plus, an oversized system can handle future expansion.

Step 1. Find airflow needs:

Start the process by determining how much air your tools need.

Find the airflow requirements of each machine you'll connect to the system, and the corresponding duct size necessary using **Table 3–2** on *page 47*.

Step 2. Lay out the ducts:

Now lay out your duct system on paper, keeping the following in mind:

● Position the air-hungry machines closest to the collector.

● The largest duct diameter required from **Table 3–2** determines the minimum size of your system's main duct. (The collector you choose in Step 3 will influence this as well.)

● Whenever possible, build branches that serve more than one machine.

● Make duct runs as short as possible, minimizing the number of bends.

Plan the shortest, straightest runs you can because every bend and foot of duct adds air resistance, known as static pressure loss, shown in **Table 3–2**.

After you've laid out a tentative duct system, determine the static pressure loss for each branch using the chart. Be sure to include the equivalent footage for 45° wyes and 90° elbows. For each branch you also need to include whatever amount of the main duct exists between that branch and the collector.

Note: Each foot of flexible hose equals 3' of rigid duct, so include

3-9. Selection Guidelines for Ductwork and Fittings

Spiral/industrial metal

Flexible Metal Duct | Spiral Duct | 45° Wye | Long-Radius 90° Elbow

- Designed specifically for dust collection, and therefore the most efficient.
- Available only through specialty suppliers, such as industrial supply catalogs and online retailers. See the sources at the end of the article.
- Costlier than HVAC-style: Ductwork: $1.80+ per foot; wye fitting: $15+ each; elbow fitting: $10+ each (prices given are for 4" ductwork).
  Note: Prices for heavy-gauge industrial tees and wyes can run much higher.
- Used in professional shops and ruggedly built.
- Fittings are designed to maximize airflow and material movement in system.
- The spiral style is very rigid and has a smooth seam to minimize resistance.
- May be available in a wider range of diameters.
- Fittings generally work with metal, PVC, or plastic pipe.
- Flexible metal duct can make gentle bends around obstructions.

HVAC-style metal

Snap-Lock Duct | Flared Tee | Adjustable Elbow

- Designed to move air only, not solid materials, such as dust and chips, so less efficient overall than the industrial components above.
- Readily available at any home center in 24- or 26-gauge steel.
- Priced economically: Ductwork: $1.50+ per foot; tee fitting: $6+ each; elbow fitting: $2+ each (prices given are for 4" ductwork).
- Easy to assemble and install using screws or rivets.
- No wye fittings available, just tee-style. Choose flared tees over straight tees.
- No long-radius elbows available, short-radius only.
- Fittings generally work with metal, PVC, or plastic pipe.

Polyvinyl Chloride (PVC)

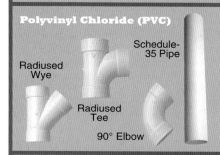

Radiused Wye | Radiused Tee | Schedule-35 Pipe | 90° Elbow

- Designed to move liquids, but capable of carrying air, dust, and chips.
- Readily available in any home center or hardware store.
- Economical in 4" sizes: Ductwork: $.35+ per foot; wye fitting: $2.25+ each; elbow fitting: $1.25+ each (prices given are for 4" pipe).
  Note: Use "schedule-35" type (drain, waste, vent), not "Schedule-40."
- Diameters larger than 4" may be difficult to find and costlier.
- Easy to cut, assemble, and install using special adhesive, screws, or rivets.
- We recommend grounding to prevent static buildup in system.
- Long-radius elbows and wye fittings are available, but only fit PVC pipe.
- PVC is quieter than metal.

Plastic

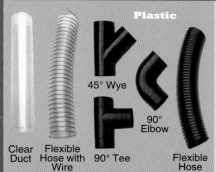

Clear Duct | Flexible Hose with Wire | 90° Tee | 45° Wye | 90° Elbow | Flexible Hose

- Designed for dust collection at lower cost less than industrial metal.
- Only available through ductwork suppliers and woodworking retailers.
- Flexible hose available in black or clear.
- Plain flexible hose is economical, type with spiral wire costs more: Black hose, no wire: $1.60+ per foot; black or clear w/wire: $3.60+ per foot (4").
- Flexible plastic hoses join rigid ducts to machines and fit around obstacles.
- Flexible hose is well-suited to temporary use and for runs that get disconnected when not in use (such as an across-the-floor run to a tablesaw).
- Spiral-wire hose provides crush-resistance and simplifies system grounding.
- Static pressure loss of flexible hose is approximately three times higher than rigid pipe, so it is not well-suited for building an entire system.
- Clear plastic rigid ductwork allows views of blockages in ducts. Usually smaller in diameter, this style requires PVC "sleeves" to connect with fittings.

Blast Gates

Plastic Gate | Aluminum Gate | "Half" Gate

- Absolutely necessary for controlling airflow in any dust-collection system serving multiple machines.
- Available through ductwork and woodworking retailers in plastic or cast aluminum styles.
- Moderately priced: Plastic gate: $5+ each; aluminum: $7+ each; aluminum half-gate: $15+ each (4").
- Blast gates allow you to close off airflow at individual machines or branches to maintain optimum airflow to the machines being used.
- Half gate can be inserted at any point in duct by cutting a narrow slit.
- Blast gates generally fit inside-diameter of any pipe or hose above, though they may require gaskets to achieve a tight seal in non-matching applications.

Table 3–2
Airflow Requirements for Machines

MACHINE	MIN. CFM REQUIRED	MIN. DUCT DIAMETER
Bandsaw (12 to 16")	350	4"
Belt-disc sander	450	5"
Drill press	350 to 400	4"
Drum sander (12 to 24")	550 to 700	5"
Floor sweep	350 to 450	4 to 5"
Jointer (up to 8")	350 to 450	4 to 5"
Jointer (over 8")	450 to 550	5"
Radial-arm saw/mitersaw	450 to 600	5"
Router, table-mounted	200	3"
Scrollsaw	200	3"
Spindle sander	400	4"
Tablesaw (10")	350	4"
Thickness planer (10 to 15")	400	4"
Thickness planer (16 to 20")	600 to 800	5 to 6

Table 3–3
Figuring System Static Pressure Loss

DUCT DIAMETER	45°	90°	SP LOSS PER FOOT
3"	3.5'	5'	.10
4"	3'	6'	.07
5"	4.5'	9'	.055
6"	6'	12'	.045

Note: The 45- and 90-degree readings are based on long-radius industrial fittings. HVAC and PVC fittings will be less efficient, yielding more static pressure loss.

these runs in the total. Also, port designs vary, so it's tough to calculate their static-pressure loss. To be safe, add 1½" for each port on the branch.

As an example, let's say you have a 6" straight main duct. At 14' from the collector, a 45° wye branches into a 4" line to the tablesaw. That branch is 12' long with two 90° bends, and connects to the saw with 3' of flexible hose. Here's the static pressure loss for that branch:

14' (6" straight)	x .045	=	.63" sp loss
3' (one 4" 45° wye)	x .07	=	.21" sp loss
12' (4" straight)	x .07	=	.84" sp loss
2' (two 90° @ 6')	x .07	=	.84" sp loss
9' (3' of flex hose)	x .07	=	.63" sp loss
One tablesaw dust port		=	1.50" sp loss
Total sp loss for branch		**=**	**4.65"**

The branch with the greatest total static pressure loss is the one that determines what your dust collector will have to overcome. If you ever run your system with two blast gates open at the same time, add both totals to get your static pressure loss.

Step 3. Choose and buy a dust collector: Each should be rated by horsepower, cfm, and a maximum static pressure. Generally, plan to eliminate any collector with fewer than 1.5 hp. Smaller, portable machines generally lack the power needed for a large built-in system.

The cfm rating will be shown prominently, but the static pressure rating is more important. The number must be higher than the highest loss in your system, calculated in Step 2, to prevent chips and dust from settling in the ducts.

You'll also need to know the highest airflow value (determined in Step 1) to find the minimum cfm rating you should purchase. Don't be surprised if the dust collector that meets your static-pressure needs is rated at about double the cfm any of your machines requires. Manufacturers often rate cfm with no ducts attached, so their ratings are higher than you'll get in real-world use. Your dust collector also needs to accommodate the largest diameter duct in your system. In fact, a good rule of thumb is to make your main duct the maximum size your dust collector can accept.

Plan the Ductwork

It makes sense to plan your ductwork before you purchase anything. That's because every inch of duct diameter, every foot of length, and every elbow adds resistance (SP loss). As the resistance increases, the volume of air the dust collector can move through the ductwork (stated in standard cubic feet per minute, or SCFM) decreases.

Once you decide on what size ducting to run and

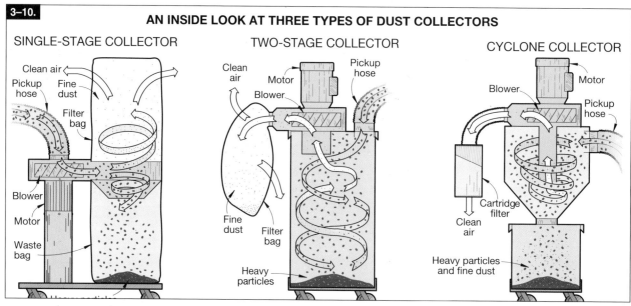

3–10.

AN INSIDE LOOK AT THREE TYPES OF DUST COLLECTORS

SINGLE-STAGE COLLECTOR

Clean air
Pickup hose
Fine dust
Filter bag
Blower
Motor
Waste bag

TWO-STAGE COLLECTOR

Clean air
Motor
Blower
Pickup hose
Fine dust
Filter bag
Heavy particles

CYCLONE COLLECTOR

Blower
Motor
Pickup hose
Cartridge filter
Clean air
Heavy particles and fine dust

With a single-stage dust collector, everything that is sucked in passes through the impeller. Heavier chips drop into the bottom bag, while the top one traps the dust. With a two-stage dust collector, heavy chips drop out first, so only fine dust passes through the impeller. The filter bag then captures the fine particles. With a cyclone dust collector, centrifugal force spins even fine dust out against the sides of the cyclone chamber. The funnel then directs it into the waste can.

where, you can calculate the total SP loss in your ductwork as pointed out earlier. To refresh you, though, here's a quick example of SP loss in 4" duct. A tablesaw requires 350 cfm of airflow. Hook it up to your dust collector with a straight 10' run of rigid duct and 5' of flexible hose, and your dust collector will have to draw 350 cfm at 1.75" of static pressure. Hook your table-saw directly to the dust collector with 10' of flexible hose, and now the dust collector has to draw the same 350 cfm, but at 2.1" of static pressure.

Keep in mind, the air through a dust-collection system must move at a rate of at least 3,500 feet per minute to keep woodworking dust and debris suspended until it reaches the collector. In order to achieve that velocity through 4" round duct, a dust collector must pull no less than 305 scfm. Of course, adding ductwork also adds resistance, which reduces the speed and volume of air moving through it. So read the cfm ratings provided by manufacturers for their machines. Here's a short rundown of the three available types, shown in **3–10**. All use impellers, so the main difference between them—in addition to price—is how they separate debris at different stages.

Single-stage machines no doubt represent the majority used in home workshops. They're nor-mally portable, 110-volt, and from 1 to 1½ hp. Typically, these lowest-priced machines have twin bags. After passing through the impeller, heavy debris drops into the bottom bag while the top bag filters the air. On the smaller portables, one bag both collects the debris and filters the air.

In a two-stage collector, the intake air flows into a collection drum first so the large, heavy particles drop out before reaching the impeller. Only fine dust passes through the impeller and into the filter bag.

Cyclone collectors capitalize on their funnel system to push both large particles and fine dust into a collection drum.

Advice for Installing the System

You've bought all the dust-collection system components, and now you're starting to scratch your head trying to figure out how they all go together. Well,

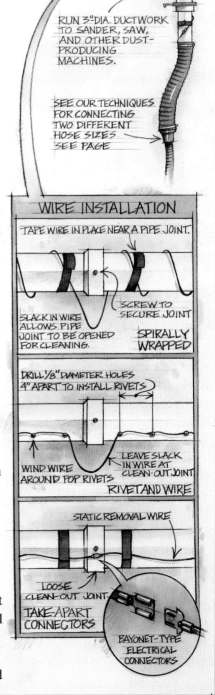

SPLICE BRANCH GROUND WIRE TO MAIN WIRE.

MAKE YOUR OWN BLAST GATES SEE PAGE

RUN 4"-DIA. DUCT WORK TO PLANER, JOINTER AND OTHER CHIP PRODUCING MACHINES.

4" DIA. 3" DIA.

RUN 3"-DIA. DUCTWORK TO SANDER, SAW, AND OTHER DUST-PRODUCING MACHINES.

SEE OUR TECHNIQUES FOR CONNECTING TWO DIFFERENT HOSE SIZES SEE PAGE

PVC DUST-COLLECTION LINE

TAPE WIRE IN PLACE

STATIC-GROUND WIRE

SPLICE GROUND WIRE AND HOSE-CORE WIRE TOGETHER

CORE WIRE FROM HOSE

HOSE CLAMP

FLEXIBLE HOSE

GROUND WIRE TO DUST COLLECTOR

WIRE INSTALLATION

TAPE WIRE IN PLACE NEAR A PIPE JOINT.

SLACK IN WIRE ALLOWS PIPE JOINT TO BE OPENED FOR CLEANING.

SCREW TO SECURE JOINT

SPIRALLY WRAPPED

DRILL ⅛" DIAMETER HOLES 4" APART TO INSTALL RIVETS

WIND WIRE AROUND POP RIVETS

LEAVE SLACK IN WIRE AT CLEAN-OUT JOINT

RIVET AND WIRE

STATIC REMOVAL WIRE

LOOSE CLEAN-OUT JOINT

TAKE-APART CONNECTORS

BAYONET-TYPE ELECTRICAL CONNECTORS

if you've purchased PV sewer and drain pipe, which is the favorite of most woodworkers, here is some advice to help you accomplish the installation more smoothly.

Illustration 3–11 shows a typical system using PVC ducting. You'll want to fasten joints with screws rather than glue so you can open the system easily to dislodge jams.

And remember to run a 4"-diameter line to your heaviest chip producers—planer, jointer, etc.—and locate the collector as close to them as possible. A 3" line will adequately serve saws, sanders, and the like.

Guard Against Static

Air and wood particles moving through the dust collection system quickly build up static electricity charges in any nonconductive hose or piping (those not made of metal). When this static buildup discharges, it could ignite the flammable wood dust particles inside the piping. If the sawdust burns fast enough, you have an explosion.

A static ground for non-conductive ductwork can prevent static-charge buildup and potential disaster. To install one, simply run a wire along or around the pipe, as shown in **3–11**. Insulated

3–11. **TYPICAL PVC DUST-COLLECTION SYSTEM**

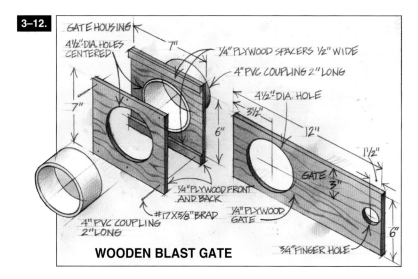

3–12.

GATE HOUSING
4½"-DIA HOLES CENTERED
7"
¼" PLYWOOD SPACERS ½" WIDE
4" PVC COUPLING 2" LONG
4½"-DIA. HOLE
3½"
6"
12"
7"
1½"
GATE
3"
¼" PLYWOOD FRONT AND BACK
#17X5⁄8" BRAD
¼" PLYWOOD GATE
4" PVC COUPLING 2" LONG
6"
¾" FINGER HOLE

WOODEN BLAST GATE

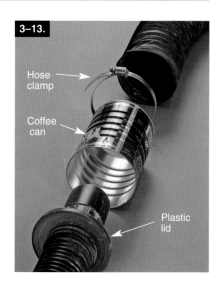

3–13.

Hose clamp

Coffee can

Plastic lid

or uninsulated 18- or 20-gauge copper wire, either solid or stranded, works fine.

At pipe joints, leave slack in the wire or install bayonet connectors (at electronics stores) to facilitate opening the system in case of a clog. Run a ground wire along each branch, and splice each into the main wire.

Connect the ground wire's conductor to the dust collector's metal frame or housing. The other end should extend to the outlet port on the tool, but it doesn't need to connect to anything, except to keep it in place. (Think of the wire as an antenna rather than as a conductor in a circuit.)

The coiled wire core in some flexible plastic tubing allows easy grounding. Simply strip the plastic away from a few inches of the wire core, and then connect the wire to your system ground, as shown in **3–11**.

Control the Flow

For maximum efficiency in a branched system, shut off the flow from all ports except the one in use. To do that, install blast gates.

You can buy metal or plastic blast gates for various pipe sizes from a number of woodworking-equipment dealers. Or, refer to **3–12** to build a wooden blast gate.

The connectors shown are made from a PVC pipe coupling cut in half and epoxy-glued into place. You can alter the dimensions and couplings shown to fit different sizes or types of pipe.

Make the Ends Meet

To mate the 3" or 4" system to most of your tools, you'll need hoses and adapters. Here are two methods for joining a big hose to a smaller hose or connector:

Coffee can connector. Cut both ends out of a 12-ounce coffee can, like the one shown in **3–13.** Slide it into the end of a 4" hose, and clamp it. Then, cut a

hole in the coffee can's plastic lid to fit the smaller connector. Snap the lid onto the can, and plug in the smaller connector. Seal the joint between the connector and the plastic lid with silicone sealer or tub and tile caulk, if you wish. For a blast gate, cap the can with an uncut plastic lid.

Wooden doughnut adapter. Another way to reduce is to make a wooden ring that inserts into the larger hoses. On ¾"-thick scrapwood, first lay out a circle the same size as the outside diameter of the small hose or connector and another concentric one that matches the inside diameter of the larger container.

Then bore the smaller hole with a circle cutter, holesaw, or similar tool. Bandsaw or scrollsaw around the larger circle to complete the doughnut. Install the adapter ring into the large port, and plug the small connector into it. Glue or caulk the joints, if you desire.

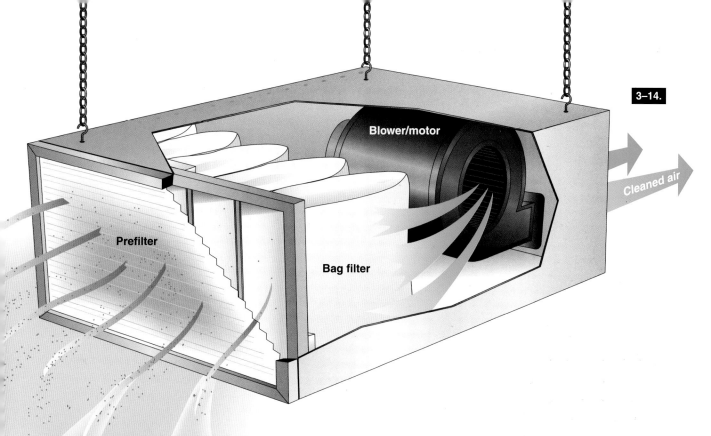

Blower/motor

Cleaned air

Prefilter

Bag filter

STEP UP TO AIR FILTRATION

Your first and best line of defense in the war against workshop dust is to collect the debris at its source before it has a chance to go airborne. But, because no method is 100-percent effective, an air-filtration system provides air support, trapping floating debris before it falls and minimizing the amount you inhale.

Keep in mind that using one of these systems doesn't negate the use of a respirator. Employ an air-filtration system regularly, though, and you'll find yourself spending less time under the mask. Just a few years ago, there were only a few models on the market. Today, you can choose from many brands in sizes ranging from a ceiling-hung, home-workshop size to near-industrial-size units and even torpedo-style benchtop units.

First and Foremost: Air Movement

As you can see from **3–14**, there's nothing overly complex about how an air-filtration system does its thing. The blower first pulls dust-laden air into the prefilter, removing larger particles, then through

a tightly woven bag filter that grabs the smaller particles. The cleaned air exhausts out the other end of the box.

In fact, these machines are so simple, your buying decision boils down to two key factors: airflow (the volume of air the blower can pull through the filter, measured in cubic feet per minute, or cfm), and the effectiveness of that filter. Let's start with airflow.

The more air an air-filtration system can move, the faster it will clean the air. At a minimum, the model you choose should change the air in your shop at least six times an hour. Because everyone's shop is a different size, every shop's cfm requirement is different. See Figuring Flow to help you calculate how much airflow you need for your shop.

Typically, the type of unit shown here will have airflow ratings in a range from 90 cfm to 490 cfm.

If you have a large shop, you may need more air-flow than this size can provide. In that case, you can either add another unit to meet the minimum or step up to a larger, more powerful system.

Although air-filtration units aren't considered loud (you can carry on normal conversation under them while they're running at full speed), a multispeed or variable-speed system allows you to quiet the

This relatively coarse prefilter allows large particles into the bag filter. Note that to take out the prefilter, you must first remove two screws and an access panel.

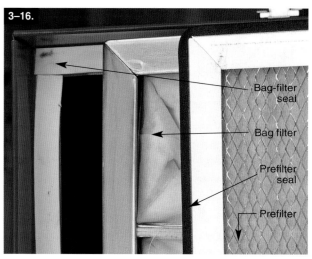

To reduce the amount of dust bypassing the filters, this manufacturer puts a foam seal around the prefilter and another behind the bag filter.

machine by slowing the motor. Obviously, doing so also reduces airflow, so lower speeds should be used only when you're not generating lots of fine dust.

Filtration Facts

If the blower and motor provide the muscle in an air-filtration system, the filters provide the finesse. After all, it takes a fine touch to handle near-microscopic-size dust particles.

The prefilter's main purpose is to protect the bag filter, which is finer, more fragile, more expensive, and more difficult to clean (**3–15**). Most of the tested models use a prefilter that can be tapped out; eventually, these need to be replaced due to wear and tear on their cardboard frames.

You can clean the durable metal-framed prefilters on Craftsman, Grizzly, and Penn models by spraying them with

water. But you must let the prefilter dry thoroughly before using it again. (To get back to work sooner, temporarily insert a pleated furnace filter while the foam prefilter dries.)

But how well do they work? A filter's efficiency is expressed as a percentage of a certain size particle removed from the air passing through it. (Dust particles from 0.5 to 10 microns in size are the most damaging to your health.)

Some manufacturers claim their units remove 95 percent of the 5-micron particles, while others are listed as removing 80 to 85 percent of smaller 1-micron particles (**3–16**). Manufacturers don't give efficiency ratings for both 1- and 5-micron particles. So which is better: a smaller percentage of small particles or a higher percentage of large particles? Truth is, there's just no way of knowing.

SHOP TIP

Figuring Flow

To find the minimum cfm requirement for your shop space, you first need to know the volume of air in that room. To do that, use this calculation (round dimensions to the nearest foot):

Length x width x ceiling height = volume

Now, multiply the volume by 6 (air exchanges per hour), and divide by 60 (minutes in an hour) to find how much airflow you need in your shop.

Volume × 6 ÷ 60 = minimum cfm require

For example, a 12 x 20' shop with a 10' ceiling has a volume of 2,400 cubic feet. 2,400 × 6 ÷ 60 = 240, so the air-filtration system for this shop should move at least 240 cfm.

Touchpad controls are sealed from the outside to reduce dust penetrating to the switches.

More Considerations

• **Power and speed controls.** Among these ceiling units, you'll find simple pull chains for turning the machine on and off, as well as smooth, dust-resistant touchpads that not only turn the machine on and off, but also control the speed (three-speeds aren't unusual) and off-timer (**3–17**).

• **Remote control.** If you mount your air-filtration unit on a high ceiling, a remote control becomes important (**3–18**). Some manufacturers include them with the unit, others charge extra. Most remotes require line-of-sight to the air-filtration unit, much like a TV remote. However, radio frequency remote control is beginning to appear.

• **Off-timers.** If you sometimes can't remember to turn things off, you'll like the timers offered by some manufacturers. They can be

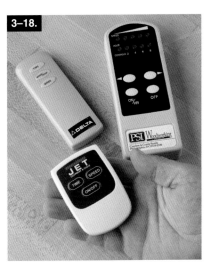

The remote control being held is easy to tuck into a pocket or misplaced. The one on the far right can be mounted to a wall, like a wireless switch.

set to scrub the air for hours at a time, then power down automatically. These can be set by either touchpad or remote control.

• **Exhaust.** Air-filtration units exhaust cleaned air. A diffused exhaust prevents stirring up already settled dust. Only some models have them; others have directable louvers.

SHOP-MADE DUST-COLLECTION SYSTEMS

Once you've decided upon a model and installed it, it's time to turn your attention to some other debris catchers—for those machines that are naturally hard to control. But you can make these dust catchers yourself! See how on the following pages.

Dust-Collection System for Your Bandsaw

To make a bandsaw more efficient, we mounted an outlet switch and connected a three-outlet adapter to one end, as shown in **3–19**. By just flipping the one switch, we can turn the bandsaw, light, and vacuum all at one time on or off.

For dust control, we've fitted the base cavity of the 14" bandsaw stand with a portable wet/dry vacuum, as shown in **3–21**. Then, we connected it to our bandsaw dust port with a 36"-long section of 2½" flexible hose and adapters (from woodworking tool suppliers). Although not necessary, we enclosed the vacuum to make it less noticeable and reduce the noise. If your bandsaw base doesn't have room for a vacuum, set the vacuum aside and still make use of the switch and three-outlet adapter.

Seal Up Your Tablesaw

Contractor's saws cost less than the cabinet style, but they spew all of the sawdust right into your workshop. Here's a simple way to set up a line of defense. Most contractor models are enclosed on three sides, but open on the back, where the motor hangs, and underneath. We used ⅛" Baltic birch plywood to make a two-piece cover for the back.

Measure the outside dimensions of the opening, and then measure to find where you need to leave gaps for the belt and the motor mount. Again, use cardboard to

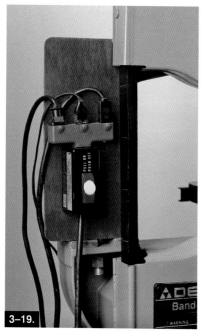

3–19.

This bandsaw has a dust-collection outlet switch mounted to it and a three-outlet adapter connected to one end.

arrive at the right shapes. Cut rectangular pieces to cover the various areas, and then tape those pieces together until you have the final shape. Use that as a pattern to cut the actual cover for the plywood.

As shown in **3–22**, one piece fits around the drive belt and another slides over to meet it. The kerf above the belt opening allows you to flex the thin plywood for installation. The lip glued onto the mating piece covers any gap.

Self-adhesive Velcro strips, available at home centers, serve to hold the dust cover to the saw. Cut them to size, and apply them where shown. You'll have to

remove the cover to swing the saw blade to any angle other than 90°. The alternative would be to cut a pathway for the motor mount to follow, which would open up an escape route for the sawdust. You can buy a bag that snaps in place from woodworking supply houses.

Tablesaw Sanding Table

This fixture does triple duty, functioning as a sanding station, table extension, and storage well for saw accessories (**3–24** and **3–25**). A built-in dust chute links the table to a shop vacuum or dust collector. Build it to fit your particular saw.

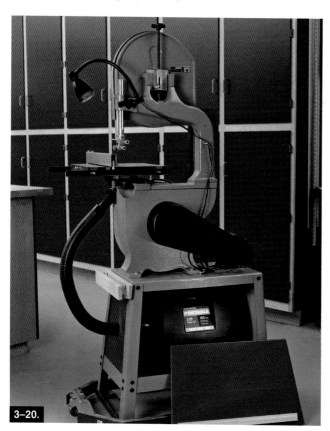

3–20.

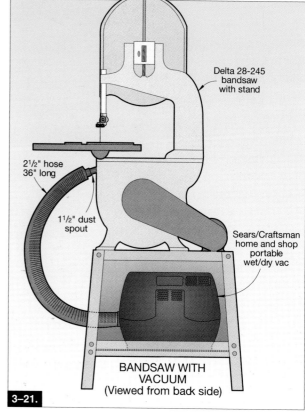

2½" hose 36" long

1½" dust spout

Delta 28-245 bandsaw with stand

Sears/Craftsman home and shop portable wet/dry vac

BANDSAW WITH VACUUM
(Viewed from back side)

3–21.

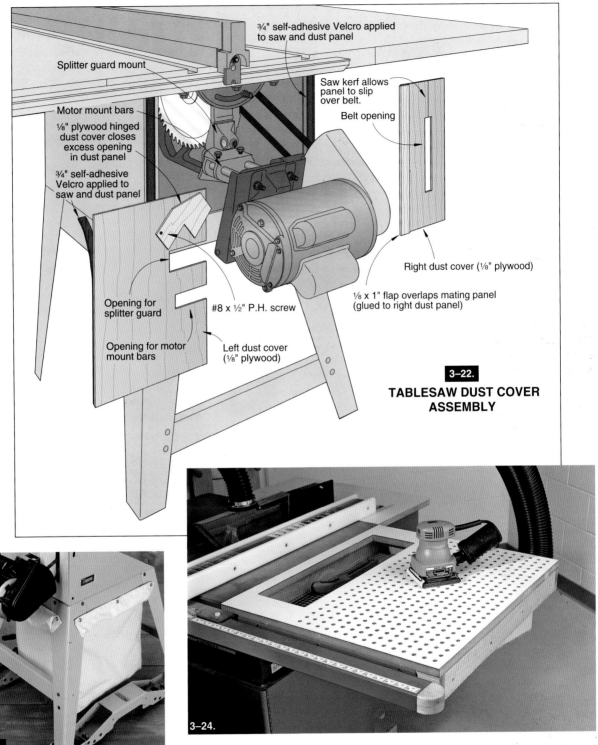

¾" self-adhesive Velcro applied to saw and dust panel

Splitter guard mount

Saw kerf allows panel to slip over belt.

Belt opening

Motor mount bars

⅛" plywood hinged dust cover closes excess opening in dust panel

¾" self-adhesive Velcro applied to saw and dust panel

Right dust cover (⅛" plywood)

⅛ x 1" flap overlaps mating panel (glued to right dust panel)

Opening for splitter guard

#8 x ½" P.H. screw

Opening for motor mount bars

Left dust cover (⅛" plywood)

3–22.

TABLESAW DUST COVER ASSEMBLY

3–23.

Shape the cover to fit tightly around the parts of your tablesaw. You'll still kick up sawdust above the saw, but a lot of the dust will fall right in the bag.

3–24.

We used a piece of perforated hardboard to mark the numerous hole locations. Drill the holes, then countersink them slightly.

Use ducting out the back of the table to fit your dust-collection system or shop vacuum.

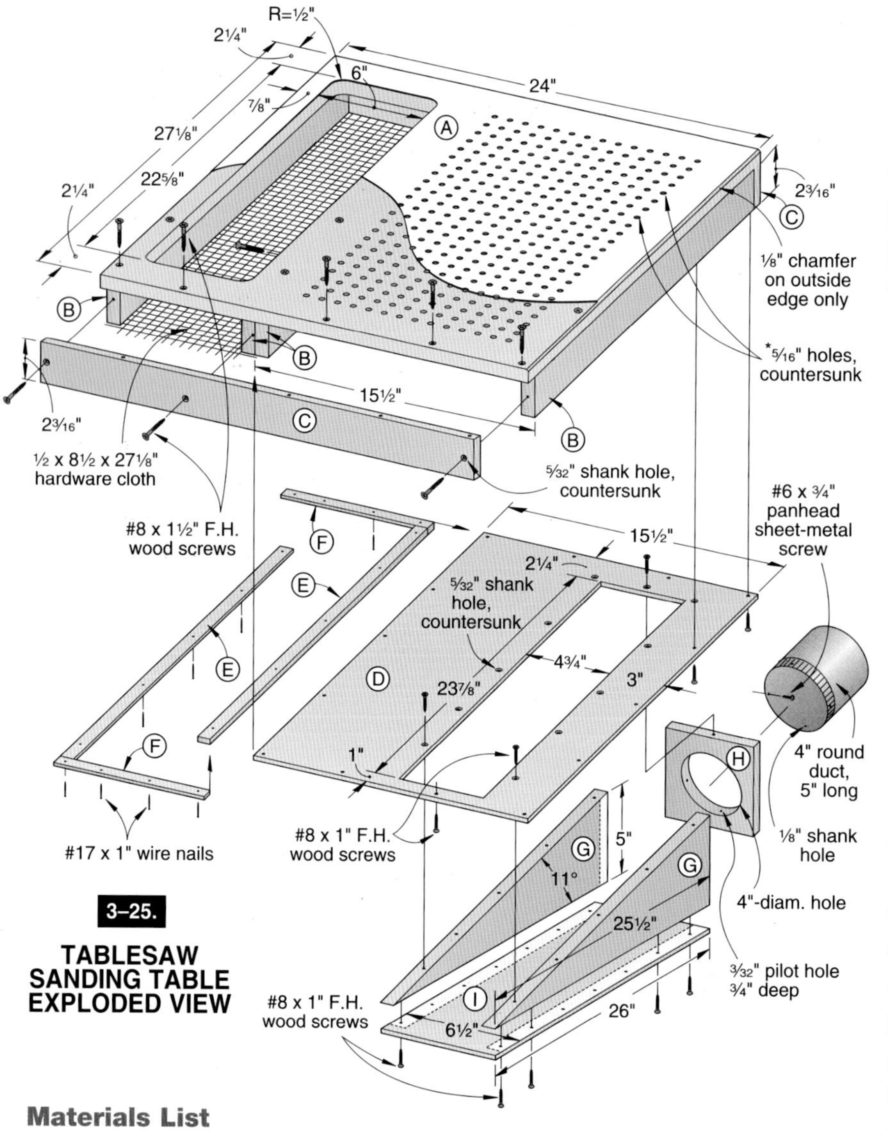
R=½"
2¼"
6"
24"
⅞"
27⅛"
22⅝"
2¼"
A
2³⁄₁₆"
C
⅛" chamfer on outside edge only
B
B
*⁵⁄₁₆" holes, countersunk
2³⁄₁₆"
15½"
C
B
½ x 8½ x 27⅛" hardware cloth
⁵⁄₃₂" shank hole, countersunk
#6 x ¾" panhead sheet-metal screw
#8 x 1½" F.H. wood screws
F
15½"
2¼"
E
⁵⁄₃₂" shank hole, countersunk
E
D
4¾"
3"
23⅞"
F
1"
H
4" round duct, 5" long
#17 x 1" wire nails
#8 x 1" F.H. wood screws
5"
G
⅛" shank hole
11°
G
4"-diam. hole
25½"
⁳⁄₃₂" pilot hole ¾" deep
#8 x 1" F.H. wood screws
I
6½"
26"

3–25.

TABLESAW SANDING TABLE EXPLODED VIEW

Materials List

PART	FINISHED SIZE T	W	L	MATL.	QTY.
		SANDING TABLE			
A tabletop	¾"	24"	27⅛"	MF	1
B spacers	¾"	2³⁄₁₆"	25⅝"	H	4
C end caps	¾"	2³⁄₁₆"	24"	H	2
D bottom	¼"	15½"	27⅛"	HB	1
E screen molding	³⁄₁₆"	¾"	25⅝"	H	2
F screen molding	³⁄₁₆"	¾"	8½"	H	2
		DUST CHUTE			
G sides	¾"	5"	25½"	H	2
H end cap	¾"	5"	5"	H	1
I bottom	¼"	6½"	26"	HB	1

Materials Key: MF = medium density fiberboard; H = hardwood (maple or birch); HB = hardboard.
Supplies: #8 x 1" flathead wood screws; #8 x 1½" flathead wood screws; ½" hardware cloth; plastic laminate; #6 x ¾" panhead sheet-metal screws; 4" round duct 5" long; #17 x 1" wire nails.

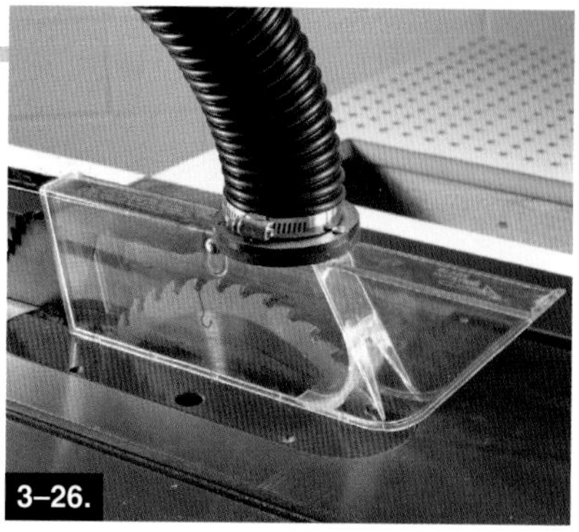

3–26.

Saw-Top Dust Collector

Catch the dust before it falls with this so-easy to make adaptor (**3–26** and **3–27**), and then create a cleaner shop environment.

This practical apparatus combines both dust collection and safety. It works so well you'll never want to remove the blade guard again. Dust and chips thrown forward from the saw blade bounce off the acrylic deflector and get pulled into the flexible-vacuum hose. Hang the hose directly above the blade and connect it to your shop vacuum or dust-collection system. Seal the joint between the blade guard and the hardboard adapter with a bead of silicone sealant. Hold the acrylic deflector in place with silicone sealant or acrylic adhesive.

Basement Window Exhaust Fan

Looking for a fast, inexpensive way to freshen up the air quality in your shop? Here's a basic setup for simple, effective shop ventilation (**3–28** and **3–29**). It allows you easy access to the sliding window in front of the fan, and it doesn't block light like a fan with a plywood support would.

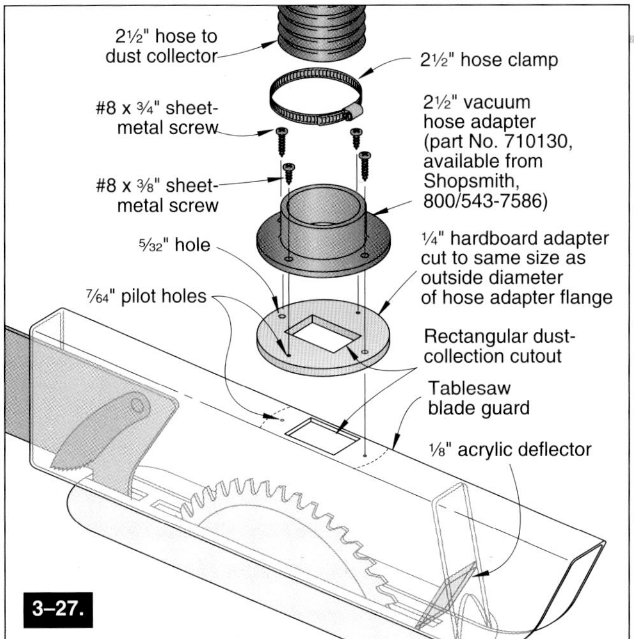

3–27.
SAW-TOP COLLECTOR EXPLODED VIEW

To build the support, we used hardware-store ⅝" aluminum U-channel screwed to our existing metal basement window frame. We pop-riveted the 12" guard-mounted exhaust fan (available at large hardware stores) to two strips of ⅛ x ¾" aluminum that in turn were riveted to the U-channel. The size of the fan will depend upon your window size. Our 12" fan with ⅓ hp delivers 700 cfm and has a totally enclosed motor suitable for dusty environments.

Warning: *Do not use the exhaust fan in an area where pushing air out of the workshop room can reverse the flow of the chimney for the furnace and possibly bring exhaust gases into the house. The structure we used the fan in was warmed with electric heat and didn't require such a precaution.*

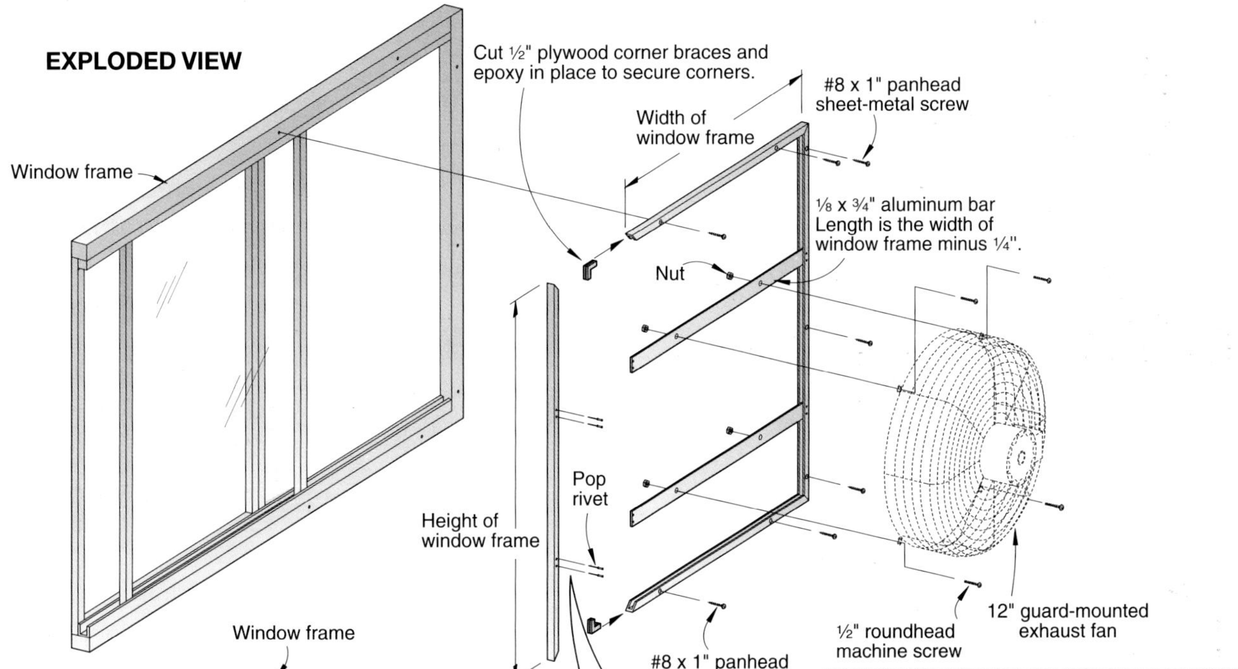

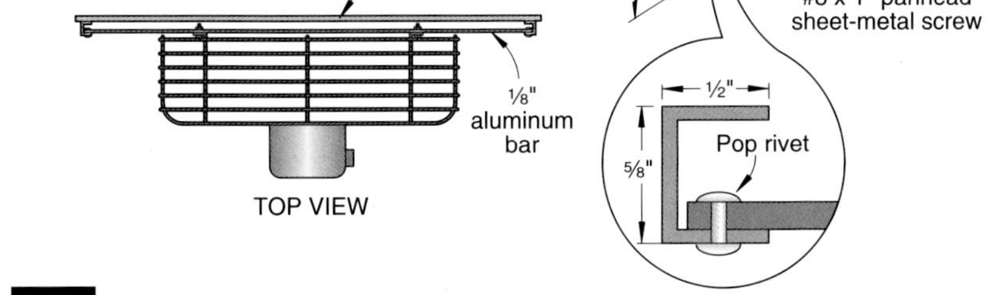

3–28.
WINDOW EXHAUST FAN EXPLODED VIEW

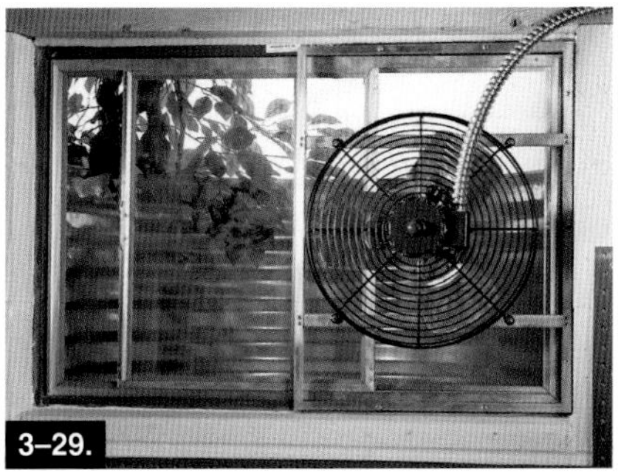
3–29.

CHAPTER 4

Shop Heating Systems at a Glance

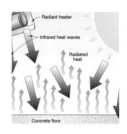

WOODWORKERS LOVE THE *winter, when outdoor chores end and nasty weather drives us indoors to where we wanted to be in the first place—our workshops (4–1). But if our shops lack heat, we could be sidelined for much of the woodworking season.*

To fix the problem, many people turn to portable space heaters for warmth. But these units often are marginally effective at best. For real comfort, consider a dedicated shop-heating system.

For about the price of a good cabinet saw, say $1,500, you can buy the components to heat a 24 × 24' workshop. You might even find a used appliance for a fraction of that cost by checking with local heating contractors. Professional installation will cost in the range of 20 to 30 percent more. Here are some smart, widely-available options.

4–1. *No matter what winter brings, your shop will be warm and comfortable with one of the practical solutions offered in this chapter.*

WORKSHOP HEATING CONSIDERATIONS

Shops share some of the heating concerns as homes, but significant differences exist as well. Keep the following in mind as you plan a heating system.

• **Insulation:** You can't bring warmth effectively into your workshop until you keep the cold out by sealing and insulating your shop. The up-front expense is small compared to what you'll save in the long run.

This difference is easy to see looking at heating requirements, figured in British Thermal Units (BTUs) per hour. For a 24 × 24' shop in the upper Midwest, where winter temperatures dip below 0 degrees Fahrenheit, manufacturers quote an average of about 25,000 BTUs per hour for an insulated shop, and more than 50,000 for one without insulation.

• **Air quality:** Fill the air in your shop with enough fine sawdust or finishing fumes, and you'll have the potential for an explosion. So stay away from use of open-flame heaters and from electric units with exposed heating elements. Choose a unit that, if electric, has shielded elements, or, if gas-powered, draws outside air for combustion rather than shop air.

• **Insurance and permits:** Before you install a heating system, check out local code requirements governing the types of heaters you can use, installation restrictions, and required permits. Talk to your insurance company, as well. Skirting these steps could lead to fines, or to denied claims if you have a fire—even one unrelated to the heating system, such as one caused by improperly discarded finishing rags. (You always lay your oily rags out to dry flat, right?)

• **Unique requirements:** Differences in climate, construction, and usage dictate different heating needs in every shop. Check out How to Determine Which Heating System to Use on *page 63*, for more individual considerations.

FORCED-AIR HEATERS

Forced-air heaters fall into a couple of categories: Self-contained heaters that mount to the wall or hang from the ceiling and the traditional ducted furnace, found in many homes.

Self-contained heaters have been standard issue in shops and garages for years. They don't eat floor space, and are relatively easy to install because they don't require ducting. These heaters produce heat from economical liquid propane (LP) or natural gas. Most circulate warmed air using a fan. Unlike older versions, some modern units draw combustion air from the outside, as shown in **4–2.**

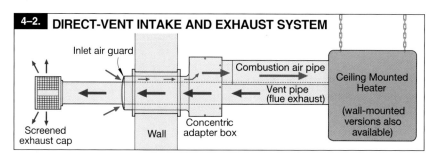

4–2. **DIRECT-VENT INTAKE AND EXHAUST SYSTEM**

Most direct-vent and separated-combustion heaters feature a 2-in-1 pipe that draws intakes air and exhausts waste through a single opening in the wall.

A traditional furnace distributes air through a series of ducts to just where you want it. Installation involves more challenges, but a furnace also accommodates central air conditioning.

If you choose a furnace that doesn't draw outside air for combustion, install it in a separate room to minimize dust and fume hazards. Illus. **4–3** shows one way to do this.

TUBE HEATERS

If you have access to gas, but don't want forced-air heating, check out a ceiling-mounted radiant tube heater, shown in **4–4** and **4–5.**

Tube heaters burn LP or natural gas, which warms the air inside a long metal pipe. Heat radiates downward, warming objects it strikes. The system has no external fan to stir up dust in the shop.

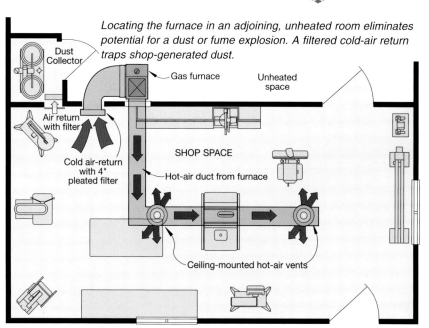

Locating the furnace in an adjoining, unheated room eliminates potential for a dust or fume explosion. A filtered cold-air return traps shop-generated dust.

4–3. **FORCED-AIR FURNACE SHOP HEATING**

4–4.

Radiant tube heaters, such as this one, are available in several sizes, and can be configured in straight, "U," and "L" shapes to fit the space.

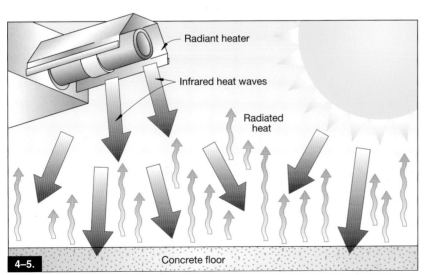

4–5.

Just like the sun, radiant heaters warm the objects in the room, such as the floor or any pieces of equipment. These, in turn radiate heat upward, warming the air.

Labels within figure 4–5: Radiant heater / Infrared heat waves / Radiated heat / Concrete floor

RADIANT HEATING

When building a new shop (oh yeah, it's a "garage," we won't tell), you might consider hydronic, in-floor radiant heating. These systems are becoming more affordable and increasingly popular in homes, shops, and buildings of all types.

As shown in **4–6**, on *page 62*, the heart of a hydronic system is a network of plastic tubes, usually imbedded in a concrete floor. Hot water pumped through the tubing heats the concrete, which acts as a giant radiator and warms everything above it.

You can power a hydronic system for a two-car-garage-sized shop with a small water heater. (If it's a gas unit, you still need to isolate the flame from shop air, of course.) You can build a "closed" system filled with antifreeze or run a water supply to the shop and let the heating system provide hot water, as well.

ELECTRIC OPTIONS

Electricity has traditionally been an expensive heat source. Even so, the setup costs compared with other systems may make it worth considering, especially if your BTU needs are low or you spend limited time in the shop during cool seasons.

4–6.

In a radiant-slab floor, heated water flows through flexible plastic tubing that gets routed and secured before pouring concrete. Electric in-floor radiant systems exist, as well.

Unless you're dedicated to very traditional woodworking, you already have electric service to your shop, so chances are you won't need anything more to run an electric heater. Units that run on 220 volts generally produce more heat.

Electric heaters come in many sizes and styles, and it's easy to add more based on need. Portables don't require special insurance, and even permanent units seldom require a permit.

Radiant panels pass electricity over a large metal plate to produce warmth. These heaters are fairly immune to dust and fume dangers.

According to a manufacturer, heating with radiant electric panels costs about the same as using a natural gas or LP forced-air system. Electricity costs more per BTU but, because radiant heating warms objects and not just the air, electric panels heat using fewer BTUs.

Other electric options include in-floor, ceiling-mounted, and simple "plug-and-play" baseboard units.

IS WOOD GOOD?

A wood-burning stove seems like the ultimate romantic source of shop heat for many woodworkers. After all, you're making fuel all the time in the form of scraps and (heaven forbid) mistakes. But wood stoves do have drawbacks.

First, those kiln-dried scraps burn up pretty quickly, so you'll need a supply of split firewood. Even with good wood, an inexpensive stove can be hard to regulate, causing wide temperature swings. And unless you make special trips to stoke the fire, you'll lose your heat when not in the shop.

Insurance companies may balk at a stove's open flame and hot surfaces. Also, some communities with strict air standards regulate the use of wood stoves and fireplaces. A stove may look great in the shop, but it isn't the safest heat source.

HOW TO DETERMINE WHICH HEATING SYSTEM TO USE

Whatever type of heating system you choose, answer these questions before you shop. Discuss the answers with a heating contractor or salesperson to ensure that you select the size and type of system that best suits your needs.

1. Does your shop stand alone or is it attached to another heated structure?

2. How many exterior walls does your shop have, and are they currently insulated?

3. Is it a dedicated shop, or a space also used for a garage or other purposes?

4. What are the shop dimensions?

5. How high are the ceilings?

6. What construction materials make up your shop (wood, brick, concrete block, etc.)?

7. Do you know the insulation values in the walls and ceiling?

8. How many windows does the shop have, and are they single-pane or high-efficiency units?

9. How many exterior doors are there?

10. Does the shop have overhead garage doors? If so, are they insulated?

11. How many hours per week do you spend in the shop during cold seasons?

12. When not in the shop, will you heat it to at least above freezing?

13. Do you have a gas line near the building, or will you have to run one?

CHAPTER 5

Workshop Safety and Security

WHETHER YOUR SHOP IS READY *to go or still in the planning stages, you don't want to bury safety and security on your "to do" list. Safety should never be taken for granted, no matter how long you've been woodworking and how well you know your tools. And just as you perform a regular schedule of tool maintenance to keep them in working order, you also need to provide for their security. How can you protect against theft? Or fire? In this chapter we'll give you some sound security advice, tell you how to guard your health in the shop, as well as how to handle some basic first aid. You'll also find a handy shop safety checklist for you to copy and then display prominently.*

TOOLS AS TARGETS

Crime probably doesn't enter your mind when you're working away in your shop. Workshops, though, can be a prime target for burglary. Juvenile burglars may not be as interested in your tools as in electronic gear that is easier to haul away and convert to cash, but professional burglars are another story. That's because a pro may have already lined up a market for tools, and if he knows that you're well stocked, he'll come for them. And it won't take long for a pro to clean you out.

One of the best lines of defense against theft is to keep your tools out of public sight. Anything you can do to hide them from the casual passersby increases your security, say law enforcement officers. If your shop is in a garage that faces a street, keep the overhead door closed when tools are out. If you must work with the door open, put tools back into cabinets or drawers between uses. Place larger equipment along walls, if possible, rather than display them in the doorway. Arrange perforated hardboard or open-shelf tool storage so that it's inconspicuous. (See **5–1** for practical applications of these suggestions.) And be sure to close and secure the door when you leave the shop for extended periods of time. Frosted glass and/or window shades will help, too. Or apply plastic film to make clear windows opaque if you don't really need them for light.

Give Your Tools Identity

Tools recovered by police often lack any identifying marks as to their owner. That makes them almost impossible to return them or to prove ownership. Serial numbers recorded on paper even may be a waste of time, because many tool serial numbers are on labels that peel off.

So what can you do? It's a good bet that your local law enforcement department has an identification system in place. With such a system, the department assigns an alpha-numeric code (letters and numbers) to you. You mark this code on your tools and other property. Sometimes, you'll be loaned an engraving tool or invisible marking pen, along with an inventory form for you to fill out and put in a safe place. If you ever have tools stolen, you simply contact the department to tell them what's missing and how it was marked. They can then trace ownership of recovered tools.

Of course, alterations to a tool can make it easier to spot, too. Adding an aftermarket accessory or something as simple as a longer replacement cord could help you pick a tool out from a stack of recovered stolen goods. Painting a logo or your name on a tool could also help.

Whatever option you take, do something, plus make records of

5–1.

Hand and portable power tools in view through an open garage door invite thieves. In the shop shown here, closed cabinets hide them, and large stationary tools back up to the sidewalls.

your tools. Note all their identifying marks and characteristics. You can also take photos of them, or use a video camera. All help you give accurate descriptions to police and your insurance company.

Ways to Deter Burglars

Burglars and amateurs alike don't want to spend a lot of time getting into a place. They don't want to make a lot of noise, and once inside, they need to exit quickly carrying armfuls of loot.

Instead of breaking windows, thieves will cut or tap out a small hole that they can reach through to unlatch a window. You can foil that with auxiliary latches. Even blocking a window sash with a dowel or nail inserted into a hole drilled into the frame will do the trick.

On basements windows, install a security grid or bars to prevent them from being kicked in. (Be sure you can remove them from inside in case of fire.)

Make sure doors are secured, too. Burglars won't be able to get in (or back out) easily if the shop door—preferably solid core without glass—has a double-cylinder deadbolt on it. To secure an overhead garage door, lock a padlock through a hole in the door track just above the rollers.

Install inexpensive, easy-to-install floodlights with motion sensors outside your shop, especially near doors and

windows. Fitted with dusk-to-dawn timers, they've proved to be effective deterrents.

To install an alarm system or add one to an existing home-security system, seek advice from a professional installer. An inadequate security system will provide you a false peace of mind, and that can be worse than no security system at all.

Adding sophisticated and expensive systems like heat sensors and sprinklers surely will protect your shop from fire. But all it really takes is some common sense and pretty ordinary equipment.

Your First Line of Defense

In industry and commercial establishments, the first line of defense always is to sound the alarm. That means, if you have a fire, you call the fire department first. Then, if you know what you're doing, you try to extinguish the fire. Accordingly, you should post the number of your fire department (or an emergency number like 911), along with other emergency numbers, close to the telephone, and consider it an essential item in your fire-protection equipment.

Smoke detectors also should be part of your alarm system. You

should install one on each level of a multi-floor structure, or in different areas of a large shop. Battery-operated smoke detectors only cost about $10 each, so don't penny-pinch.

What about extinguishing the fire? Of the several types of fire extinguishers available, two types should be used in the home workshop. A Class A pressurized-water extinguisher (about $65 for one of 2½-gallon capacity) works well for fighting fires in cloth, wood, and other ordinary non-liquid, non-electrical combustibles. The second type does it all: A 10-lb. dry-chemical, A-B-C-rated fire extinguisher (about $50) will put out a paint or a varnish fire and it works on other flammable liquids, grease, electrical fires—such as with a motor—and wood, paper, and cloth as well. And remember to always place the fire extinguisher at the entrance to the room because you don't want to put your life in jeopardy by going to a location in the center or the far corner of the shop to grab it. Never cut yourself off from escape.

There's still another fire-fighting item, though, and an inexpensive one. Depending on where the shop is located, you could have a hose with a nozzle connected to a valve tap on a cold-water line. Normally, a basement shop would be close to a utility room, with valve taps readily available. Then, an appropriate length of garden hose to reach your shop area is all you need.

Additional Practices That Help Prevent Fires

Although statistics on the causes of fire in the home don't pinpoint workshops, there's a good chance that flammable liquids, such as acetone and lacquer thinner, are often the culprits. With them, proper storage is the best insurance.

You can buy nonflammable storage units and waste receptacles, but they're expensive (cabinets cost $300 and up; a six-gallon capacity waste container, about $50). But if you normally work with flammable finishing materials, they're worth the investment. In lieu of them, always store flammable liquids in tightly closed, metal containers. Never keep even dirty solvent, such as turpentine or brush cleaner, in an open one.

Some finishing oils, including linseed oil and tung oil, have the tendency to self-combust under the right conditions, if left open. So you want to keep all of your oily cloths in a lidded metal container, too. The alternative is to hang the cloths outside until they completely dry.

But, even if you put all of this advice into practice, don't expect big reductions in your homeowner's insurance premium. That's because most of your premium payment goes for liability, not for fire coverage. Although having fire protection equipment does knock something off the cost, it

5–2.

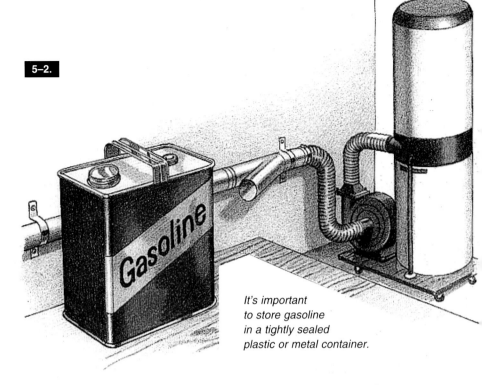

It's important to store gasoline in a tightly sealed plastic or metal container.

won't be a lot. Your payback, more than anything, is really peace of mind.

How to Deal with Gasoline

Many woodworkers share their garage workshop with an automobile, a power mower, chainsaw, and possibly other internal-combustion tools and equipment. That means there's gasoline around, and along with it, the danger of fire.

Sure, gasoline fumes smell strong. They're also highly flammable as well as invisible. But did you know that fumes from gasoline are heavier than air? Because of that weight, they sink through the air to the floor like water running downhill.

Fumes don't always stay low, though. You can stir them up just by walking into a room. And the fumes can rise all the way to the

ceiling before they settle again. Either on the floor or in an unsettled state in the air, it's possible they'll ignite from a glowing ember, high heat, open flame, or spark.

Of course, you can avoid catastrophe. First of all, always store gasoline in a sealable container intended for that use (**5–2**). It doesn't matter if it's metal or plastic (never glass), just so you can keep it tightly sealed. Then, place that container away from a heat source or flame, such as a furnace, electric heater, or gas water heater. This advice also applies to gasoline-powered tools and equipment. Keep their gas tanks fully closed and store them properly.

How to Retire a Fire

Knowing how to use a fire-extinguisher properly to fight a fire can mean the difference between saving your shop or watching it go up in smoke. A lack of know-how, on the other hand, could lead to personal injury, while a lack of preparedness could doom your workshop.

Equip your shop with an A-B-C rated fire extinguisher, and keep it by the door leading into your shop. It's also a good idea to have a phone and a flashlight installed in the shop, again, by the door. Now, here's what to do when the temperature in your shop rises:

1 First, call the fire department. Then, if it looks feasible to fight the fire, remove the extinguisher from its mount and check to make sure it is full.

2 Stand back 8' from the fire, pull the extinguisher's string, and remove the hose from the holder on the container.

3 Aim the hose at the absolute base of the blaze, and then squeeze the handle and lever together to start the chemical stream flowing from the nozzle.

4 Sweep the fire with the stream in a side-to-side motion, advancing as the fire extinguishes.

5 When the fire is out, overhaul the area of the fire, searching for hot spots to put out.

6 Stand back from the fire area and watch the debris carefully until you're absolutely positive that it won't reignite.

SAFETY EQUIPMENT

Shop safety includes a whole lot more than leaving the guard in place on your tablesaw. There's also your eyes, ears, and lungs to think about. To protect these, select the right gear, and then make a habit of using it everytime you're in the shop.

Protect Your Eyes

All safety eyewear must be officially approved. That approval comes from the American National Standards Institute (ANSI), a voluntary organization that looks after the development of standards used in business, industry, government, and educational institutions through-out the United States. ANSI wrote the standard for safety eyewear for the industrial workplace, but your workshop differs only in size. The possible hazards to your eyes, such as flying chips of metal or wood, dust, or contact with harmful liquids, remain the same.

ANSI Sets the Standard for Protection

All quality protective eyewear, including face shields, complies with ANSI standard Z87.1 (of 1989) and will have that letter and numbers stamped or molded into the frame or shield. Lenses (usually of hard polycarbonate) that comply will bear the manufacturer's initials (AO for American Optical, X for UVEX, etc.) somewhere out of the line of sight. Any manufactured protective eyewear you consider purchasing should bear both of those inscriptions.

What does the ANSI standard mean to you? For one thing, the frames and lenses work together for protection. Industrial safety glasses have lenses that withstand nearly four times the impact of regular impact-resistant lenses. Compliant frames have inner retention lips that keep unshat-tered lenses from being driven into your eyes under the force of heavy impact. They also meet the

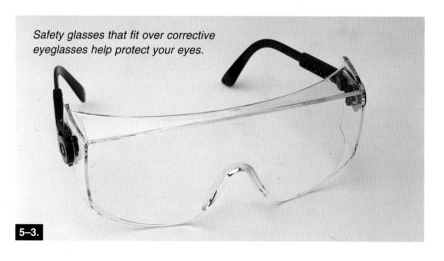

Safety glasses that fit over corrective eyeglasses help protect your eyes.

5–3.

standards for pressure and impact that regular frames do not. And for complete protection, all industrial safety glasses should have side shields.

You'll find, though, some contemporary styles of one-piece, wraparound safety glasses that might not carry the ANSI Z87.1 imprint or the initials of the maker. They may meet or exceed the standard, but due to their one-piece construction, they don't comply with ANSI's lens-and-frame stipulations.

How To Determine Your Eyewear Options

Depending on your eyesight, you have several options in safety eyewear. If you don't require corrective lenses (or wear contact lenses), you may select prefabricated safety glasses with clear lenses in place (5–3), safety frames in which safety lenses are inserted, or goggles. (There's also a combination eyeglass/goggle available.) If you have to wear corrective lenses while woodworking, you can don prefabricated safety glasses or goggles over your normal eyewear; use a flip-up face shield, as shown in 5–4; or have prescription safety glasses made.

How much does ANSI-compliant eyewear cost? Not much, considering what you're protecting. You can buy a pair of prefabricated Z87-level safety glasses for as little as $5 or as much as $30. (Lens quality gets better as the price goes up.)

A flip-up face shield protects your face as well as your eyes.

5–4.

Goggles run from about $8 to $20; face shields, $15 to $20. The cost of prescription safety glasses varies with the fashion and quality of the frame, as it does with regular corrective lenses.

With safety glasses, always check (or ask) for ones with scratch-resistant lenses. And to prevent them from fogging when you're wearing a dust mask, have them treated with an antifog coating. Many companies offer permanent antifog coating on nonprescription safety glasses.

Hearing Protection

If there is one thing to remember from reading this, it's that hearing loss is cumulative and permanent. Hearing protection can't restore what you've already lost, but it can halt further deterioration.

If you value your hearing, you'll want to wear ear protection for any noise greater than 85 dB. For very loud noise, such as that made by a chain saw, you'll need added protection, such as earplugs under earmuffs. (See 5–5 for tool loudness ratings.) Permanent

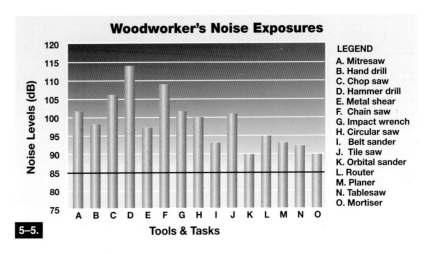

Woodworker's Noise Exposures

LEGEND
A. Mitresaw
B. Hand drill
C. Chop saw
D. Hammer drill
E. Metal shear
F. Chain saw
G. Impact wrench
H. Circular saw
I. Belt sander
J. Tile saw
K. Orbital sander
L. Router
M. Planer
N. Tablesaw
O. Mortiser

5–5. Tools & Tasks

damage to your hearing ability can result from exposure to 100 dB or more for two hours or less.

Note: The decibel scale by which sound is measured happens to be logarithmic, not linear. As demonstrated in 5–6, that means that a 100 dB noise is 10 times as loud as a 90 dB noise. And 90 dB is 10 times as loud as 80 dB.

In industry, the federal Occupational Safety and Health Administration (OSHA) helps reduce noise at its source by doing site-specific studies and giving recommendations to manufacturers for quieting operations. You can do a similar thing in your home shop by purchasing low-noise power tools and equipment.

Even with quieter tools, though, you'll still need hearing protection when noise exceeds dangerous exposure levels, such as when you're routing. So how much do you need? You first must understand how hearing protection is rated.

Manufacturers of hearing protectors assign each of their products a laboratory-based Noise Reduction Rating (NRR), and by law, it must be shown on the label of each hearing protector sold. The NRR supposedly equals the drop in decibels (attenuation)

Foam earplugs

5–7.

provided by the device. For example, an NRR of 20 would reduce a 100 dB noise to an audible 80 dB. In the real world of your shop, however, the actual NRR proves to be somewhat less. That's why you should select hearing protection with an NRR of at least 25.

Wear the Proper Hearing Protection

According to a 1997 study by the National Institute of Occupational Safety and Health (NIOSH), laboratory data show that earmuffs provide the highest real-world noise attenuation values, followed by foam earplugs. However, other data from OSHA and industrial sources, such as 3M, find that properly fitted foam or flexible plastic earplugs offer the greatest protection—from an NRR rating of about 29 to 33. NIOSH, more

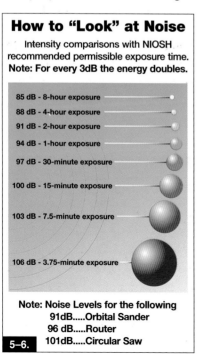

How to "Look" at Noise

Intensity comparisons with NIOSH recommended permissible exposure time.
Note: For every 3dB the energy doubles.

85 dB - 8-hour exposure
88 dB - 4-hour exposure
91 dB - 2-hour exposure
94 dB - 1-hour exposure
97 dB - 30-minute exposure
100 dB - 15-minute exposure
103 dB - 7.5-minute exposure
106 dB - 3.75-minute exposure

Note: Noise Levels for the following
91dB.....Orbital Sander
96 dB.....Router
101dB.....Circular Saw

5–6.

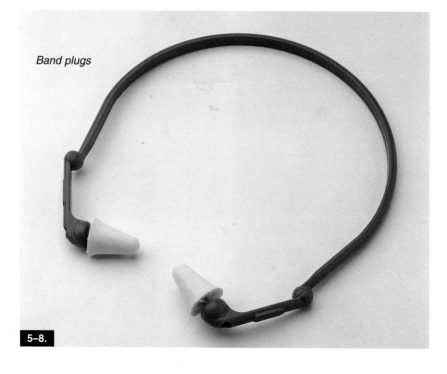

Band plugs

5–8.

generally speaking, states that "the best hearing protector is the one that the worker will wear."

Basically, you'll find three types of hearing protectors. Foam earplugs that mold to fit your ear canal offer the highest NRR and cost the least (about 15 cents a pair) (**5–7**). Band plugs (**5–8**), similar to foam ones but made of flexible plastic and joined with a head/neck band, come next, and cost a bit more. Earmuffs (**5–9**), usually with the lowest NRR (17 to 23), have prices around $15. Top-of-the-line models can have an NRR as high as 29, but cost as much as $25 a pair.

Why Hearing Protection Fails

Researchers at 3M, which manufactures several styles of hearing protectors, have studied why hearing protection frequently fails in the industrial workplace. Some of their findings follow:

• **Improper sizing and insertion.** The wearer tends to fit plugs too loosely, even though they're available in various sizes. If fit too tightly, they are a discomfort and the wearer removes them. Also, a person can have different size ear canals, so each must be sized separately.

• **Incompatibility with other protective equipment.** Earmuffs often don't seal properly over safety glasses. Long hair also interferes.

• **Poor communication.** Hearing protection tends to attenuate high pitches, typical of voices. Wearers loosen, alter, or remove protectors to hear others.

• **Wear and tear.** Seals wear down on muffs. Foam plugs become less flexible and unable to properly mold to the ear canal. Premolded plugs shrink. Ear wax and perspiration also build up on them. Earplugs should be checked frequently and pushed in. Even chewing gum can shift them out of position.

Finally, here's a test to see if earplugs fit properly: After inserting the plugs, cup your hands over your ears, then take them away. If you hear a difference, they're not being worn correctly. Remove and refit the plugs, and then try again. And don't forget to wash them in mild soap and water after wearing them a few times.

THE DANGER OF DUST

Exposure to wood dust in excess of 5 milligrams per cubic meter of air is hazardous to your health, says OSHA. It's even more so from western red cedar. Because that very common wood has been linked to respiratory problems, OSHA limits its dust to 2.5 milligrams per cubic meter.

How much is 5 milligrams? It's actually less than two ten-thousandths of an ounce. (A dime weighs eight-hundredths of an ounce!) So according to OSHA standards, a woodshop measuring 15 × 30' with a 10' ceiling would reach the exposure limit when there are two-hundredths of an ounce of wood dust in the air.

Earmuffs

5–9.

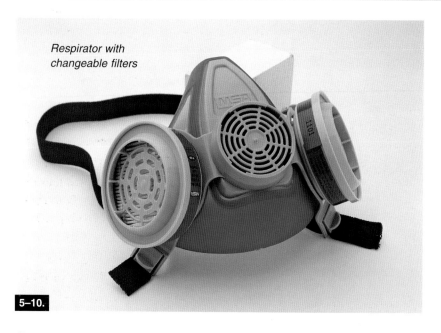

Respirator with
changeable filters

5–10.

Granted, that's not much dust. But OSHA cares about it because exposure to wood dust has been associated with a variety of adverse health effects that include dermatitis, nonallergenic and mucosal respiratory effects, allergic respiratory ailments, as well as cancer. You and your home shop don't fall under OSHA's scrutiny, but for your own well-being, you'll want to do all that you can to cut down your exposure to dust.

How to Protect Yourself Against All Dust

The highest degree of dust control consists of a three-pronged approach. Of primary importance is the installation of a dust-collection system that captures it at the source. The second prong is an air-filtration system that pulls out air-borne particles. And the third is the use of personal dust protection. Of course, most woodworkers typically start with the latter, then add the rest as their shop activity and hobby grow. So we'll look at dust masks as your first line of lung protection.

Protection for your respiratory system has two categories: nuisance protection masks and respirators. For occasional light sanding that won't generate heavy dust, you can opt for the common paper throwaways with the single elastic band and the metal nose clip. These nuisance masks run about 30 cents each. The next step up is the woven cloth or polyester mask with two elastic bands and an exhaust port. Although they're still disposable, they're NIOSH approved and can cost around $2 apiece. Costlier variations of these are washable for years of wear.

For prolonged exposure to fine dust, mist, and dangerous fumes you'll need an air-purifying respirator with changeable filters that remove specific, unhealthy contaminants from ambient air (**5–10**). These half masks, because they're made of rubber or silicone, are flexible to fit your facial contours. Several straps ensure a tight fit. This type of protection can run you $30 or more. And you'll have a choice of filters, depending on the kind of protection needed.

High-efficiency particulate air (HEPA) filters are 99.97 percent efficient in removing particles of 0.3 micrometer in diameter. A set of them may cost as much as the respirator itself. NIOSH (standard 42 CFR84) designates them as follows:

N100	Not resistant to oil particulates
R100	Resistant to oil particulates
P100	Oilproof

Here's a tip concerning air-purifying respirators: Never simply store one on a shelf. Keep it in a sealed plastic bag; otherwise it will filter the ambient air and clog the filters while it just sits there unused.

To test the respirator, put it on and cover the air outlet with one hand. Then blow gently. Anywhere your other hand can feel air escaping around the mask is where it will leak when you inhale, so readjust it for fit.

FIRST AID FOR THE SHOP

Every shop, no matter its size, needs a first-aid kit to handle medical emergencies—from a splinter to a cut. You'll find two general types of pre-assembled kits available in a variety of sizes at most drug stores. Unit-type kits contain dressings, ointments, and other needs packaged in one-treatment units of from 16 to 32 in quantity. Cabinet-type kits, on the other hand, have all the same items, but they are packaged in quantities for more than one treatment, such as a box of pressure dressings rather than just one.

And because shop accidents do happen, you should know what to do when they occur. The following first-aid procedures were developed and are advocated by the American Red Cross.

For Scrapes, Cuts, and Punctures

Step 1: Stop the bleeding by holding a sterile gauze dressing (or clean cloth) over the wound. If necessary, add more layers but don't remove the first one. Elevate the wounded part of the body above the heart; gravity should help slow down the bleeding in the wound.

Note: Shock impairs your ability to think clearly. If the bleeding or the wound is more severe than you have coped with in the past, don't hesitate to call for help.

Step 2: After bleeding is controlled, wash your hands. Then wash in and around the wound. Rinse thoroughly. Dry the wound by blotting gently with a sterile gauze pad or clean cloth. Cover with a sterile dressing.

Watch carefully for signs of infection. Consult your doctor about the need for a tetanus shot.

For Splinters

Step 1: Remove splinters in surface tissues with tweezers sterilized in boiling water or over an open flame.

Step 2: Splinters just below the skin are worked out with the tip of a sterilized needle and then removed with a tweezers. Keep an eye on the area for infection. Often, a small broken-off piece will cause the area to fester. If it is too deep to work out, consult a doctor.

For Blisters

Step 1: Small blisters are best left unbroken. If the pressure does not fade, however, wash the area with soap and water, and then use a sterilized needle to make a small hole at the base of the blister to drain it.

Step 2: Apply a sterile dressing to protect the area from further irritation. Watch for infection.

For Eye Injury, Penetrating Object

Step 1: If a splinter or other object penetrates the eye area, do not attempt to remove the object or wash the eye. Call for help.

Step 2: Cover both eyes loosely with a clean dressing. (Both eyes must be covered so the injured eye does not move.)

Step 3: Stay calm and call a doctor or hospital for instructions.

For Poisons, Swallowing

If you believe someone has swallowed a poison, such as paint remover, stain, or varnish, call the local poison control center or your doctor immediately. Check the container label for ingredients.

For Poisons, Splashed in the Eyes or on the Skin

Step 1: For the eyes, pour lukewarm water gently into the affected eye, directing it away from the other eye. Continue flushing from 2–3" above the eye for five minutes.

For the skin, remove all clothing around the area and flush with generous amounts of water for several minutes.

Step 2: Follow further instruction on the container label. Call poison control or your doctor.

From the Power Tool Institute, Inc.

SHOP SAFETY CHECK

1. Do you know exactly what you're going to do, and feel like doing it?

 Think through the operation and the moves you must make before you make them. And don't do anything with power tools if you're tired, angry, anxious, or in a hurry.

2. Is your work area clean?

 Keep your work area uncluttered, swept, and well lighted. The work space around equipment must be adequate to safely perform the job you're going to do.

3. What are you wearing?

 Don't wear loose clothing, work gloves, neck ties, rings, bracelets, or wristwatches. They can become entangled with moving parts. Tie back long hair or wear a cap.

4. Do you have the right blade or cutter or the job?

 Be sure that any blade or cutter you're going to use is clean and sharp so it will cut freely without being forced.

5. Are all power tool guards in place?

 Guards—and anti-kickback devices—also must work. Check to see that they're in good condition and in position before operating the equipment.

6. Where are the start/stop switches?

 Ensure that all the woodworking machines you'll use have working start/stop buttons or switches within your easy reach.

7. Are the power cords in good shape?

 Don't use tools with signs of power-cord damage; replace them. Only work with an extension cord that's the proper size for the job (see the chart), and route it so that it won't be underfoot.

Minimum Extension Cord Wire Gauge Size				
Nameplate Amps	Wire Gauge Size by Cord Length			
	25'	50'	100'	150'
0-6	18	16	16	14
6-10	18	16	14	12
10-12	16	16	14	12
12-16	14	12	(not recommended)	

8. Do you have the power tools properly grounded?

 Tools other than double-insulated ones come with three-wire grounding systems that must be plugged into three-hole, grounded receptacles. Never remove the grounding prong from the plug.

9. Do you know what safety equipment you need for the job?

 Around cutting tools, always wear safety glasses, goggles, or a face shield. Add a dust mask when sanding. Wear hearing protection when required. (If you can't hear someone from 3' away, the machine is too loud and hearing damage may occur.)

10. Where are the chuck keys and wrenches?

 Check that all chuck keys, adjusting wrenches, and other small tools have been removed from the machine so they won't interfere with the operation.

11. Have you checked your stock?

 Inspect the wood you're going to use for nails, loose knots, and other materials. They can be hidden "bombs" that possibly may injure you or damage your equipment.

12. Where's your pushstick?

 Keep a pushstick or pushblock within reach before beginning any cut or machining operation. And avoid getting into awkward stances where a sudden slip could cause a hand to move into the blade or cutter.

CHAPTER 6

The Workbench: Heart of the Shop

C AN'T WAIT TO START BUILDING *something? Look over the workbenches presented in this chapter. They range from one you can easily create in a single day to a clever dust-catching model that would be the centerpiece of anyone's shop. You'll find five in all, and one of them is sure to fit your needs as well as your wallet.*

On the last few pages, you'll learn how to build a nifty tool crib you can attach to your workbench. Then, there's a proven method to flatten a battered and bruised workbench top, if you have one that's already seen years of use.

So whether you're still planning your shop or ready to outfit or upgrade it, there's plenty here to keep you busy.

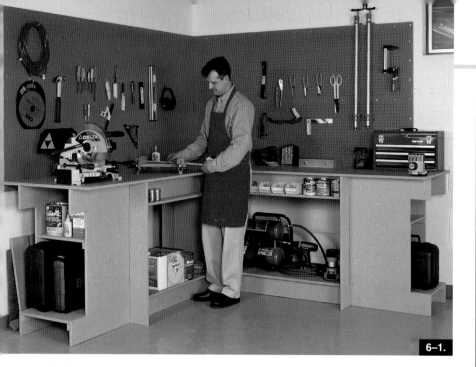

6–1.

ONE-DAY WORKBENCH

Inexpensive material and an ingenious design work together in this sturdy bench project (**6–1**) that's a cinch to build. Also see (**6–2**).

Cut Out the Bench Parts

1 Temporarily fasten 4'-long 2 × 4 extensions to a pair of sawhorses, and lay a sheet of ¾" particleboard across them. The extensions will support both the workbench parts and the cutoffs as you work. Referring to the Materials List, mark the width of the right base (A) on the sheet.

2 Measure the distance from the right edge of your portable circular saw's shoe to the right side of the blade. Lay a second sheet of particleboard on the first, offsetting it from part A's marked width the distance you measured

on your saw. Clamp the second sheet to the first, and cut part A to width, as shown in **6–3**.

3 Using the cutoff from the first part A as a straightedge, clamp it to the other sheets, offsetting it your measured distance from the parts' dimensions. Cut the rest of the parts A, B, C, D, and E to size.

4 Lay out the end and center cutouts on the right bases (A) and left bases (B), where shown on **6–4**. Drill blade-start holes and jigsaw the cutouts, or see Making Inside Cuts with a Circular Saw on *page 79*.

5 True up one edge of the four base center cutouts. Cut the upper and lower long base shelves (F) from the cutouts from the right bases (A), and

Using the edge of a second sheet to guide your saw, cut the first part A to width. Save the cutoff for a straightedge for cutting the other parts.

Corner or Straight Bench: It's Your Choice

We'll show you how to make the corner bench shown in 6–1. If you prefer a straight bench (6–2), follow the same instructions but make two right bases (A), two ends (C), and the long top (D). Omit the biscuit slots in (D). Use the cutouts from the right bases (A) to make two long base shelves (F). Fit your straight bench with optional shelves at each end by making the long and short cleats (H, I) and the long and short shelves (J, K) from the extra particleboard.

6–2.

the upper and lower short base shelves (G) from the cutouts from the left bases (B).

Form Interlocking Slots

1 Make the slot template, shown on **6–4**, from leftover particleboard. To cut the slot as accurately as possible, use a jigsaw and straightedge. Mark the outside corner, where shown.

6–3.

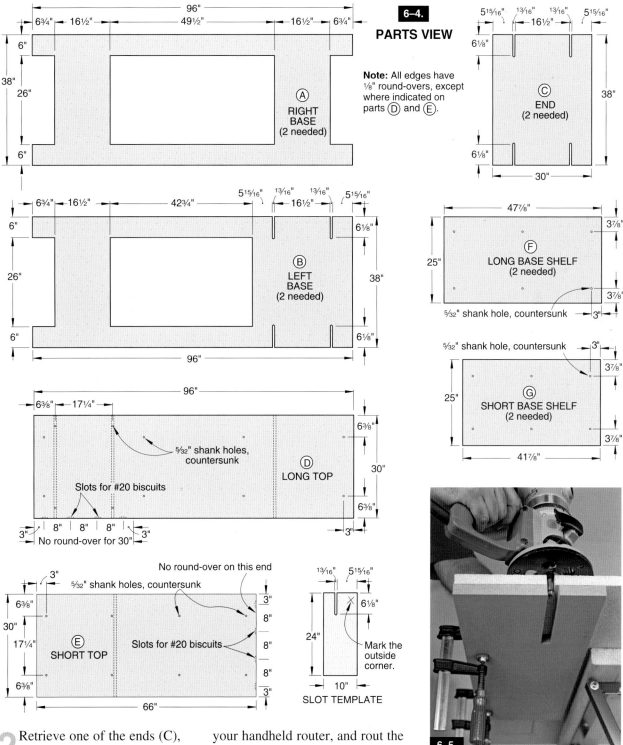

6–4.
PARTS VIEW

Note: All edges have ⅛" round-overs, except where indicated on parts (D) and (E).

96"
6¾" — 16½" — 49½" — 16½" — 6¾"
6"
38"
26"
6"
Ⓐ
RIGHT
BASE
(2 needed)

5¹⁵⁄₁₆" ¹³⁄₁₆" ¹³⁄₁₆" 5¹⁵⁄₁₆"
16½"
6⅛"
Ⓒ
END
(2 needed)
38"
6⅛"
30"

6¾" — 16½" — 42¾" — 5¹⁵⁄₁₆" ¹³⁄₁₆" ¹³⁄₁₆" 5¹⁵⁄₁₆"
16½"
6"
6⅛"
26"
Ⓑ
LEFT
BASE
(2 needed)
38"
6"
6⅛"
96"

47⅞"
3⅞"
25"
Ⓕ
LONG BASE SHELF
(2 needed)
3⅞"
⁵⁄₃₂" shank hole, countersunk — 3"

⁵⁄₃₂" shank hole, countersunk — 3"
3⅞"
25"
Ⓖ
SHORT BASE SHELF
(2 needed)
3⅞"
41⅞"

96"
6⅜" — 17¼"
⁵⁄₃₂" shank holes, countersunk
6⅜"
Slots for #20 biscuits
Ⓓ
LONG TOP
30"
6⅜"
3" 8" 8" 8" 3"
No round-over for 30" — 3"

3"
No round-over on this end
⁵⁄₃₂" shank holes, countersunk
6⅜"
3"
8"
30"
17¼"
Ⓔ
SHORT TOP
Slots for #20 biscuits
8"
8"
6⅜"
3"
66"

¹³⁄₁₆" 5¹⁵⁄₁₆"
6⅛"
24"
Mark the outside corner.
10"
SLOT TEMPLATE

2 Retrieve one of the ends (C), and clamp the template to it, aligning the template's marked corner with one of the part's corners. Chuck a ½" flush-trim bit in your handheld router, and rout the slot, as shown in **6–5**. In the same manner, rout the other three slots, where shown on **6–4**, flipping the template as needed.

6–5.

With the slot template (painted green for clarity in this photo) guiding the flush-trim bit's pilot bearing, rout the slot in the end (C).

3 Using it as a template, clamp the slotted end (C) in turn to the other part C and the two left bases (B), and rout the slots in those parts, where dimensioned on **6–4**.

Rout, Drill, Slot, Finish, and Assemble

1 Chuck a ⅛" round-over bit in your handheld router, and rout all edges of parts A, B, C, F, and G. On the long top (D) and short top (E), rout all the edges except where the two tops butt together, where indicated on **6–4** and **6–6**.

2 Drill ⁵⁄₃₂" countersunk shank holes in parts D, E, F, and G, where shown on **6–4**.

3 Adjust your biscuit cutter to center a slot in the thickness of the particleboard. Plunge mating slots in parts D and E, where shown. Biscuits inserted

without glue in the slots keep the tops aligned.

4 To protect the bench from moisture and grime, apply two coats of finish to all the parts before assembly. Apply a third coat to the top after assembly. To keep the cost down, you can use any house paint or finish you have around. (We used satin polyurethane.)

5 With the finish dry, enlist a helper and assemble the bench, as shown on **6–6** and in **6–7**. Take care in handling the bases (A, B). The 6"-wide "rails" are somewhat fragile until the bases are interlocked and supported by the ends (C) and shelves (F, G). Using the previously drilled

shank holes in parts D, E, F, and G as guides, drill pilot holes in their mating parts, and drive the screws.

Add the Optional Shelves

1 To add shelves to the ends of your workbench, where shown on **6–6**, cut the long and short cleats (H, I) and shelves (J, K) to size.

2 Rout ⅛" round-overs on the front edges of the shelves. Drill ⁵⁄₃₂" countersunk shank holes through the cleats. Locate the holes 1" in from both ends of each cleat; then evenly space two more holes between these. Apply two coats of finish to the shelves and cleats.

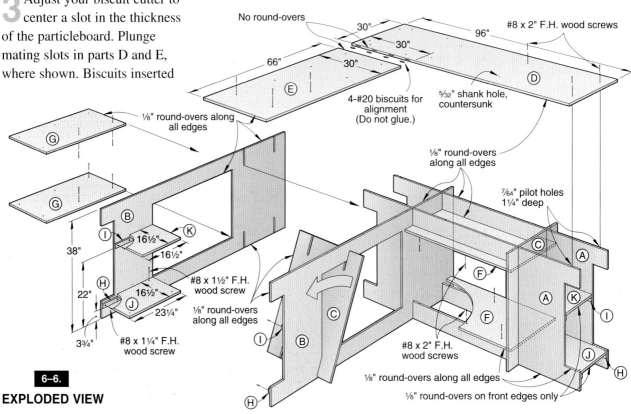

No round-overs

30"
30"
96" — #8 x 2" F.H. wood screws
66"
30"
E
4-#20 biscuits for alignment (Do not glue.)
⁵⁄₃₂" shank hole, countersunk
D

⅛" round-overs along all edges
G
G

⅛" round-overs along all edges

⁷⁄₆₄" pilot holes 1¼" deep

38"
B
I
16½"
K
C
A
22"
H
16½"
F
A
K
#8 x 1½" F.H. wood screw
J
16½"
23¼"
⅛" round-overs along all edges
A
F
I
C
3¾"
#8 x 1¼" F.H. wood screw
I
B
#8 x 2" F.H. wood screws
J
H
H
⅛" round-overs along all edges
⅛" round-overs on front edges only

6–6.
EXPLODED VIEW

6–7. | With a Helper, Assembly Is a Snap

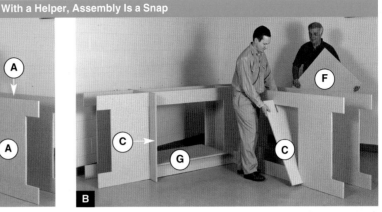

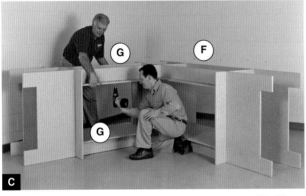

A

Interlock the slots in the left bases (B) with the right bases (A).

B

Interlock the slots in the ends (C) with the bases. Add the lower base shelves (F, G).

C

Drill pilot holes, and screw the upper base shelves (F, G) in place.

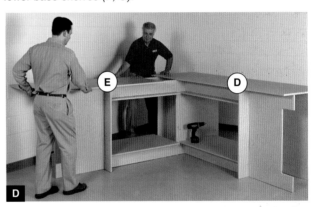

D

Position the long and short tops (D, E). Drill pilot holes, and screw them in place.

SHOP TIP

Making Inside Cuts with a Circular Saw

Get arrow-straight edges when making inside cuts like the ones needed to remove the cutouts in the workbench bases by using your circular saw, a straight-edge, and a technique called plunge cutting.

Measure the offset from the edge of your circular saw's shoe to the side of the blade. Clamp a straightedge to the part, offsetting it that distance from the cutoff's layout line. With the saw's depth adjustment loose, raise the blade above the part's surface. Starting at one inside corner, hold the shoe against the straightedge, and swivel the blade guard out of the way. Then switch on the saw, and lower the blade through the part, as shown in 6–8.

Keeping the saw's shoe against the straightedge, continue your cut, stopping at the next corner's intersecting layout line. Repeat with the other three sides of the cutout. Finish the cuts at the corners with a handsaw.

6–8.

Materials List for One-Day Workbench

PART	FINISHED SIZE			Mtl.	Qty.
	T	W	L		
A right bases	¾"	38"	96"	P	2
B left bases	¾"	38"	96"	P	2
C ends	¾"	30"	38"	P	2
D long top	¾"	30"	96"	P	1
E short top	¾"	30"	66"	P	1
F long base shelves	¾"	25"	47½"	P	2
G short base shelves	¾"	25"	41½"	P	2
H long cleats	¾"	1½"	23¼"	P	4
I short cleats	¾"	1½"	16½"	P	4
J long shelves	¾"	16½"	23¼"	P	2
K short shelves	¾"	16½"	16½"	P	2

Material Key: P = particleboard.
Supplies: #8 x 1¼" flathead wood screws; #8 x 1½" flathead wood screws; #8 x 2" flathead wood screws; #20 biscuits.
Router Bits: ½" flush-trim; ½" round-over.

Cutting Diagram

¾ x 48 x 96" Particleboard (2 needed)

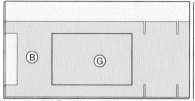

¾ x 48 x 96" Particleboard (2 needed)

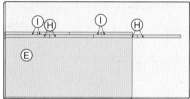

¾ x 48 x 96" Particleboard

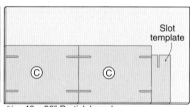

Slot template

¾ x 48 x 96" Particleboard

¾ x 48 x 96" Particleboard

3 Using the shank holes in the cleats as guides, drill pilot holes in the bases (A, B), and screw the cleats in place. Position the long cleats so the long shelves are flush with the bottom of the bases' end cutouts. Then position the short cleats for the height of the items you plan to store. The long shelves screw down as shown. Add the short shelves.

6–9.

This workbench is easy to build and extremely strong.

DURABLE WORKBENCH

Sure, this workbench (**6–9**) is simple to build, but it's also super strong. And to make this dandy you employ rugged mortise-and-tenon joinery and inexpensive lumberyard stock for the base. For the benchtop, we laminated hard maple to handle a lifetime of shop action. Bench dogs and a bench vise expand its usefulness.

Design Notes

To keep costs down on this project, we hand-picked straight-grained pine 2 × 10s for the workbench base at a local lumberyard (**6–10**). In addition, we checked each 2 × 10 for twist and bow, and chose the straightest and driest pieces available. (If you have a moisture meter, take it with you when you shop.)

After getting the stock back to the *WOOD*® magazine shop, we stickered the boards, and let them acclimate to our indoor environment for several weeks before cutting the parts (A, B, C, D) from along the edges, where shown in **6–10**. This allowed us to use the straightest grain and achieve the best results.

First, Build the Sturdy legs

1 From 1½"-thick, straight-grained pine, rip and crosscut eight pieces 3¼" wide by 33¼" long for the leg blanks. Plane the edges of the stock before ripping it to finished width to remove the rounded corners. (See Design Notes for our method of obtaining straight-grained piccces from common lumberyard 2 × 10 stock.)

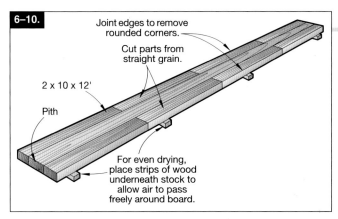

6–10.

Joint edges to remove rounded corners.

Cut parts from straight grain.

2 x 10 x 12'

Pith

For even drying, place strips of wood underneath stock to allow air to pass freely around board.

2 Cut a 3" dado ½" deep 18¼" from the bottom end of each leg blank, where shown on the Mortise Detail accompanying the End-Frame Assembly drawing in **6–12**.

3 Cut a 1 x 3 x 6" spacer to temporarily fit in the mating dadoes of two leg blanks, where shown on **6–11**. With the spacer between the pair of dadoes and the edges of the leg blanks flush, glue and clamp the pieces together. Then, remove the spacer before the glue dries. (We used pieces of scrapwood stock between the clamp jaws and legs to prevent the metal jaws from denting the softwood.) Repeat the clamping process for each leg.

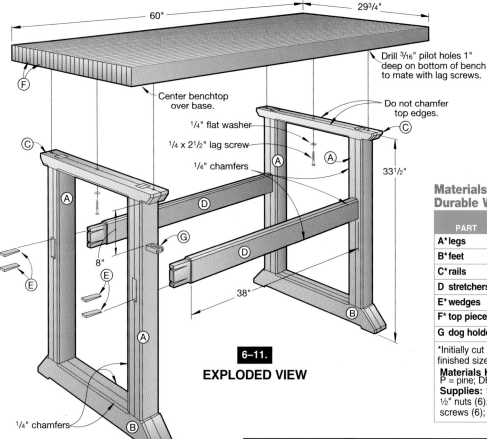

60"

29¾"

Drill 3/16" pilot holes 1" deep on bottom of bench to mate with lag screws.

F

Center benchtop over base.

Do not chamfer top edges.

¼" flat washer

¼ x 2½" lag screw

¼" chamfers

C

A

A

C

33½"

A

D

G

8"

D

E

E

B

A

38"

¼" chamfers

B

6–11.
EXPLODED VIEW

Materials List for Durable Workbench

PART	FINISHED SIZE			Mtl.	Qty.
	T	W	L		
A* legs	3"	3"	33¼"	LP	4
B* feet	3"	3¼"	29½"	LP	2
C* rails	3"	1½"	28"	LP	2
D stretchers	1½"	3½"	44"	P	2
E* wedges	⅜"	1"	3¼"	DH	8
F* top pieces	1 1/16"	2¼"	60"	M	28
G dog holder	1 1/16"	1¾"	2⅝"	M	1

*Initially cut parts oversize in width. Trim to finished size according to the instructions
Materials Key: LP = laminated pine; P = pine; DH = dark hardwood; M = maple.
Supplies: ½" all-thread rods 27¼" long (3); ½" nuts (6); ½" flat washers (6); ¼ x 2½" lag screws (6); ¼" flat washers; clear finish.

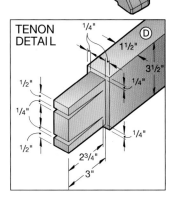

TENON DETAIL

¼"

D

1½"

3½"

¼"

½"

¼"

¼"

½"

¼"

2¾"

3"

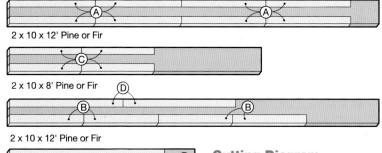

A

A

2 x 10 x 12' Pine or Fir

C

2 x 10 x 8' Pine or Fir

D

B

B

2 x 10 x 12' Pine or Fir

Cutting Diagram

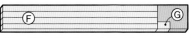

F

G

1 1/16" x 9¼" x 72" Maple (7 pieces) (6.25 bd. ft.)

Remove the clamps, scrape the glue from one edge, and plane ⅛" from the scraped edge to get it flat. Rip the opposite edge for a 3¹⁄₁₆" width.

Next, plane ¹⁄₁₆" from the cut edge to remove the saw marks and to obtain the 3" finished width. Repeat this process for each leg.

Add the Feet and Rails for a Wobble-Free Base

For the feet (B) and the rails (C), see the End-Frame Assembly in **6–12** and Parts View drawings in **6–13**, and follow the same method described to form the legs (A). Cut the pieces oversized in width, cut the dadoes, glue the pieces together with the dadoes and edges of the boards aligned, and then trim to finished width.

Clamp the two feet (B) bottom edge to bottom edge. Mark a centerpoint 3¼" from each end of the clamped-together feet. Now, use a compass to mark a ½" hole (¼" radius) at each centerpoint. Draw straight lines to connect the edges of each circle, where shown in **6–14**.

Mark a 45° cutline across the end of each leg, where shown on the Parts View drawing in **6–13**. Do the same thing to the end of the rails, where shown on the End-Frame Assembly drawing in **6–12**.

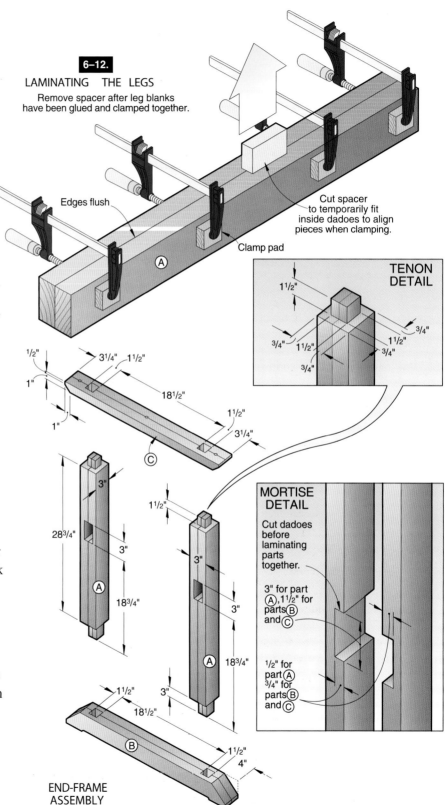

6–12.

LAMINATING THE LEGS
Remove spacer after leg blanks have been glued and clamped together.

Edges flush

Cut spacer to temporarily fit inside dadoes to align pieces when clamping.

Clamp pad

(A)

TENON DETAIL

1½"
¾" 1½"
¾" 1½"
¾"
¾"
¾"

½"
1"
3¼" 1½"
18½"
1"
1½"
3¼"
(C)

MORTISE DETAIL

Cut dadoes before laminating parts together.

3" for part (A), 1½" for parts (B) and (C)

½" for part (A)
¾" for parts (B) and (C)

3"
28¾"
(A)
18¾"

1½"
3"
(A)
3"
18¾"
(A)

1½"
3"
18½"
(B)

1½"
4"

END-FRAME ASSEMBLY

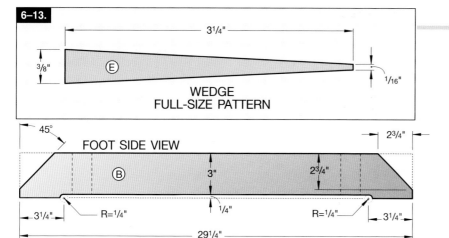

6–13.

3¼"

³⁄₈"

Ⓔ

¹⁄₁₆"

WEDGE
FULL-SIZE PATTERN

45°

FOOT SIDE VIEW

2¾"

Ⓑ

3"

2¾"

3¼"

R=¼"

¼"

R=¼"

3¼"

29¼"

PARTS VIEWS AND PATTERN

RAIL TOP VIEW

³⁄₈" holes

Ⓒ

³⁄₈" hole

1½"

1½"

1½"

12½"

12½"

1½"

28"

6–14.

Clamp the feet together, and drill a ½" hole at the marked centerpoints to form the radiused bottoms.

4 As shown in **6–14**, drill a ½" hole at each marked center-point. Remove the clamps, and bandsaw between the holes along the inside edge of the marked line. Sand to the line to remove the saw marks.

5 Using the dimensions on the End-Frame Assembly drawing in **6–12** and Parts View drawings in **6–13**, miter-cut (we used a bandsaw) both ends of each foot (B) and both ends of each rail (C). Sand smooth.

6 Drill a trio of ³⁄₈" holes in each rail (C), where shown on the Parts View drawing in **6–13**.

Assemble the Base

1 Mount an auxiliary wood fence to your miter gauge and a dado blade to your tablesaw. Cut tenons to the sizes shown on the End-Frame Assembly drawing

and accompanying Tenon Detail in **6–12**.

2 Glue and clamp each end frame together, checking for square.

3 Route ¼" chamfers along the edges of the end frames, where shown on **6–11**.

4 Cut the stretchers (D) to size. Cut a 3"-long tenon at each end of each stretcher to fit snugly through the leg mortises.

5 Rout a ¼" chamfer along the edges of the stretchers between the tenons.

6 Using the Tenon Detail accompanying the Exploded View drawing in **6–11**, bandsaw a pair of V-shaped notches in each tenon.

7 Cut eight wedges (E) to the size shown on the Parts View drawing in **6–13**. (For contrast

6–15.

Tap the hardwood wedges into the notches. After the glue dries, trim the wedges flush with the legs.

against the light pine, use a dark-colored hardwood for the wedges; we choose genuine mahogany.)

8 Glue and clamp the stretchers in place between the end frame assemblies. Inject a bit of glue in each notch, and using a mallet, tap the wedges into the notches, and check for square.

9 Being careful not to mar the surface of the leg, trim the wedges flush, as shown in **6–15**.

Build a Top That Can Take a Pounding

Note: You either can laminate your own maple top as described below or substitute a solid-core door from a local lumberyard or home center. Ask to find out if the company has any doors that customers have rejected because of mistakes in staining or cutting. You can purchase these for a fraction of their retail cost. Avoid doors returned to the store because of warpage.

1 Cut 28 pieces of 1 1/16"-thick maple (F) to 2 1/4 × 61" for the laminated top. For reference when drilling and laminating later, mark an **X** on the best (defect-free) edge (not face) of each strip.

2 Using **6–16** for reference, construct and attach a long fence to your drill press to ensure consistently spaced holes. Add a support to each end. Mark the reference marks on the fence, where shown on the drawing.

3 With the marked edge facing out, align the ends with the reference marks on the fence, and drill three 5/8" holes in 24 of the 28 benchtop pieces (F).

4 Still using the fence and your marks, drill three 1/2" holes 3/4" deep with a 1/2" hole centered inside each 1 1/2" hole in two of the remaining four pieces.

5 Glue and clamp eight of the predrilled pieces (F) face-to-face, with the edges and ends

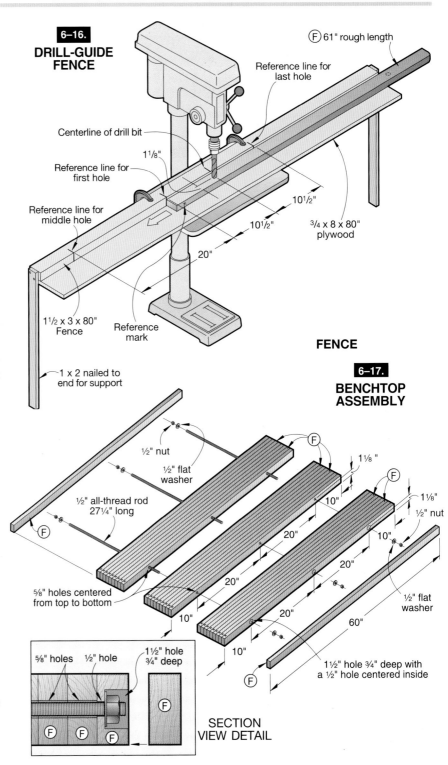

6–16.
DRILL-GUIDE FENCE

F 61" rough length
Reference line for last hole
Centerline of drill bit
1 1/8"
Reference line for first hole
Reference line for middle hole
10 1/2"
10 1/2"
3/4 × 8 × 80" plywood
20"
1 1/2 × 3 × 80" Fence
Reference mark
1 × 2 nailed to end for support

FENCE

6–17.
BENCHTOP ASSEMBLY

1/2" nut
1/2" flat washer
1/2" all-thread rod 27 1/4" long
F
5/8" holes centered from top to bottom
F
1 1/8"
F
1 1/8"
1/2" nut
10"
20"
10"
20"
20"
10"
20"
F
1/2" flat washer
60"
1 1/2" hole 3/4" deep with a 1/2" hole centered inside
F

5/8" holes   1/2" hole   1 1/2" hole 3/4" deep
F   F   F   F

SECTION VIEW DETAIL

flush, the 5/8" holes aligned, and the **X**s facing up. Next, glue and clamp two nine-piece sections together in the same manner.

Each of the nine-piece sections should have a strip with the 1 1/2" holes on one outside edge. (See **6–17** for reference.) (We found it

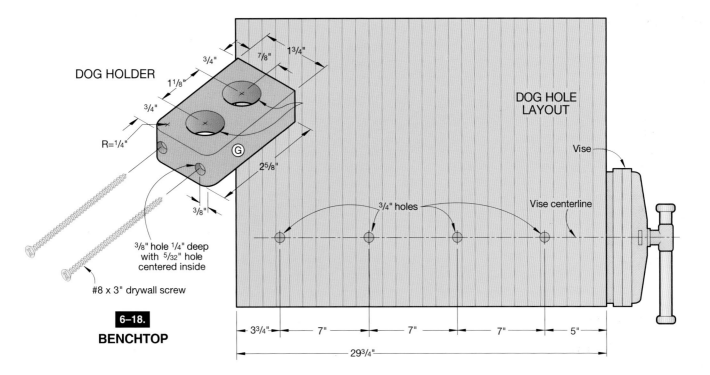

DOG HOLDER

3/4"
7/8"
1 3/4"
1 1/8"
3/4"
R=1/4"
G
2 5/8"
3/8"

3/8" hole 1/4" deep
with 5/32" hole
centered inside

#8 x 3" drywall screw

6–18.
BENCHTOP

DOG HOLE
LAYOUT

Vise

Vise centerline

3/4" holes

3 3/4"   7"   7"   7"   5"

29 3/4"

easier to laminate three sections; then glue and clamp the three sections together to form the top.) You should still have two maple strips (F) with no holes in them.

6 Using a hacksaw, cut three pieces of ½"-diameter all-thread rod to 27¼" long.

7 Spread glue on the mating edges, and clamp the three sections edge-to-edge using pipe clamps and the all-thread rod with nuts and flat washers attached. Check that the surfaces are flush. (We used a ratchet to tighten the ½" nuts on the all-thread rod.) Alternate back and forth between the clamps and the nuts on the threaded rods to achieve even clamping pressure.

8 Glue the remaining two top pieces (F) to the edges of the top assembly to hide the holes and threaded rods.

9 Scrape off the excess glue, and then belt-sand both surfaces of the benchtop flat.

10 Fit your portable circular saw with a carbide-tipped blade. Clamp a straightedge to the benchtop, and trim ½" off one end of the benchtop. Repeat at the other end.

Finishing Up

1 Finish-sand the base and top. Wipe off the dust.

2 Center the benchtop assembly on the base. Clamp the top to the base. Using the previously drilled holes in the rails (C) as guides, drill six 3/16" pilot holes 1" deep into the bottom side of the benchtop assembly. The holes in the rail are slightly oversized to allow the lag screws to move with the expansion and contraction of the benchtop. Using ¼" lag

screws and flat washers, fasten the base to the top.

3 Add the finish to all surfaces. (We applied three coats of Danish oil.)

4 Drill the mounting holes, and add a vise using the instructions provided with the vise.

5 Mark and drill ¾" dog holes through the benchtop, where shown in **6–18**.

6 If you use the same type of round bench dogs we did, mark the layout for the dog holder (G) on a piece of 1 1/16" maple. Mark the centerpoints for the dogs and mounting screws. Bore the holes for the dogs; then cut the dog holder to shape. Drill the mounting holes, sand smooth, and apply the finish. Finally, screw the dog holder to the leg nearest the vise.

6–19.

DROP-LEAF MOBILE WORKBENCH

Yes, a mobile workbench can be rock solid. This baby (6–19) not only gets out of traffic when cars come into the garage workshop, but it's super sturdy and loaded with storage. Its fold-down leaves will give you the most efficient against-the-wall storage you'll ever find.

Build a Solid Foundation

1 Cut the two base side panels (A) to the size listed in the Materials List from ¾" birch plywood.

2 Cut the base banding pieces (B, C) to size from ¾" maple. Glue the banding pieces to the panels where shown on 6–20. Sand the banding flush.

3 Using the same drawing for reference, mark the location of and cut or rout ¾" dadoes and grooves ¼" deep into the mating inside surface of each of the plywood side panels.

4 Mark and cut a ¾ × 5" taper on the bottom rear corner of each side panel. When assembling the base later, make certain that the tapers are directly across from each other. The tapers allow the casters to come in contact with

the floor without the corners of the side panels rubbing against the floor when moving the workbench around.

5 Cut the shelves (D) and dividers (E, F) to size from ¾" plywood. Now, cut a pair of notches in the upper corners of the top divider (E), where shown on 6–21.

6 Cut the banding strips (G), and band both ends of each shelf (D).

7 Glue and clamp the shelves and dividers between the side panels. Be sure to check that all items are square.

8 Cut the toe kick (H) and cleats (I) to size. Drill mounting holes through the cleats, attach them, and screw the toe kick in place so the outside face of the toe kick is recessed ¼" from the outside surface of the banding pieces (C, G).

Add the Pull-Out Handles and Casters

1 Cut the handles (J) to size. Joint or plane each to ¹¹⁄₁₆" thick. Transfer **6–31** (the full-size Handle Pattern), on *page 92*, to one end of each handle blank. Cut the contoured ends to shape. Then, rout or sand ⅛" round-overs along all of the handle edges.

2 Drill the ½" and ⅜" holes in the handle, where dimensioned on **6–31** (the Handle Pattern). Cut four ⅜" dowels to 1½" long, and glue them into the holes. The dowels act as stops when the handles are pushed and pulled in and out of the cabinet.

3 Cut the handle retainer cleats (K, L) and retainers (M) to size. Glue and screw the bottom cleats (L) to the retainers (M).

4 Turn the base cabinet upside down, and position the rigid casters flush with the inside face of the side panels (A) and outside face of the shelf banding (G). Drill pilot holes, and screw the casters in place.

Build the Workbench's Plywood Top

1 From ¾" birch plywood, cut the benchtop center panels (N) and outer leaf panels (O) to the sizes listed in the Materials List shown on *page 89*, plus ½" in length and width.

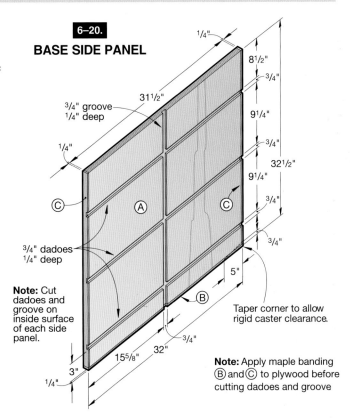

6–20.
BASE SIDE PANEL

Note: Cut dadoes and groove on inside surface of each side panel.

Taper corner to allow rigid caster clearance.

Note: Apply maple banding B and C to plywood before cutting dadoes and groove

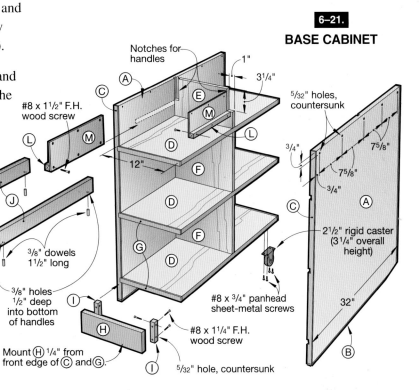

6–21.
BASE CABINET

Notches for handles

#8 x 1½" F.H. wood screw

⅛" round-overs

⅜" dowels 1½" long

⅜" holes ½" deep into bottom of handles

Mount H ¼" from front edge of C and G.

5⁄32" hole, countersunk

#8 x 1¼" F.H. wood screw

#8 x ¾" panhead sheet-metal screws

2½" rigid caster (3¼" overall height)

5⁄32" holes, countersunk

Glue and clamp the benchtop panels together in pairs with the edges and ends flush. See **6–25** for reference. (We used sliding-head-type clamps around the perimeter. Then, we drilled and countersunk holes from the bottom side and drilled screws in the middle to pull the panels tightly together.)

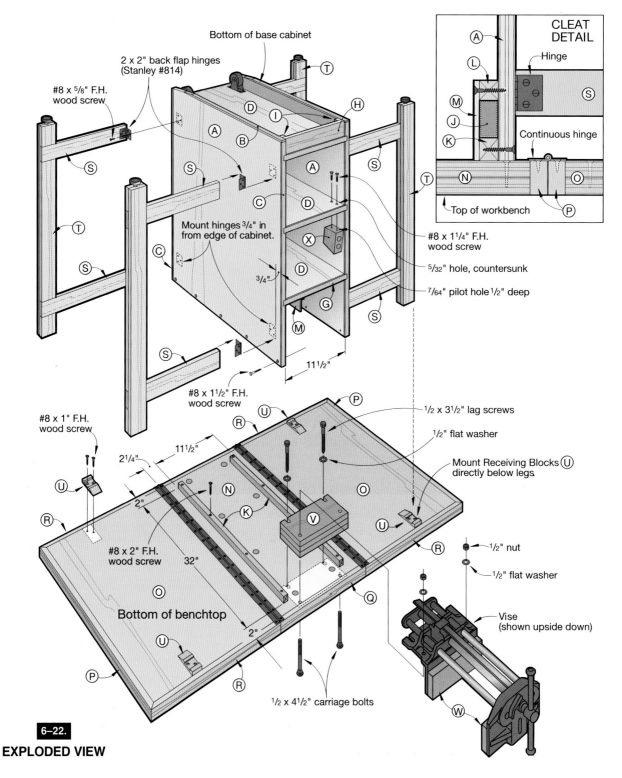

Bottom of base cabinet

2 x 2" back flap hinges
(Stanley #814)

#8 x ⁵⁄₈" F.H.
wood screw

CLEAT
DETAIL

Hinge

Continuous hinge

Top of workbench

Mount hinges ³⁄₄" in
from edge of cabinet.

3/4"

#8 x 1¹⁄₄" F.H.
wood screw

⁵⁄₃₂" hole, countersunk

⁷⁄₆₄" pilot hole ¹⁄₂" deep

11¹⁄₂"

#8 x 1¹⁄₂" F.H.
wood screw

#8 x 1" F.H.
wood screw

11¹⁄₂"

2¹⁄₄"

2"

#8 x 2" F.H.
wood screw   32"

Bottom of benchtop

2"

½ x 3¹⁄₂" lag screws

¹⁄₂" flat washer

Mount Receiving Blocks (U)
directly below legs.

¹⁄₂" nut

¹⁄₂" flat washer

Vise
(shown upside down)

½ x 4¹⁄₂" carriage bolts

6–22.
EXPLODED VIEW

3 Trim the laminated benchtop panels to the finished sizes listed in the Materials List.

4 Rip to width, then miter-cut the banding strips (P, Q, R) to size. Drill countersunk mounting holes, and screw (no glue) them to the laminated benchtops.

Cutting Diagram

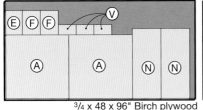

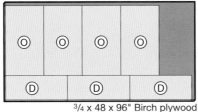

3/4 x 48 x 96" Birch plywood

3/4 x 48 x 96" Birch plywood

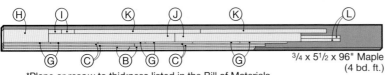

3/4 x 5 1/2 x 96" Maple (4 bd. ft.)

*Plane or resaw to thickness listed in the Bill of Materials.

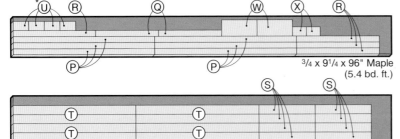

3/4 x 9 1/4 x 96" Maple (5.4 bd. ft.)

3/4 x 11 1/4 x 96" Maple (8 bd. ft.)

1/4 x 12 x 12" Plywood

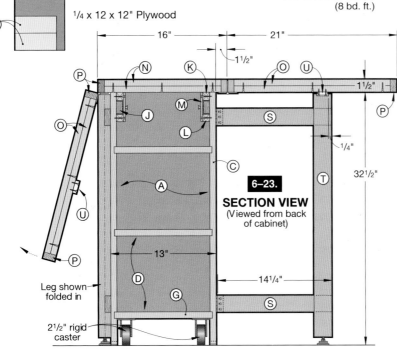

6-23.

SECTION VIEW
(Viewed from back of cabinet)

Materials List for Drop-Leaf Mobile Workbench

PART		FINISHED SIZE*		Mtl.	Qty.
	T	W	L		
BASE					
A sides	3/4"	31 1/2"	32 1/4"	BP	2
B btm. bands	1/4"	3/4"	32"	M	2
C side bands	1/4"	3/4"	32 1/4"	M	4
D shelves	3/4"	12"	31 1/2"	BP	1
E divider	3/4"	12"	8 1/2"	BP	2
F dividers	3/4"	12"	9 1/4"	BP	2
G shelf bands	1/4"	3/4"	12"	M	6
H toe kick	3/4"	3 1/4"	11 1/2"	M	1
I cleats	3/4"	1"	3 1/4"	M	2
HANDLE ASSEMBLY					
J handles	1 1/16"	2 1/8"	32"	M	2
K cleats	3/4"	1"	32"	M	2
L cleats	3/4"	3/4"	10"	M	2
M retainers	1/4"	4"	10"	P	2
BENCHTOP					
N* center	3/4"	34 1/2"	14 1/2"	BP	2
O* leaves	3/4"	34 1/2"	19 1/2"	BP	4
P banding	3/4"	1 1/2"	36"	M	6
Q banding	3/4"	1 1/2"	16"	M	2
R banding	3/4"	1 1/2"	21"	M	4
GATE LEGS AND BLOCKS					
S rails	3/4"	2 1/4"	14 1/4"	M	8
T* legs	1 1/2"	2 1/4"	31 1/4"	LM	8
U rec. blocks	1/2"	2 1/4"	3 7/8"	M	4
VISE MOUNT AND JAWS					
V spacer	2 1/4"	5"	9"	LBP	1
W vise jaws	3/4"	4"	9"	M	2
X dog holder block	1 1/2"	2 1/4"	3 1/2"	LM	1

*Initially cut parts oversized. Then, trim each to finished size according to the instructions. Note that we measure length with the grain.

Materials Key: BP = birch plywood; M = maple; P = plywood; LM = laminated maple; LBP = laminated birch plywood.
Supplies: #8 x 1/2", #8 x 1", #8 x 1 1/4", #8 x 1 1/2", and #8 x 2" wood screws; 2 x 2" heavy-duty hinges (8); 2 sections of 1 1/2" continuous hinge 36" long with #6 x 3/4" flathead wood screws; 3/8" dowel stock; 1/2 x 4 1/2" carriage bolts with flat washers and nuts (2); 2 1/2" (3 1/4" overall height) rigid casters (2) with #8 x 3/4" panhead sheet-metal screws (8); 3/8" T nuts (4); 3/8" nuts (4); adjustable floor glides (4); clear finish; paint.
Buying Guide: Record 52 1/2 ED vise, 9" jaw width, 13" jaw opening, with quick-release mechanism.

Add the Gate-Leg Assemblies Next

1 Using the dimensions in the Materials List and on **6–24**, cut the rails (S) to size.

2 To form the legs (T), cut eight pieces of ¾" maple to 2½" wide by 32" long. Glue four pairs of two pieces each face-to-face, with the edges and ends flush. After the glue dries, scrape it from one edge of each leg. Next, joint or plane that edge flat. Rip the opposite edge of each leg on your tablesaw for a 2¼" finished width. Cross-cut both ends of each leg to a 31¼" finished length.

3 Using a dado blade in your tablesaw or radial-arm saw, cut a pair of 2¼" dadoes ¾" deep, where shown on **6–24**.

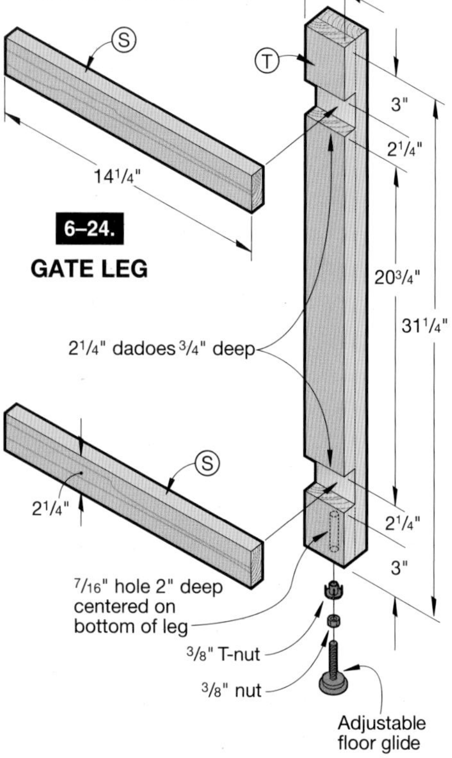
2¼"
S
T
3"
2¼"
14¼"
20¾"
31¼"
2¼" dadoes ¾" deep
S
2¼"
2¼"
3"
7/16" hole 2" deep centered on bottom of leg
3/8" T-nut
3/8" nut
Adjustable floor glide

6–24.
GATE LEG

6–25. BENCHTOP

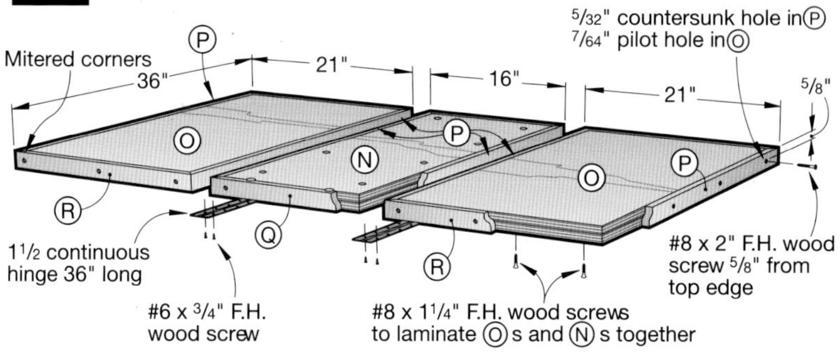
Mitered corners
P
36"
21"
16"
21"
5/32" countersunk hole in P
7/64" pilot hole in O
5/8"
O
N
P
O
P
R
Q
1½ continuous hinge 36" long
#6 x 3/4" F.H. wood screw
R
#8 x 1¼" F.H. wood screws to laminate O s and N s together
#8 x 2" F.H. wood screw 5/8" from top edge

6–26. DOG-HOLE LAYOUT

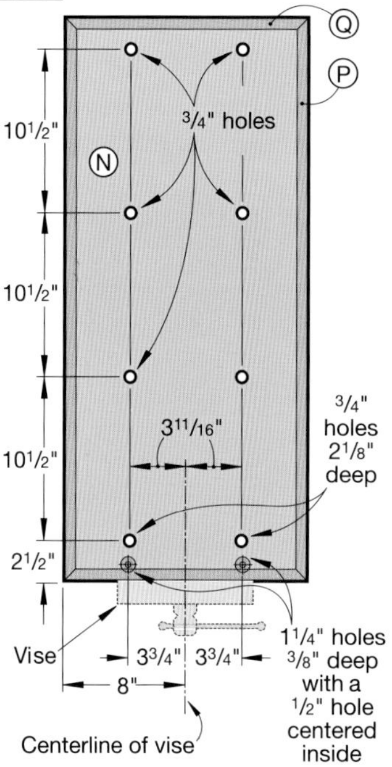
Q
P
10½"
3/4" holes
N
10½"
10½"
3 11/16"
3/4" holes 2 1/8" deep
2½"
Vise
3¾"
3¾"
8"
1¼" holes 3/8" deep with a ½" hole centered inside
Centerline of vise

4 Checking for square, glue a pair of rails (S) to each of the legs (T).

5 On the bottom of each leg, draw diagonals to find the center. Drill a 7⁄16" hole 2" deep at each centerpoint. (We used a brad-point bit to prevent wander.)

6 Tap a ⅜" T-nut into the bottom of each leg, centered over the holes you just drilled. Pry the nuts loose. Mix about an ounce of epoxy, and use an ear swab to coat the portion of each leg bottom and hole that comes into contact with the T-nuts. Tap the T-nuts back into position. When the epoxy has cured, thread a ⅜" adjustable floor glide into each of the T-nuts.

7 For locking the leg tops to the bottom surface of the benchtop leaves later, cut the receiver blocks (U) to the size listed in the Materials List and the shape shown on **6–28**. Drill the mounting holes. Set the blocks aside for now; we'll add them later.

Now Add the Spacer Block and the Vise

1 Using **6–22** and **6–27** for reference, laminate three pieces of ¾" plywood to form the spacer block (V). (We used a woodworker's vise with a 9" jaw width. The size of your spacer and its location may vary with different brands of vises.)

2 Mark the centerpoints, and bore ½" and ⅝" holes through the block, where marked on **6–27**. Cut from the edge into the ⅝" holes to form ⅝"-wide slots, where shown.

9"
1 1/16" **6 7/8"** **1 1/16"**
7/8"
5/8" slots 7/8" long

5/8" hole

6–27.
SPACER BLOCK
Ⓥ

5"

3/4" **7 1/2"** 3/4"
1/2" holes

1"

Final Assembly

1 Position a pair of saw horses about 4' apart. Set a pair of 2 x 4s on the sawhorses, and center the benchtop center section upside down on the 2 x 4s. Position the benchtop leaves next to the center section, also upside down. Using two 36"-long sections of 1 1/2" continuous (piano) hinge, drill pilot holes and screw the leaves to the center section.

2 Position the base (also upside down) on the bottom side of the benchtop center section, where shown in **6–29**, centering the base from side-to-side and end-to-end. Place the long cleats (K) on the inside of the side panels (A). (See **6–22** for reference.) Next, clamp the cleats in place to the benchtop. Lift the base off the benchtop, and drive screws through the previously drilled holes in the cleats and into the bottom side of the benchtop.

3 Reposition the base on the benchtop. Drill and countersink mounting holes through the outside face of the base cabinet into the cleats (K). Next, screw the base to the benchtop cleats.

3 Clamp the spacer block (V) to the bottom side of the benchtop center section (N), where shown on **6–22**. The outside edge of the block should be flush with the outside edge of the banding strip (Q).

4 Position the vise on the spacer block to determine the block's exact location, and adjust if necessary. With the vise clamped securely to the benchtop, use the existing holes as guides to drill mounting holes through your benchtop. (We clamped scrap stock to the top surface of our benchtop top to prevent chip-out.)

Flip the center benchtop section (minus the vise) over, and counterbore the mounting holes.

5 Cut a pair of wood jaws (W) to size, and attach them to the metal vise jaws.

6 Mark the dog-hole center-points on the center benchtop section, where shown on **6–26**. Drill the holes. Now you're ready for the final assembly.

4 Drill the mounting holes and attach the hinges to the ends of the leg rails. Attach the hinge/leg assemblies to the outside face of the side panels. (For proper spacing between the top ends of the legs and the benchtop bottoms, we slipped a receiver block between the two.)

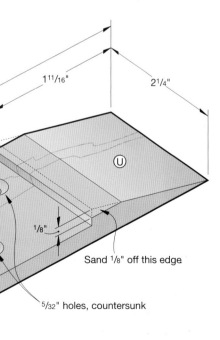

1 11/16" 2 1/4"
3 7/8"
1 9/16"
5/8"
Ⓤ
1/8"
Sand 1/8" off this edge
1/4"
1/2"
5/32" holes, countersunk

6–28.
RECEIVER BLOCK

5 Swing the legs 90° from the base sides, and slide the receiver blocks (U) in place under the legs. Use a pencil to trace their location on the benchtop bottom surface. Swing the legs against the base, and screw the receiver blocks in place.

6 Position the handles (upside down) on the cleats (K). Screw the handle retainer pieces (L, M) in place.

7 Remove all hardware and remove the legs from the base and the benchtop assemblies from the base.

8 Cut the dog holder (X) (**6–30**) to shape, and drill a pair of holes in it to house your dogs when not in use.

9 Sand all assemblies. Apply a clear finish to the benchtops and legs (polyurethane works well). Prime and paint the base and dog holder.

10 Attach the benchtops, legs, and dog holder to the base. Position the handles, and screw the handle retainers (L, M) to the base, where shown on **6–21**. Mount the vise and casters.

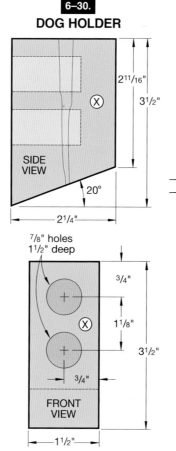

6–30.
DOG HOLDER

SIDE VIEW

2 11/16"
3 1/2"
20°
2 1/4"
Ⓧ

7/8" holes 1 1/2" deep
3/4"
1 1/8"
3 1/2"
3/4"
1 1/2"
Ⓧ

FRONT VIEW

6–29.

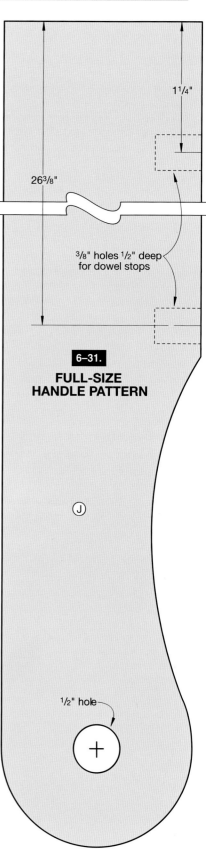

1 1/4"
26 3/8"
3/8" holes 1/2" deep for dowel stops

6–31.
FULL-SIZE HANDLE PATTERN

Ⓙ

1/2" hole

TOUGH-STUFF WORKBENCH

Here's a no-nonsense workbench (6–32) that incorporates the essentials: simple base cabinets with deep enclosed drawer and shelf storage and a flat, stable 3 × 8' top offering plenty of room for a beefy woodworking vise. Plus, it's a cinch to build.

Build Two Basic Cabinets

1 For the workbench base, start by building two basic floor cabinets with legs, using the instructions for Build the Heart of a Floor-Cabinet System: The Basic Cabinet in Chapter 7 (*page 136*). Hinge the doors so they open facing each other, as shown on **6–33**.

2 Cut the upper front rails (A) to the size shown on the Materials List. With a dado blade in your tablesaw, cut ½ × ¾" notches in the rails' ends. Clamp the rails in place, and drill angled countersunk screw holes through the rails and into the cabinets' upper side rails. Drive the screws.

Add Drawers and Shelves

1 To add drawers to one cabinet, make two pairs of drawer cleats and two drawers, referring to the drawings and instructions on *page 143* of Chapter 7, "Make Drawers Quickly." Use the

6–32.

Materials List on *page 144* of Chapter 7 only as a guide and build the cleats and drawers to the size needed for your unit. Fasten the cleats to the cabinet, where shown on **6–32**.

2 Install the drawer slides' drawer and cabinet members, where shown. Drill pilot holes, and drive the screws.

3 Position shelf standards in the other cabinet, where shown on the Shelf Standard Detail drawing in **6–33**. Drill pilot holes, and fasten them with #5 × ⅝" flathead wood screws.

4 Cut the shelf (B) and shelf edges (C) to the sizes listed. Cut slots for #20 biscuits, where

shown on **6–33**. Glue, biscuit, and clamp the edges to the shelf. Sand the edges flush with the shelf's top surface, and rout ⅛" round-overs on the front edges.

Make the Benchtop

1 Cut two 34½ × 94½" pieces of ¼" MDF for the top (D). Spread glue using a foam or short-nap roller, and clamp the two pieces together, keeping their ends and edges flush. Drill screw holes, and drive the screws.

2 Cut the end bands (E) to size, and glue and clamp them to the top (D). For help with this operation, see Making Do with Short Clamps on *page 95*. Then cut the side bands (F) to size.

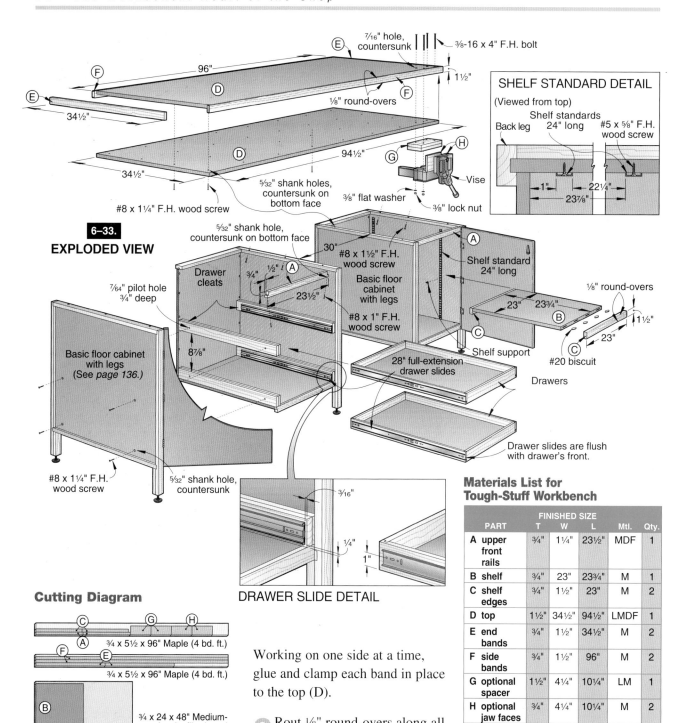

SHELF STANDARD DETAIL
(Viewed from top)

6-33.
EXPLODED VIEW

DRAWER SLIDE DETAIL

Cutting Diagram

Materials List for Tough-Stuff Workbench

PART	FINISHED SIZE			Mtl.	Qty.
	T	W	L		
A upper front rails	3/4"	1 1/4"	23 1/2"	MDF	1
B shelf	3/4"	23"	23 3/4"	M	1
C shelf edges	3/4"	1 1/2"	23"	M	2
D top	1 1/2"	34 1/2"	94 1/2"	LMDF	1
E end bands	3/4"	1 1/2"	34 1/2"	M	2
F side bands	3/4"	1 1/2"	96"	M	2
G optional spacer	1 1/2"	4 1/4"	10 1/4"	LM	1
H optional jaw faces	3/4"	4 1/4"	10 1/4"	M	2

Materials Key: M = maple; MDF = medium-density fiberboard; LMDF = laminated medium-density fiberboard; LM = laminated maple.
Supplies: #8 x 1", #8 x 1 1/4", #8 x 1 1/2", and #5 x 1/2" flathead wood screws; 24"-long shelf standards (4); shelf supports (4); #20 biscuits. Fasteners as necessary to mount your vise.
Blades and bits: Stack dado set; 1/2" round-over router bit.

Working on one side at a time, glue and clamp each band in place to the top (D).

3 Rout 1/8" round-overs along all the top's ends and edges, and sand it to 180 grit. For an easily renewed finish, apply two coats of penetrating oil finish, letting each one dry without wiping it down. Lightly sand the second coat with

SHOP TIP

Making Do With Short Clamps

To glue the end bands (E) to the top (D), you might think you'll need either 8'-long bar clamps or special edge-gluing clamps. But here's an easy method that uses the bar clamps you already have.

Screw a cleat to the bottom face of the top (D) within reach of your longest clamps and apply moderate pressure to hold the end band (E) in place, flush with the top's ends and top and bottom surfaces. Hook your long clamps on the cleat, as shown, and apply pressure to the band.

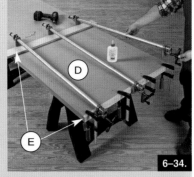

6–34.

220-grit sandpaper, and then apply a third coat. Let this coat dry for five minutes, and then wipe it with a soft cloth.

4 Remove all the hardware except the T-nuts. Apply three coats of satin polyurethane to the shelf and doors, and two coats to all the other parts, sanding between coats with 180-grit sandpaper. To seal the doors' edges, double-coat them as you apply each coat of finish.

Now Put It All Together

1 Position the cabinets 30" apart. When placing the workbench against a wall, push the cabinets as close to it as possible. Level them individually and in relation to each other. Lay the benchtop on the cabinets, leaving (for a right-hander) 3" overhanging the left-hand one. This leaves 12" overhanging the right-hand cabinet for mounting a vise. If used against a wall, push the top tight against it. Placed away from the wall,

center the top front-to-back on the cabinets.

2 Clamp the top to the cabinets, and drill angled countersunk screw holes through the cabinets' upper side rails and into the benchtop, where shown on **6–33**. Drive the screws.

3 Reinstall all the hardware and the drawers. Clip in the shelf supports, and install the shelf.

4 Referring to the instructions that come with it, install your vise. Some vises may require a thicker benchtop for mounting. If yours does, install a spacer (G) to the underside of the benchtop, as shown on **6–33**. Then bolt the vise in place. To fit your vise with wood jaw faces that are flush with the benchtop, cut the optional jaw faces (H) to size. Secure them to the vise jaws with #12 × ¾" flathead wood screws and ¼–20 × 1" flathead bolts.

FULL-SERVICE WORKBENCH WITH SURPRISES

A great workbench is the heart of a woodworking shop, and this one ticks with a different beat (**6–35** to **6–37**). On the following pages you'll learn how to build the end cabinets, drawers, and laminated benchtop. Then come the surprises! How about a center cabinet with a lift-up router table? Or, the same cabinet with a rotating top for mounting your benchtop scrollsaw? You'll see it all.

Start with the Two Basic End Cabinets

1 Cut the cabinet tops and bottoms (A), sides (B), backs (C), and shelf (D) to the sizes listed in the Materials List from ¾" medium-density fiberboard (MDF). As noted on the Cutting Diagram, MDF measures 1" wider and longer than regular 4 × 8' sheet goods. Birch plywood would also work fine for these pieces.

2 Cut the rabbets and dadoes in the sides (B), where shown on **6–38**. Note that only the left-hand cabinet is dadoed for the shelf (D).

3 Glue and clamp the cabinets (A/B/C/D) together in the configuration shown on **6–38**, checking for square.

4 Cut the base front and back (E), sides (F), and cleats (G) to size.

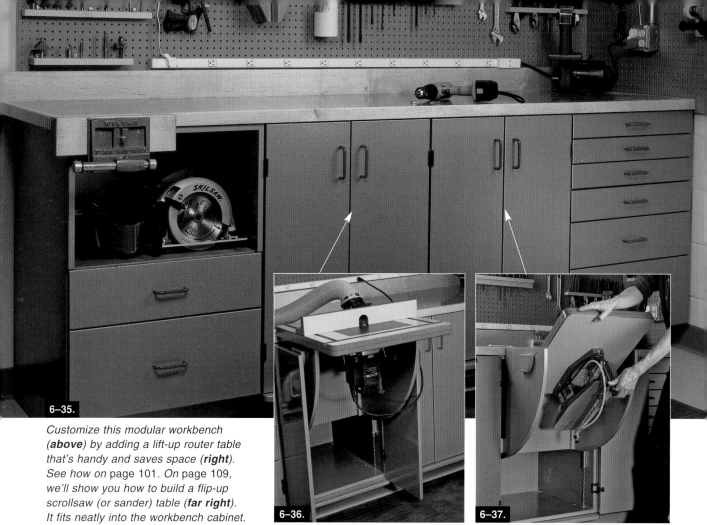

6–35.

*Customize this modular workbench (**above**) by adding a lift-up router table that's handy and saves space (**right**). See how on page 101. On page 109, we'll show you how to build a flip-up scrollsaw (or sander) table (**far right**). It fits neatly into the workbench cabinet.*

6–36.

6–37.

5 Cut the rabbets in parts E and F, where shown on **6–38**. Glue and screw the bases (E, F, G) together as shown. Fill the counterbored screw holes with wood filler, and later sand it smooth with 220-grit sandpaper.

6 Cut the left-hand cabinet front banding strips (H, I) to size. The thickness of the strips needs to be the same as the thickness of the MDF or plywood. Glue the strips to the front of the cabinet, where shown on **6–38**.

7 Use **6–39** and **6–40** for reference to drill the mounting holes and screw the bases to the bottoms of the cabinets.

Note: the back edge of the cabinet overhangs the base by ¾".

Add the Drawers for the Cabinets

1 From ½" stock (we used birch plywood), cut all the drawer fronts (J, L, N, P, R) and sides (K, M, O, Q, S) to the sizes listed in the Materials List. Cut the drawer bottoms (T) from ¼" hardboard and false fronts (U, V, W, X) from ¾" MDF.

2 Cut rabbets in the drawer sides, as shown in **6–41**.

3 Drill countersunk mounting holes, and glue and screw each drawer (except the drawer

fronts) together, measuring diagonally to check for square.

4 Using the instructions supplied with the drawer guides and the dimensions on **6–39** and **6–40**, screw the guides to the bottom side edge of each drawer. Make sure each guide is flush with the front of the drawer box. Mount the mating guide hardware to the inside of the end cabinets.

5 Drill counterbored shank holes on the inside face of the false drawer fronts. Attach a wire pull to each false front. Fit the drawers in place in the cabinets. Apply two pieces of double-faced tape to the back side of each false

Cutting Diagram

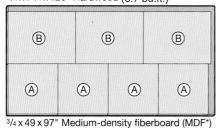

³/₄ x 5¹/₂ x 72" Hardwood (3 bd.ft.)

³/₄ x 7¹/₄ x 120" Hardwood (6.7 bd.ft.)

³/₄ x 49 x 97" Medium-density fiberboard (MDF*)

³/₄ x 49 x 97" Medium-density fiberboard (MDF*)

¹/₄ x 48 x 96" Hardboard

³/₄ x 49 x 97" Medium-density fiberboard (MDF*)

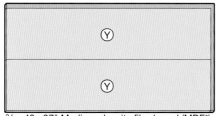

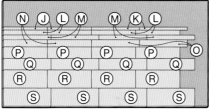

¹/₂ x 48 x 96" Birch plywood

³/₄ x 49 x 48¹/₂"
Medium-density
fiberboard (MDF*)

*Standard 4x8' sheet
of particleboard or
plywood will work
in place of MDF.

Materials List for Full-Service Workbench

PART		T	FINISHED SIZE W	L	Mtl.	Qty
CARCASES AND BASES						
A	top & bottom	³/₄"	23"	23¹/₂"	MF	4
B	sides	³/₄"	23³/₄"	31³/₄"	MF	4
C	back	³/₄"	23¹/₂"	31³/₄"	MF	1
D	shelf	³/₄"	22³/₄"	23"	MF	1
E	front & back	³/₄"	4"	23¹/₂"	MF	4
F	sides	³/₄"	4"	19"	MF	4
G	cleats	³/₄"	4"	23¹/₂"	MF	4
H	banding	³/₄"	³/₄"	22¹/₂"	MF	1
I	banding	³/₄"	³/₄"	15"	H	2
DRAWER 1						
J	front & back	¹/₂"	1³/₈"	20³/₄"	BP	2
K	sides	¹/₂"	1³/₈"	22¹/₂"	BP	4
DRAWERS 2 AND 3						
L	fronts and backs	¹/₂"	2¹/₄"	20³/₄"	BP	4
M	sides	¹/₂"	2¹/₄"	22¹/₂"	BP	4
DRAWER 4						
N	front and back	¹/₂"	3¹/₄"	20³/₄"	BP	2
O	sides	¹/₂"	3¹/₄"	22¹/₂"	BP	2
DRAWER 5						
P	front & back	¹/₂"	4¹/₂"	20³/₄"	BP	4
Q	sides	¹/₂"	4¹/₂"	22¹/₂"	BP	4
DRAWER 6						
R	front & back	¹/₂"	8¹/₄"	20³/₄"	BP	4
S	sides	¹/₂"	8¹/₄"	22¹/₂"	BP	4
DRAWER BOTTOMS AND FRONTS						
T	bottoms	¹/₄"	21¹/₂"	22¹/₂"	HB	8
U	drawers 1, 2, & 3	³/₄"	3³/₈"	24"	MF	3
V	drawer 4	³/₄"	4³/₈"	24"	MF	3
W	drawer 5	³/₄"	6³/₈"	24"	MF	2
X	drawer 6	³/₄"	10"	24"	MF	2
BENCHTOP						
Y	top	1¹/₂"	23³/₈"	96"	LMF	1
Z	banding	³/₄"	1¹/₂"	97¹/₂"	H	1
AA	banding	³/₄"	1¹/₂"	24¹/₈"	H	2
BB	backboard	³/₄"	5¹/₄"	97¹/₂"	H	1
CC	spacer	³/₄"	4¹/₂"	7"	H	1

Materials Key: MF = medium-density fiberboard; H = hardwood; BP = birch plywood; HB = hardboard; LMF = laminated medium-density fiberboard.
Supplies: #8 x 1", #8 x 1¹/₄", #8 x 1¹/₂", #8 x 2" flathead wood screws; 3³/₄" pulls (8); 22" bottom-mount drawer guides (8 pair); primer; paint; polyurethane. For the bench vise, ³/₈ x 3" lag screws (4) with flat washers.
Buying Guide: Pivot-jaw woodworking vise.

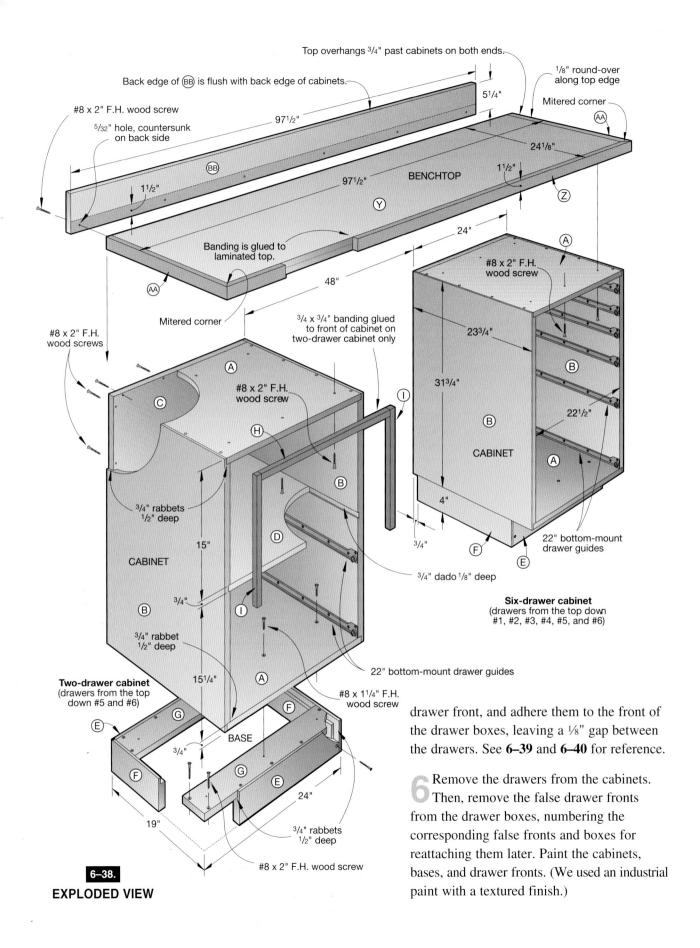

Top overhangs ³/₄" past cabinets on both ends.

Back edge of (BB) is flush with back edge of cabinets.

¹/₈" round-over along top edge

Mitered corner

#8 x 2" F.H. wood screw

⁵/₃₂" hole, countersunk on back side

5¹/₄"

97¹/₂"

(AA)

24¹/₈"

(BB)

97¹/₂"

BENCHTOP

1¹/₂"

(Z)

1¹/₂"

(Y)

Banding is glued to laminated top.

24"

#8 x 2" F.H. wood screw

(A)

48"

(AA)

23³/₄"

Mitered corner

³/₄ x ³/₄" banding glued to front of cabinet on two-drawer cabinet only

31³/₄"

(B)

#8 x 2" F.H. wood screws

(A)

(I)

22¹/₂"

(C)

#8 x 2" F.H. wood screw

(B)

(H)

(B)

(A)

(B)

CABINET

³/₄" rabbets ¹/₂" deep

15"

4"

³/₄"

(D)

³/₄"

(F)

(E)

CABINET

³/₄" dado ¹/₈" deep

22" bottom-mount drawer guides

(B)

(I)

Six-drawer cabinet
(drawers from the top down #1, #2, #3, #4, #5, and #6)

³/₄" rabbet ¹/₂" deep

15¹/₄"

(A)

22" bottom-mount drawer guides

Two-drawer cabinet
(drawers from the top down #5 and #6)

#8 x 1¹/₄" F.H. wood screw

(E)

(G)

(F)

BASE

³/₄"

drawer front, and adhere them to the front of the drawer boxes, leaving a ⅛" gap between the drawers. See **6–39** and **6–40** for reference.

(F)

(G)

(E)

24"

6 Remove the drawers from the cabinets. Then, remove the false drawer fronts from the drawer boxes, numbering the corresponding false fronts and boxes for reattaching them later. Paint the cabinets, bases, and drawer fronts. (We used an industrial paint with a textured finish.)

19"

³/₄" rabbets ¹/₂" deep

6–38.
EXPLODED VIEW

#8 x 2" F.H. wood screw

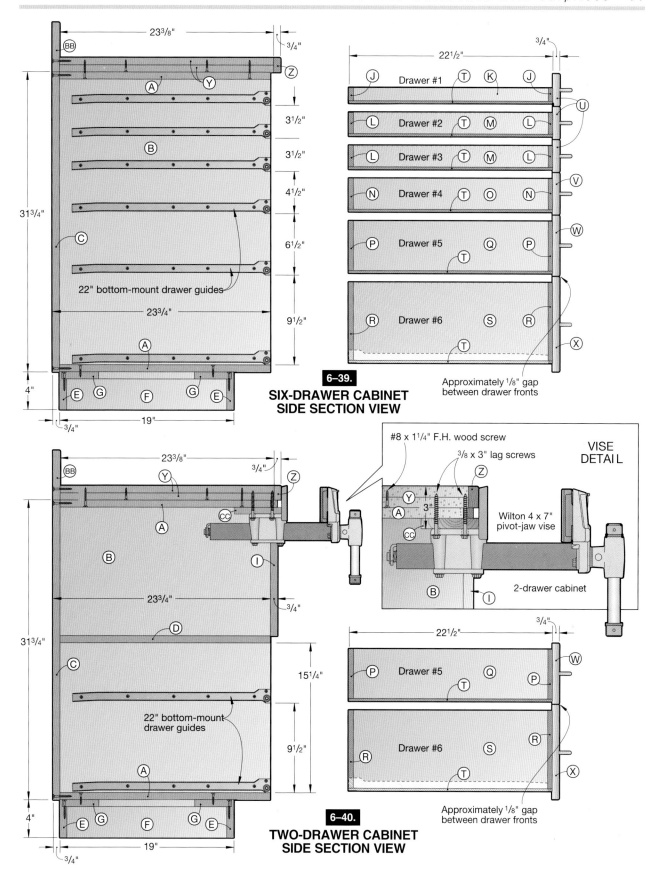

6–39.
SIX-DRAWER CABINET
SIDE SECTION VIEW

22" bottom-mount drawer guides

Drawer #1
Drawer #2
Drawer #3
Drawer #4
Drawer #5
Drawer #6

Approximately 1/8" gap
between drawer fronts

#8 x 1 1/4" F.H. wood screw
3/8 x 3" lag screws
VISE
DETAIL

Wilton 4 x 7"
pivot-jaw vise

2-drawer cabinet

6–40.
TWO-DRAWER CABINET
SIDE SECTION VIEW

22" bottom-mount
drawer guides

Drawer #5
Drawer #6

Approximately 1/8" gap
between drawer fronts

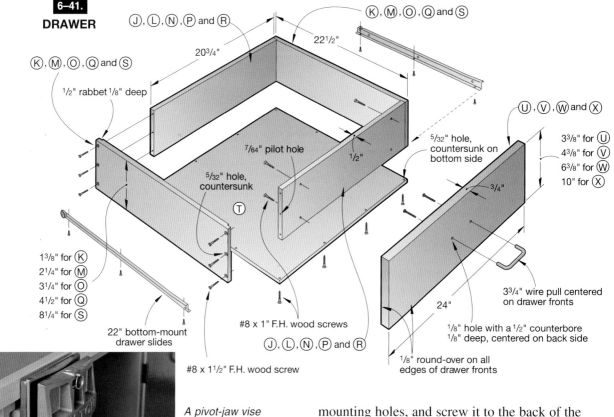

6–41.
DRAWER

(J), (L), (N), (P) and (R)

(K), (M), (O), (Q) and (S)

20³⁄₄"

22¹⁄₂"

(K), (M), (O), (Q) and (S)

¹⁄₂" rabbet ¹⁄₈" deep

⁷⁄₆₄" pilot hole

¹⁄₂"

⁵⁄₃₂" hole, countersunk

(T)

⁵⁄₃₂" hole, countersunk on bottom side

(U), (V), (W) and (X)

3³⁄₈" for (U)
4³⁄₈" for (V)
6³⁄₈" for (W)
10" for (X)

³⁄₄"

1³⁄₈" for (K)
2¹⁄₄" for (M)
3¹⁄₄" for (O)
4¹⁄₂" for (Q)
8¹⁄₄" for (S)

3³⁄₄" wire pull centered on drawer fronts

24"

22" bottom-mount drawer slides

#8 x 1" F.H. wood screws
(J), (L), (N), (P) and (R)

¹⁄₈" hole with a ¹⁄₂" counterbore ¹⁄₈" deep, centered on back side

¹⁄₈" round-over on all edges of drawer fronts

#8 x 1¹⁄₂" F.H. wood screw

A pivot-jaw vise mounts easily to the bottom side of the benchtop.

6–42.

Laminate a Benchtop to Take Plenty of Abuse

1 Cut two pieces of ³⁄₄" MDF for the top (Y) to the size listed in the Materials List plus 1" in length and width. Spread glue over the mating surfaces, and clamp the two pieces together, keeping the edges and ends flush. (Because it's difficult to apply clamping pressure in the middle of this large panel, we drilled countersunk mounting holes, and drove 1¹⁄₄" flathead wood screws from the bottom side of the panel to hold the two sheets together.)

2 Trim the 1¹⁄₂"-thick top (Y) to the finished size (23³⁄₈ x 96") listed in the Materials List.

3 Cut and miter-cut the banding pieces (Z, AA) to size, and glue them to the top, where shown on **6–38**. Next, cut the backboard (BB) to size. Drill

mounting holes, and screw it to the back of the laminated top, where shown on the drawing.

4 Remove the guides from the drawers. Apply a clear finish of your choice (we used polyurethane) to the benchtop and drawers. Later, reattach the painted false drawer fronts to the clear-coated drawers. Then reattach the drawer guides.

Add the Vise for Plenty of Holding Power

1 With the aid of a helper, position the end cabinets and laminated top in position in your shop. Working from inside the cabinets, drill mounting holes and secure the laminated top to the top of the cabinets with #8 x 2" wood screws.

2 To add a vise of your choice, cut the spacer (CC) to size, and attach it to the bottom of part A, where shown on the Vise Detail on **6–40**. Using the same drawing for reference, drill the mounting holes, and use lag screws to secure the vise to the workbench assembly (**6–42**).

❖

LIFT-UP ROUTER TABLE

On previous pages, we showed you how to build the end cabinets and the sturdy laminated top for the cabinet-style workbench. Now, you'll find it easy to add another pair of cabinets—one with a lift-up router table as described here (**6–43**), and another following it with a flip-over top that accommodates a benchtop power tool.

Begin with the Basic Carcase Assembly

1. Rip and crosscut the carcase top and bottom (A), sides (B), back (C), and doors (D) to the sizes listed in the Materials List from ¾" medium-density fiberboard (MDF) or birch plywood.

2. Mark the locations, and cut or rout the rabbets on the inside face of the sides (B), where dimensioned on **6–45** and **6–46**.

3. Drill the ½" chain-access holes through the bottom (A), where dimensioned on **6–46**.

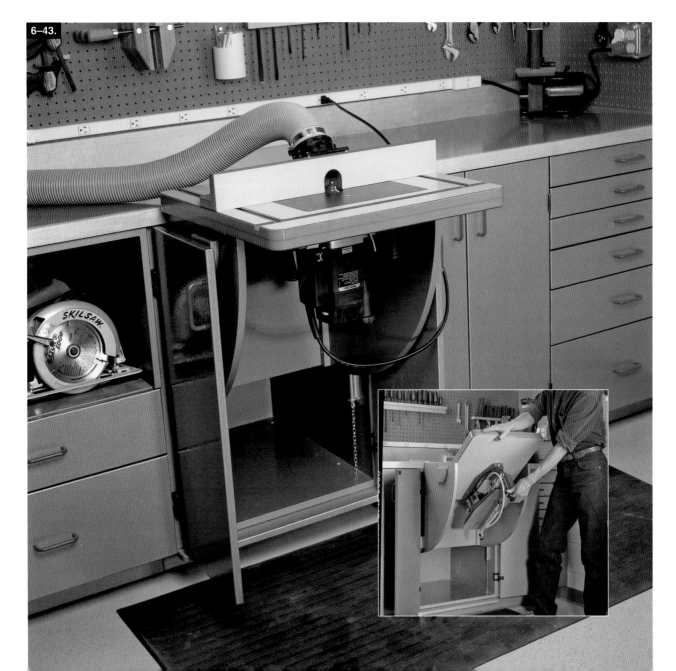

6–43.

6–44.

SIDE SECTION VIEW

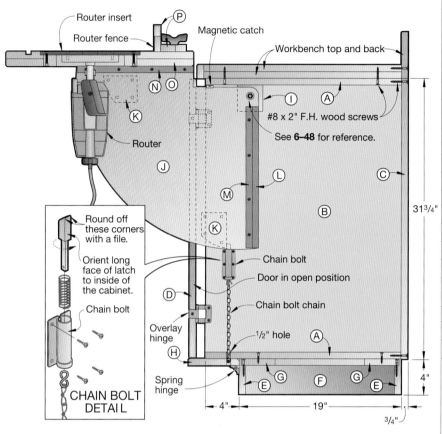

Router insert
Router fence
Magnetic catch
Workbench top and back
#8 x 2" F.H. wood screws
See **6–48** for reference.
Router
Router

Round off these corners with a file.
Orient long face of latch to inside of the cabinet.
Chain bolt

CHAIN BOLT DETAIL

Chain bolt
Door in open position
Chain bolt chain
Overlay hinge
½" hole
Spring hinge

31¾"

4"

4"   19"   ¾"

Cutting Diagram

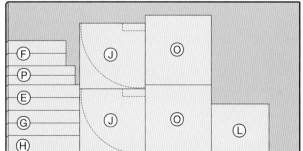

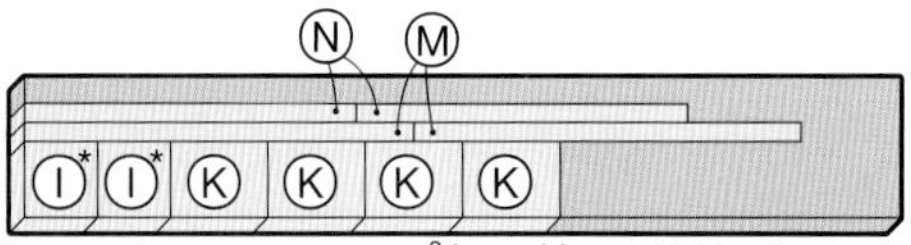

¾ x 5½ x 36" Hardwood
(1.5 bd.ft.)

Materials List for Lift-Up Router Table

| PART | | FINISHED SIZE | | | | |
		T	W	L	Mtl.	Qty.
CARCASE AND BASE						
A	top & bottom	¾"	23"	23½"	MF	2
B	sides	¾"	23¾"	31¾"	MF	2
C	back	¾"	23½"	31¾"	MF	1
D	doors	¾"	11¾"	31½"	MF	2
E	front & back	¾"	4"	23½"	MF	2
F	sides	¾"	4"	19"	MF	2
G	cleats	¾"	4"	23½"	MF	2
H	foot pedal	¾"	6"	23½"	MF	1
I	spacers	½"	3"	3"	H	2
PULL-UP						
J	support arms	¾"	20¾"	21½"	MF	2
K	latch blocks	¾"	3"	4"	H	4
L	panel	¾"	16"	19"	MF	1
M	panel cleats	¾"	¾"	16"	H	2
N	router top cleats	¾"	¾"	13⅝"	H	2
TOP AND FENCE						
O*	router top	1½"	22"	22"	LMF	1
P	fence & base	¾"	3"	22"	MF	2

*Cut parts oversized. Trim to finished size according to the instructions.

Materials Key: MF = medium-density fiberboard; H = hardwood; LMF = laminated medium-density fiberboard.

Supplies: #6 x 1", #8 x 1¼", #8 x 1½", #8 x 2", and #8 x 2½" flathead wood screws; two pair 270° no-mortise hinges; 3¾" pulls (2); double magnetic catch and strike plates; ⅜ x 3½" flathead machine screws with nuts and flat washers (2); ⅜" locknuts; ⅜" ID x ½" OD x 1" bronze or nylon bushings (2); 4" chain bolts (2); pair of 2⁷⁄₁₆ x 1½" spring hinges; plastic laminate; contact cement; ¼-20 x 1½" flathead machine screws; ¼" prongless T-nuts (4); mini-channel;plastic wing nuts with ⅜" flat washers (2); ⅜" x 1½" carriage bolts; primer (2); paint; polyurethane.

*Plane or resaw to thickness listed in the Materials List.

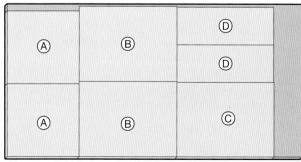

¾ x 49 x 97" Medium-density fiberboard (MDF)

¾ x 49 x 97" Medium-density fiberboard (MDF)

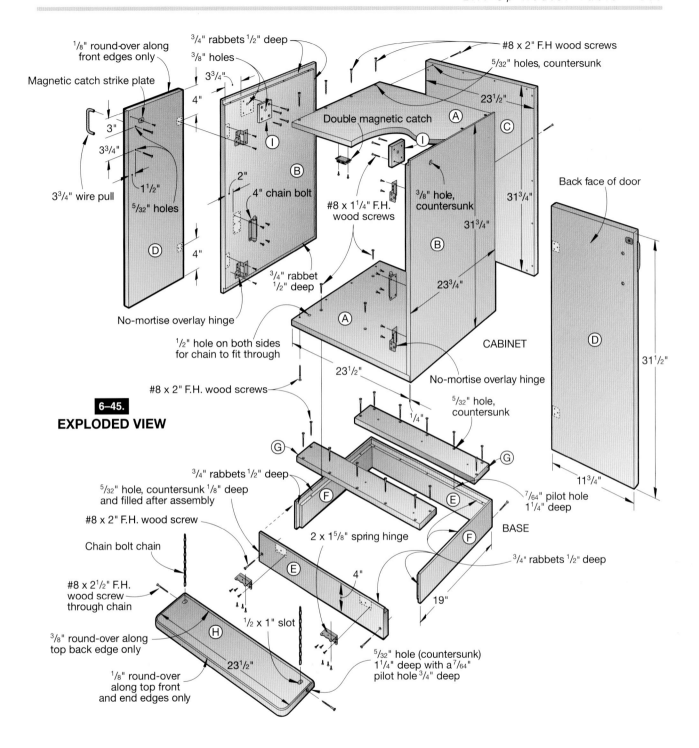

1/8" round-over along front edges only

Magnetic catch strike plate

3/4" rabbets 1/2" deep

3/8" holes

3 3/4"

4"

3"

3 3/4"

3 3/4" wire pull

1 1/2"

5/32" holes

D

4"

No-mortise overlay hinge

1/2" hole on both sides for chain to fit through

2"

4" chain bolt

3/4" rabbet 1/2" deep

No-mortise overlay hinge

Double magnetic catch

#8 x 1 1/4" F.H. wood screws

3/8" hole, countersunk

#8 x 2" F.H wood screws

5/32" holes, countersunk

23 1/2"

A

C

I

31 3/4"

B

31 3/4"

B

23 3/4"

23 3/4"

A

CABINET

Back face of door

D

31 1/2"

11 3/4"

23 1/2"

#8 x 2" F.H. wood screws

6–45.
EXPLODED VIEW

5/32" hole, countersunk
23 1/2"

1/4"

G

G

5/32" hole, countersunk

7/64" pilot hole 1 1/4" deep

3/4" rabbets 1/2" deep

5/32" hole, countersunk 1/8" deep and filled after assembly

#8 x 2" F.H. wood screw

Chain bolt chain

#8 x 2 1/2" F.H. wood screw through chain

3/8" round-over along top back edge only

1/8" round-over along top front and end edges only

F

E

2 x 1 5/8" spring hinge

F

E

4"

H

1/2 x 1" slot

23 1/2"

19"

BASE

3/4" rabbets 1/2" deep

5/32" hole (countersunk) 1 1/4" deep with a 7/64" pilot hole 3/4" deep

4   Cut the spacers (I) to size from 1/2" solid stock. Glue and screw a spacer to the inside face of each side (B), where dimensioned on **6–46** on *page*

105. Next, drill a 3/8" hole through each spacer and through the mating side (B), where dimensioned on **6–48**. (To keep our 3/8" holes perpendicular to the sides, we did

this on a drill press.) Countersink the holes on the outside surface of each carcase side.

5 Glue and clamp the basic cabinet pieces (A, B, C) together, checking for square. Drill the mounting holes, and strengthen the assembly with wood screws.

6 Drill a pair of holes through each door (D) for the pulls. Next, rout a 1/8" round-over on the front edges of each door.

7 Mount a pair of no-mortise overlay hinges on each door, where shown on **6–45**. Screw the doors to the cabinet, leaving a 1/16" gap between the door bottoms and bottom surface of the bottom (A). (See **6–44**.)

Construct the Base

1 Cut the base front and back (E), sides (F), and cleats (G) to size.

2 Cut the rabbets in parts E and F, where shown on **6–45**. Glue and screw the base (E, F, G) together in the configuration shown. Later, fill the counterbored screw holes with wood filler, and sand smooth.

3 Using **6–46** on *page 105* for reference, cut the foot pedal (H) to shape. Form a pair of 1/2" slots for the chains in the pedal, where shown on the drawing. Drill the holes for the 2½"screws. Rout 1/8" round-overs on the top front and ends. Then, rout a 3/8" round-over along the top back edge of the pedal.

Add the Pull-Up Assembly

1 Cut the two support arms (J) to size (20¾ × 21½"). With the edges and ends flush, stick the two pieces together face-to-face. Mark the centerpoints, where shown on **6–46**. Drill the 1/2" hole. Lay out the 19" radius and notch. Cut and sand the support arms to shape. Separate the pieces, and remove the tape.

2 Cut four latch-block blanks (K) to 3 × 4" from 3/4" stock, and stick them together in pairs with double-faced tape. Make two copies of the full-size latch-block pattern (K) in **6–46**, and adhere one to each set of blanks. Cut and sand the pieces to shape, drill the 5/32" holes, countersink the holes on opposite sides of each pair, separate the pieces, and remove the paper patterns.

3 Screw (no glue) the latch blocks (K) to the outside face of each support arm (J), where dimensioned on the Router Table Support Arms drawing in **6–46**.

4 Using **6–48** for reference, drive a 3/8" bushing (1/2" O.D.) into the 1/2" hole in each support arm (J). Drive the bushings flush with one surface, and trim them flush with the other.

5 Test the workings of the support arms (J) by mounting the chain bolts on the inside face of each cabinet side, where dimensioned on the Side drawing (B) in **6–46**. Remove the bolts

from the metal housing, and round the corners of the bolts with a file so they don't dig into the latch blocks (K). See the Chain Bolt Detail in **6–44** for reference. Reassemble the chain bolts so that the long edge of the bolts faces inward.

6 Using **6–48** for reference, bolt the support arms to the cabinet sides and spacers. Swing the support arms up and down, checking to see the latch blocks clear the tops of the chain bolts.

7 Cut the pull-up back panel (L) and cleats (M, N) to size. Drill countersunk holes in the cleats, and glue and screw the M cleats to the back panel. Screw the N cleats to the arms (J). See the Pull-Up Exploded View in **6–47** for reference.

Add the Router Tabletop

1 Cut two pieces of 3/4" MDF for the router top (O) to size plus 1" in length and width. Glue and clamp them together face-to-face. Later, trim the top to finished size, radiusing the corners, where shown on the Router Tabletop drawing (O) in **6–46**.

2 Using contact cement, apply plastic laminate cut slightly oversized to the top and bottom surfaces of the router tabletop (O). Trim the laminate with a flush trim bit, and then rout a 1/8" chamfer along the top and bottom edges.

6–46. **PARTS VIEWS**

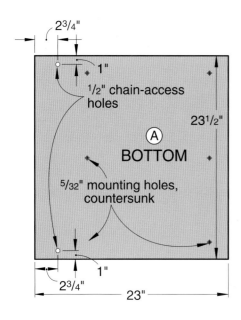

2³/₄"
1"
1/2" chain-access holes
23¹/₂"
Ⓐ
BOTTOM
5/32" mounting holes, countersunk
1"
2³/₄"
23"

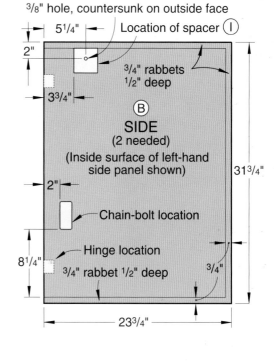

3/8" hole, countersunk on outside face
5¹/₄"
Location of spacer Ⓘ
2"
3/4" rabbets 1/2" deep
3³/₄"
Ⓑ
SIDE
(2 needed)
(Inside surface of left-hand side panel shown)
31³/₄"
2"
Chain-bolt location
Hinge location
8¹/₄"
3/4" rabbet 1/2" deep
3/4"
23³/₄"

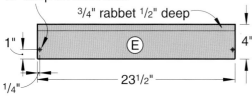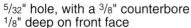

5/32" hole, with a 3/8" counterbore 1/8" deep on front face
3/4" rabbet 1/2" deep
1"
Ⓔ
4"
23¹/₂"
1/4"
BASE FRONT AND BACK

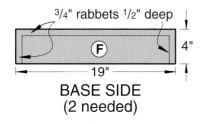

3/4" rabbets 1/2" deep
Ⓕ
4"
19"
BASE SIDE
(2 needed)

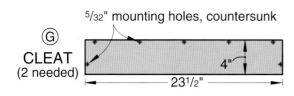

5/32" mounting holes, countersunk
Ⓖ
CLEAT
(2 needed)
4"
23¹/₂"

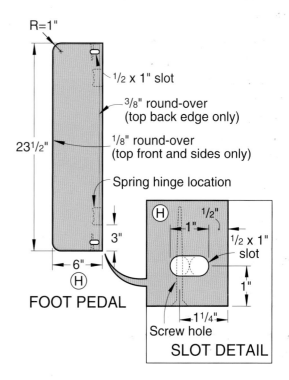

R=1"
1/2 x 1" slot
3/8" round-over (top back edge only)
1/8" round-over (top front and sides only)
Spring hinge location
23¹/₂"
3"
6"
Ⓗ
FOOT PEDAL
Ⓗ
1/2"
1"
1/2 x 1" slot
1"
1¹/₄"
Screw hole
SLOT DETAIL

6–46 continued. **PARTS VIEWS**

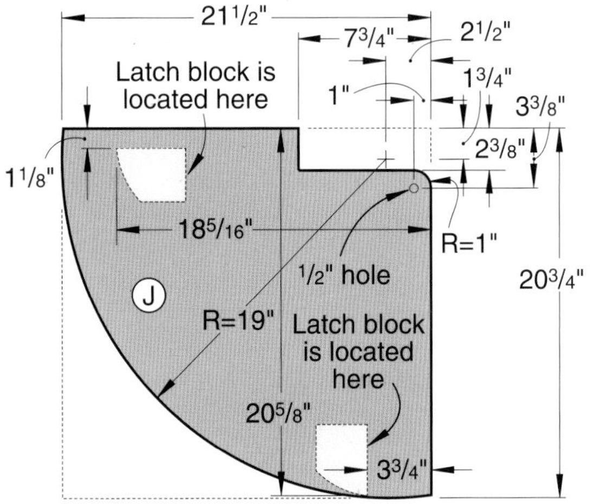
21 1/2"
7 3/4"
2 1/2"
Latch block is located here
1"
1 3/4"
3 3/8"
2 3/8"
1 1/8"
18 5/16"
R=1"
1/2" hole
20 3/4"
J
R=19"
Latch block is located here
20 5/8"
3 3/4"

ROUTER TABLE SUPPORT ARM (2 needed)

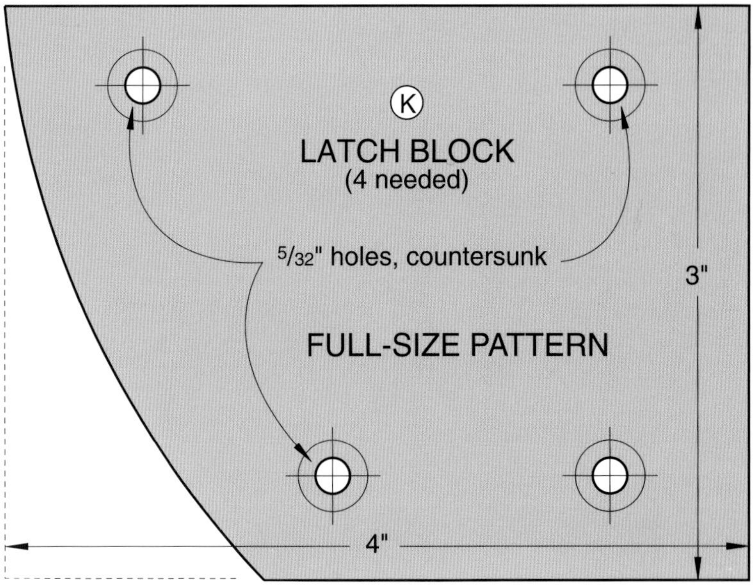
K
LATCH BLOCK
(4 needed)
5/32" holes, countersunk
3"
FULL-SIZE PATTERN
4"

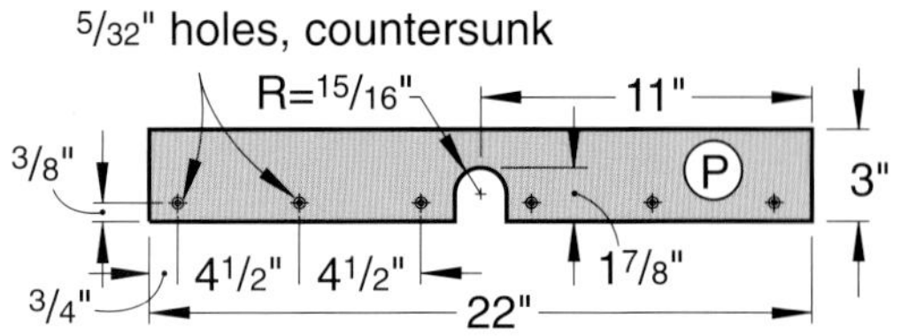
5/32" holes, countersunk
R=15/16"
11"
3/8"
P
3"
3/4"
4 1/2"
4 1/2"
1 7/8"
22"

FENCE

Note: Plastic laminate is applied to top and bottom surfaces before cutting any holes or dadoes.

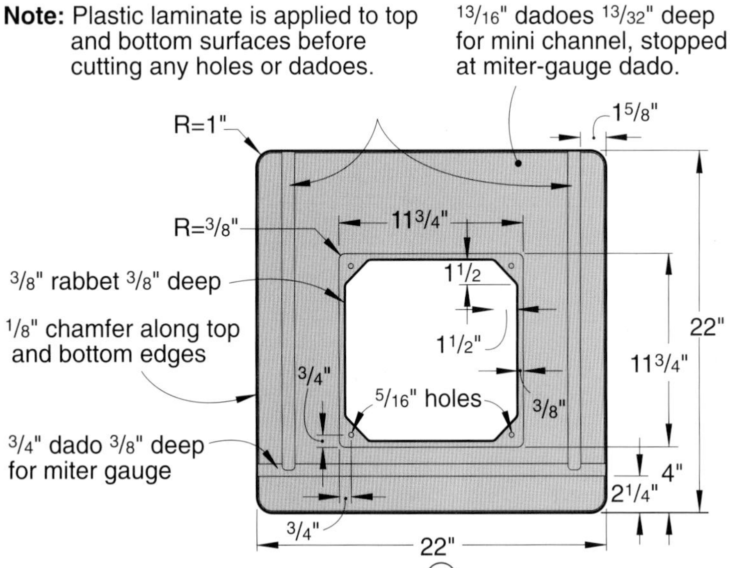
13/16" dadoes 13/32" deep for mini channel, stopped at miter-gauge dado.
R=1"
1 5/8"
R=3/8"
11 3/4"
3/8" rabbet 3/8" deep
1 1/2
1/8" chamfer along top and bottom edges
1 1/2"
22"
11 3/4"
3/4"
5/16" holes
3/8"
3/4" dado 3/8" deep for miter gauge
4"
2 1/4"
3/4"
22"

O
ROUTER TABLETOP

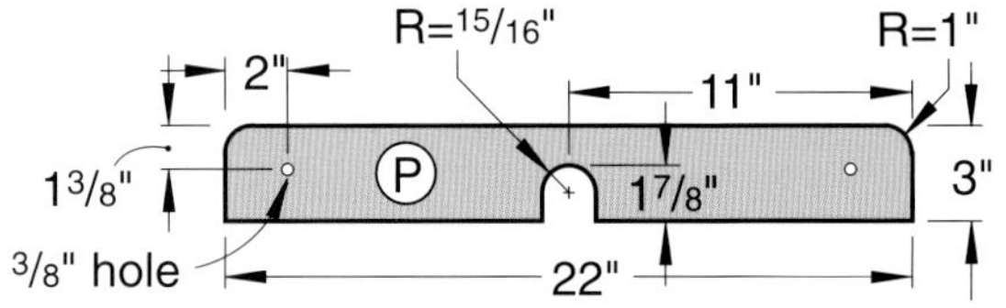
R=15/16"
R=1"
2"
11"
1 3/8"
P
1 7/8"
3"
3/8" hole
22"

FENCE BASE

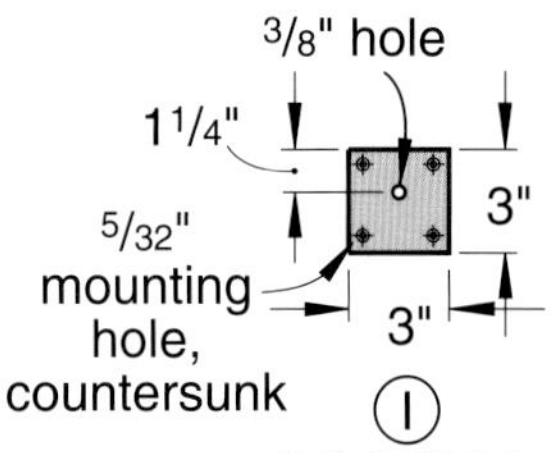
3/8" hole
1 1/4"
3"
5/32" mounting hole, countersunk
3"

I
SPACER
(2 needed)

3 Cut or rout the miter-gauge and mini-channel grooves in the tabletop, where dimensioned on the drawing. Cut the grooves to match the width of your miter gauge and mini-channel. (We used mini-channel available at most electrical supply outlets.)

Note: When tucking the router tabletop under the workbench, it's

important to have the router plate (with the router attached) secured to the tabletop so it doesn't fall out. For that reason, we secured the plate to the tabletop with screws and T-nuts.

4 The phenolic ⅜ × 12 × 12" router plate we bought and used actually measured a bit less. So at the tablesaw, we cut

adjoining edges, trimming the plate to 11¾" square.

5 With your router, form a router-plate recess in the table-top to accept a mounting plate.

6 Mark the centerpoints, and use your drill press to drill ⁵⁄₁₆" holes in the corners of the router plate, ¾" in from the

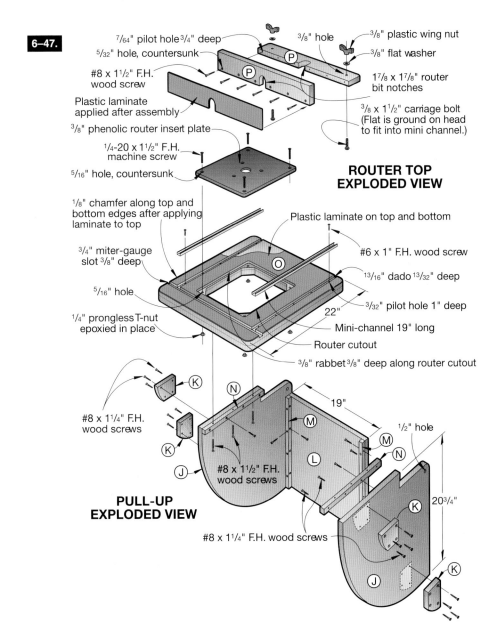

6–47.

7/64" pilot hole 3/4" deep
5/32" hole, countersunk
#8 x 1 1/2" F.H. wood screw
Plastic laminate applied after assembly
3/8" phenolic router insert plate
1/4-20 x 1 1/2" F.H. machine screw
5/16" hole, countersunk

3/8" hole
3/8" plastic wing nut
3/8" flat washer
1 7/8 x 1 7/8" router bit notches
3/8 x 1 1/2" carriage bolt (Flat is ground on head to fit into mini channel.)

ROUTER TOP EXPLODED VIEW

1/8" chamfer along top and bottom edges after applying laminate to top
3/4" miter-gauge slot 3/8" deep
5/16" hole
1/4" prongless T-nut epoxied in place

Plastic laminate on top and bottom
#6 x 1" F.H. wood screw
13/16" dado 13/32" deep
3/32" pilot hole 1" deep
22"
Mini-channel 19" long
Router cutout
3/8" rabbet 3/8" deep along router cutout

#8 x 1 1/4" F.H. wood screws
#8 x 1 1/2" F.H. wood screws

PULL-UP EXPLODED VIEW

19"
1/2" hole
20 3/4"

#8 x 1 1/4" F.H. wood screws

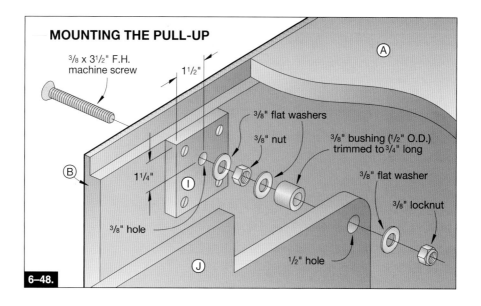

MOUNTING THE PULL-UP

3/8 x 3 1/2" F.H. machine screw

1 1/2"

3/8" flat washers

3/8" nut

3/8" bushing (1/2" O.D.) trimmed to 3/4" long

3/8" flat washer

3/8" locknut

1 1/4"

3/8" hole

1/2" hole

6–48.

corners. Center the router plate in the recess, and use the holes as guides to drill 5/16" holes through the corners of the recess in the tabletop (O).

7 Using epoxy, adhere a 1/4" prongless T-nut into each 5/16" hole on the bottom side of the tabletop.

8 Mark the centerpoint on the router plate, and bore a 1 1/2" hole (or to a diameter just slightly larger than the diameter of your largest-diameter router bit) through the center of the plate.

9 Countersink the 5/16" holes in the router plate to house the 1/4 x 20 x 1 1/2" flathead machine screws.

10 Remove the base plate from your router. Using the holes in your base plate as guides, drill corresponding holes in the router plate. Countersink the holes. Fasten your router to the router

plate. Insert the router and plate into the router-table opening, and secure the unit in place with the machine screws.

Add a Fence to the Router Tabletop

1 Using **6–46,** *page 105,* and the Router Top Exploded View in **6–47**, cut the fence and fence base pieces (P) to size. Cut a U-shaped notch in each, and drill the holes in each, where dimensioned.

2 Glue and screw the two pieces together in the configuration shown in **6–47** to form the fence.

3 Use contact cement to apply plastic laminate to the front of the fence. Trim the laminate flush.

4 Cut two 19" lengths of mini channel, and drill countersunk holes for #6 x 1" flathead wood screws. Clamp the channel in place, drill pilot holes into the top (O), and drive the screws.

Finish It Up, and Add the Router

1 Remove the doors and all previously attached hardware from the carcase. Prime the carcase, doors, base, foot pedal, and pull-up parts with an oil-based primer. Apply the paint. Finish the exposed MDF surfaces of the router top and fence with polyurethane.

2 Drill the mounting holes, and screw the carcase assembly to the base, locating the back edge of the base 3/4" in from the back edge of the carcase. See **6–44**.

3 Mount spring hinges to the foot pedal (H) and footpedal/ spring hinges to the base front (E).

4 Mount the chain bolts, and feed the chains through the 1/2" holes in the bottom (A) and into the slots in the foot pedal. Let the chains hang slack, and run the #8 x 2 1/2" wood screws into the pre-drilled holes in the edges

of the foot pedal, securing the chains to the pedal. Check to see that both chain bolts operate simultaneously. Slight differences in the lengths of the chains can be equalized by removing the short chain, clamping one end of the chain in a vise and stretching the chain by pulling on the other end with pliers. Make sure the chain bolts are extended fully when the foot pedal is in the up position. With the bolts working properly, cut the excess chain off below the foot pedal.

5 Using **6–48** for reference, install the ⅜ × 3½" flathead machine screws through the sides (B) and spacers (I). Secure each with a washer and nut. Hang the support arms (J) on the bolts with additional washers and nuts, where shown on the drawing. Rotate the arms into the up position, and attach the back (L) between the arms. Step on the foot pedal to release the pull-up assembly to check the movement. A little grease applied to the curved edge of each latch block (K) will smooth the operation of the pull-up assembly.

Note: With the carcase unsecured, installation of the router tabletop will cause the whole assembly to tip forward. The carcase must be temporarily weighted or secured to prevent this.

6 Clamp the router tabletop (O) to the support arms (J), centered side-to-side. Screw the router tabletop in place.

7 File two flat areas opposite each other on the heads of two ⅜ × 1½" carriage bolts. Slide the bolts into the mini-channel in the router top and into the bottom of the fence base (P). Secure them loosely with a washer and wing nut. Check the fit of the fence assembly on the router top.

8 Position the cabinet under the workbench top, and screw it to the wall and the workbench top.

FLIP-UP BENCHTOP TOOL TABLE

Add a benchtop tool (a scrollsaw is shown in **6–49**) to this flip-out-of-the-way table that fits right into the support arms used on the router-table cabinet shown on the previous pages. When you're done with the tool, stow it easily and quickly out of sight.

In the last several pages, we explained how to construct the space-saving lift-up router table and cabinet. Now, use **6–50** to build a swivel top to support your favorite benchtop tool. To do this,

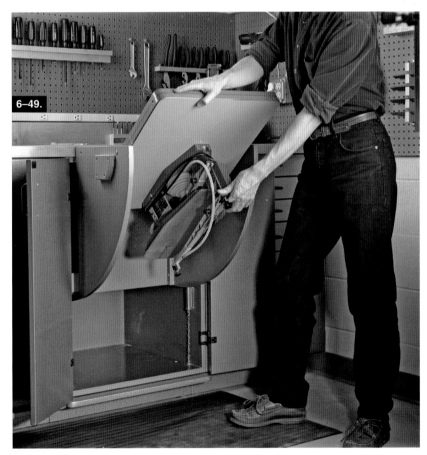

6–49.

A scrollsaw is the tool of choice on this workbench table that flips out of the way.

6–50. **FLIP-UP BENCHTOP**

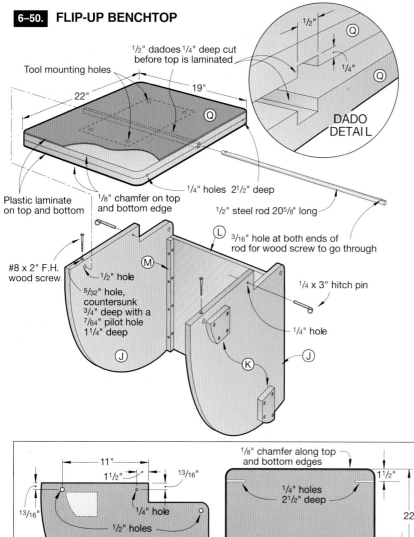

Tool mounting holes

22"

19"

$\frac{1}{2}$" dadoes $\frac{1}{4}$" deep cut before top is laminated

$\frac{1}{2}$"

$\frac{1}{4}$"

Ⓠ

Ⓠ

Ⓠ

DADO DETAIL

Plastic laminate on top and bottom

$\frac{1}{8}$" chamfer on top and bottom edge

$\frac{1}{4}$" holes 2$\frac{1}{2}$" deep

$\frac{1}{2}$" steel rod 20$\frac{5}{8}$" long

$\frac{3}{16}$" hole at both ends of rod for wood screw to go through

#8 x 2" F.H. wood screw

$\frac{1}{2}$" hole

$\frac{5}{32}$" hole, countersunk $\frac{3}{4}$" deep with a $\frac{7}{64}$" pilot hole 1$\frac{1}{4}$" deep

Ⓛ

Ⓜ

$\frac{1}{4}$ x 3" hitch pin

$\frac{1}{4}$" hole

Ⓙ

Ⓙ

Ⓚ

PARTS VIEW

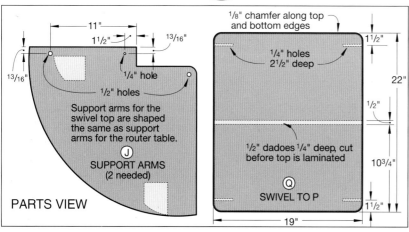

11"

1$\frac{1}{2}$"

13$\frac{1}{16}$"

13$\frac{1}{16}$"

$\frac{1}{4}$" hole

$\frac{1}{2}$" holes

Support arms for the swivel top are shaped the same as support arms for the router table.

Ⓙ

SUPPORT ARMS (2 needed)

$\frac{1}{8}$" chamfer along top and bottom edges

$\frac{1}{4}$" holes 2$\frac{1}{2}$" deep

1$\frac{1}{2}$"

22"

$\frac{1}{2}$"

$\frac{1}{2}$" dadoes $\frac{1}{4}$" deep, cut before top is laminated

10$\frac{3}{4}$"

Ⓠ

SWIVEL TO P

1$\frac{1}{2}$"

19"

build the router-table support arms (J) like you did for the router-table version. Then, add the $\frac{1}{4}$" and $\frac{1}{2}$" holes in each arm, where dimensioned on the Parts View drawing in **6–50**. Cut two pieces of $\frac{3}{4}$" sheet goods for the swivel top (Q), cut a dado down

the center of each top section, and glue them together, aligning the dadoes. Later, add plastic laminate to the top and bottom of the top (Q). The top swivels on a $\frac{1}{2}$" steel rod and locks in place with a pair of $\frac{1}{4}$ x 3" hitch pins.

❖

WORKBENCH WITH SIDE-DRAFT DUST-COLLECTION SYSTEM

Most woodworkers can't spare the shop space needed for a dedicated sanding table with at-the-source dust collection and still have a workbench. That's what is great about this project: You get both!

The workbench shown here (**6–51**) has a side-draft dust collection system, a durable top, storage, electrical outlets, a metal track system, and a router mat. Here's how it deals with dirt: The heart of the system is a furnace blower with a squirrel-cage fan. When you want to sand a project, raise the side-draft hood at the end of the bench, and switch on the motor. This starts a flow that pulls air and dust across the top of the bench. After the air moves through the hood and an opening in the left drawer support, dust particles hit the furnace filter and stay there. The cleaned air continues through the furnace blower. The forced air then exhausts through a hole in the bottom, moving through the vents cut in the base. With the hood lowered for an unobstructed work surface, you can leave the fan running to remove dust from your shop's air, just like a typical air-filtration system.

Note: We usually identify the parts of a project in alphabetical order determined by the sequence

6–51.

in which you'll cut them. But in this project, we lettered the parts in their assembly sequence. It's important that you follow this order of assembly so you'll have the clearance you'll need to attach clamps and drive screws. To keep the parts organized, use a permanent marker to write the identifying letter on the end of each part.

Let's Start with the Carcase

1 Lock your tablesaw's rip fence 30½" from the blade, and rip ¾"-thick medium-density fiberboard (MDF) for the left divider (A), the right divider (D), and the bottom (G). Without moving the fence, rip ¾"-thick birch plywood for the left and right end panels (H, I).

2 Move the rip fence to 26½" from the inner edge of the blade, and rip ¾"-thick MDF for the left and right back (C, F), the right drawer support (E), and the left drawer support (B).

3 Adjust the rip fence, and cut two stretchers (J) and two kickers (K) from MDF. Move it again to cut the outer drawer support (L). Finally, rip a pair of shelves (T) from the same material.

4 Crosscut all of these parts to the lengths shown in the Materials List. In addition to the top stretchers (J), kickers (K), and the shelves (T), there are other pairs of pieces with identical lengths. These are the left and right dividers (A, D), the left and right end panels (H, I), the right and left backs (C, F), and left and right drawer supports (B, E).

Cut Dadoes and a Rabbet

1 Put a ¾"-wide dado set in your tablesaw, and raise it for a ¼"-deep cut. Make a test cut in scrap stock to make certain that the dado's width matches the thickness of your MDF. Check that the test dado's depth is exactly ¼" deep.

2 Referring to **6–53** (*page 114*) adjust your rip fence to cut the lower horizontal dado in the right and left end panels (H, I). Move the rip fence, and then cut the upper horizontal dadoes in H and I.

Cutting Diagram

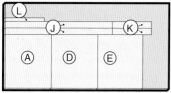

¾ x 48 x 96" MDF

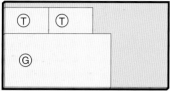

¾ x 48 x 96" MDF

¾ x 48 x 96" Birch plywood

¾ x 48 x 96" MDF

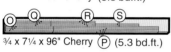

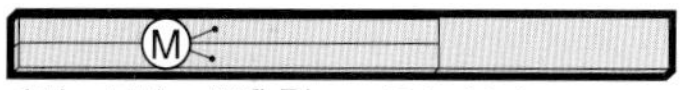

1½ x 7¼ x 96" Pine (6 bd.ft.)

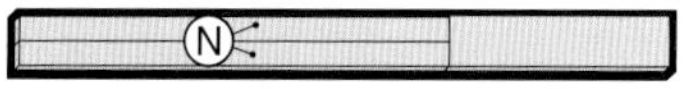

¾ x 7¼ x 96" Cherry (5.3 bd.ft.)

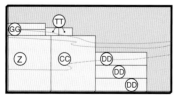

¾ x 7¼ x 96" Cherry (P) (5.3 bd.ft.)

¾ x 7¼ x 96" Maple (5.3 bd.ft.)

¾ x 7¼ x 96" Maple (5.3 bd.ft.)

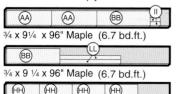

¾ x 9¼ x 96" Maple (6.7 bd.ft.)

¾ x 9 ¼ x 96" Maple (6.7 bd.ft.)

¾ x 7¼ x 96" Maple (4 needed) (5.3 bd.ft.)

Materials List for Workbench with Side-Draft Dust-Collection System

PART		FINISHED SIZE T	W	L	Mtl.	Qty.	PART		FINISHED SIZE T	W	L	Mtl.	Qty.
A	left divider	¾"	30½"	27¼"	MDF	1	X	drawer front/back	¾"	5⅞"	25⅞"	M	2
B	left drawer support	¾"	26½"	30½"	MDF	1	Y	drawer sides	¾"	5⅞"	31½"	M	2
C	left back	¾"	26½"	20¼"	MDF	1	Z	drawer bottom	½"	31½"	25⅜"	BP	1
D	right divider	¾"	30½"	27¼"	MDF	1	AA	pull-out front/back	¾"	8⅜"	26⅝"	M	2
E	right drawer support	¾"	26½"	30½"	MDF	1	BB	pull-out top/bottom	¾"	8⅜"	31½"	M	2
F	right back	¾"	26½"	20¼"	MDF	1	CC	pull-out side	½"	31½"	26⅛"	BP	1
G	bottom	¾"	30½"	62½"	MDF	1	DD	pull-out shelves	½"	7⅛"	30⁵⁄₁₆"	BP	3
H	left end panel	¾"	30½"	31¾"	BP	1	EE	pull-out shelf edges	½"	1½"	30⁵⁄₁₆"	M	6
I	right end panel	¾"	30½"	31¾"	BP	1	FF	drawer false fronts	¾"	6"	26"	M	2
J	stretchers	¾"	3¾"	63"	MDF	2	GG	backers	½"	3½"	22"	BP	2
K	kickers	¾"	3¾"	23"	MDF	2	HH	doors	¾"	12⅞"	19⅜"	EM	8
L	outer drawer support	¾"	2"	23"	MDF	1	II*	filter-holding block	1½"	2¼"	6"	M	1
M	sub-bases	1½"	3½"	62"	P	2	JJ	top	¾"	34½"	76½"	MDF	3
N	bases	¾"	3½"	63½"	C	2	KK	edge bands	¾"	2¼"	78"	M	2
O	end panel bands	¾"	¾"	28¼"	C	4	LL	end bands	¾"	2¼"	36"	M	2
P	top/bottom bands	¾"	¾"	62"	C	2	MM	hood top	¾"	4¾"	22¹⁵⁄₁₆"	M	1
Q	center bands	¾"	¾"	26¾"		4	NN	hood back	¾"	9½"	22¹⁵⁄₁₆"	M	1
R	rail bands	¾"	¾"	26"	C	4	OO	hood bottom	¾"	2¼"	21⁷⁄₁₆"	M	1
S	shelf bands	¾"	¾"	25⅞"	C	2	PP	hood sides	¾"	4¾"	9½"	M	2
T	shelves	¾"	14¾"	25⅞"	MDF	2	QQ	hood bracket	¾"	3½"	21⁷⁄₁₆"	M	1
U	doorstop	¾"	¾"	3¾"	M	1	RR	stop blocks	¾"	¾"	5"	M	2
V	false drawer cleats	¾"	1"	6"	M	4	SS	sanding-mat strips	³⁄₁₆"	⅜"	25"	C	4
W	laminate strips	¹⁄₁₆"	1"	30½"	PL	16	TT	spacers for vise	½"	4½"	8"	BP	2

*Laminate from two pieces of ¾" stock.

Materials Key: MDF = medium-density fiberboard; BP = birch plywood; P = pine; C = cherry; M = maple; EM = Edge-glued maple; PL = plastic laminate.

Supplies: Items listed in the Buying Guide plus #10 plate-joiner biscuits; ½" brads; #8 x 1", #8 x 1¼", #8x 1½" and #8 x 2" flathead wood screws; #6 x ½" panhead sheet-metal screws; #6 x ¾" and #6 x 1¼" flathead sheet-metal screws; #12 x 1" panhead wood screws; #12 x 1¼" sheet-metal screws; 24" length of ½"-dia. dowel rod; orange and yellow wire nuts; #19 x ½" brads; six-outlet, 36"-long surface metal power strip (we used Wiremold brand); 2½" wood mushroom knobs (12); 25' 12-gauge power cord; 20 x 25" furnace filters (2); octagonal junction boxes with plain cover plates (3); single-gauge switch box; 15-amp single-pole switch; switch plate; electrical staples; clamp connectors (8); router/sanding pad; sash lock; wood and foam door-jamb weatherstrip; ¼" bracket-style shelf supports (20); 10' pieces of galvanized mini-channel B-Line brand model B-72 or equivalent (2) (available from your local electrical-supply house).

Buying Guide: Woodworking vise (your choice to fit); Blum (or similar) 35mm European clip hinges, 125° inset (workbench requires eight pair); squirrel-cage blower unit.

*Plane or resaw to thickness listed in the Materials List

Plastic laminate

¾ x 48 x 96" MDF (3 needed)

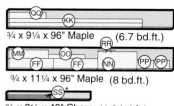

¾ x 9¼ x 96" Maple (6.7 bd.ft.)

¾ x 11¼ x 96" Maple (8 bd.ft.)

⅜ x 3½ x 48" Cherry (1.3 bd.ft.)

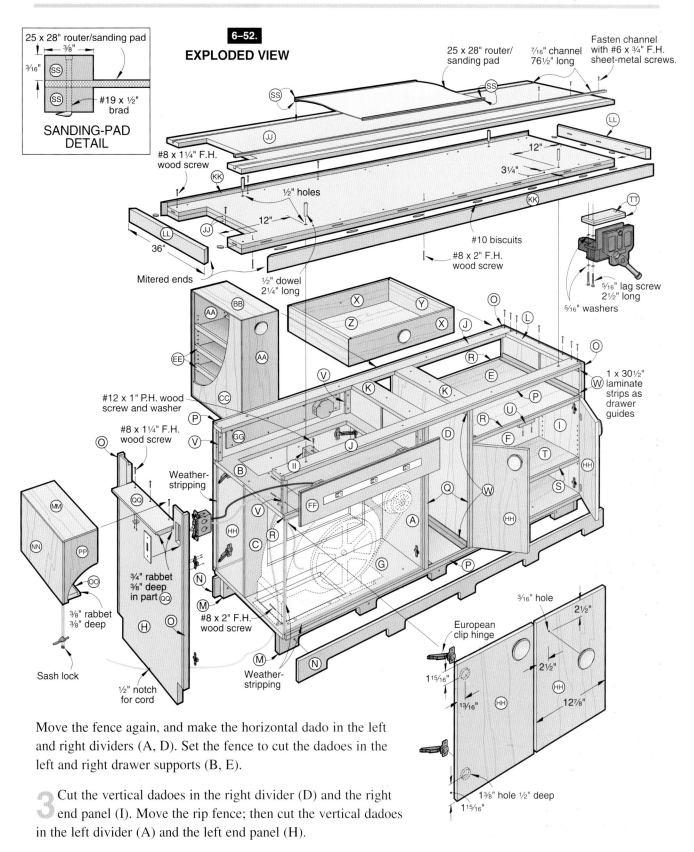

SANDING-PAD DETAIL

25 x 28" router/sanding pad
3/8"
3/16"
SS
SS
#19 x 1/2" brad

6–52.
EXPLODED VIEW

25 x 28" router/ sanding pad
7/16" channel 76 1/2" long
Fasten channel with #6 x 3/4" F.H. sheet-metal screws.

#8 x 1 1/4" F.H. wood screw
1/2" holes
12"
3 1/4"
1/2" dowel 2 1/4" long
36"
Mitered ends
12"
#10 biscuits
#8 x 2" F.H. wood screw
5/16" lag screw 2 1/2" long
5/16" washers

#12 x 1" P.H. wood screw and washer
#8 x 1 1/4" F.H. wood screw
Weather-stripping
3/4" rabbet 3/8" deep in part
3/8" rabbet 3/8" deep
Sash lock
1/2" notch for cord
#8 x 2" F.H. wood screw
Weather-stripping

1 x 30 1/2" laminate strips as drawer guides

European clip hinge
3/16" hole
2 1/2"
2 1/2"
12 7/8"
1 15/16"
13/16"
1 3/8" hole 1/2" deep
1 15/16"

Move the fence again, and make the horizontal dado in the left and right dividers (A, D). Set the fence to cut the dadoes in the left and right drawer supports (B, E).

3 Cut the vertical dadoes in the right divider (D) and the right end panel (I). Move the rip fence; then cut the vertical dadoes in the left divider (A) and the left end panel (H).

4 Referring to **6–53** (*page 115*), lay out the location of the dadoes on the bottom (G) and the stretchers (J). The stretchers are longer than the bottom, but the dadoes in both parts (G, J) must produce an 8½" space centered end-to-end. Cut these dadoes. The dado in each kicker (K) is centered in its width.

5 Raise the dado blade ½" above the saw's table. Attach a 6"-high scrapwood face to your rip fence, and then lock the fence in position with the scrapwood face just touching the edge of the dado blade. Make a test cut in scrap stock to check the setup for a ¾" wide × ½" deep dado. Cut the rabbets along the top edge of the left and right end panels (H, I). Leave the scrapwood face on the fence because you'll need it for several other cuts later.

Create the Cutouts in the Carcase

1 Use a pencil and framing square to lay out the cutout in the bottom (G). Drill a ½" hole to create an entry hole for your jigsaw's blade, and cut the rectangular opening. Drill the hole for the power cord next to this opening, and then set this part aside for the time being.

2 Chuck a 1" bit into your drill, and bore the hole in the left drawer support (B). Following

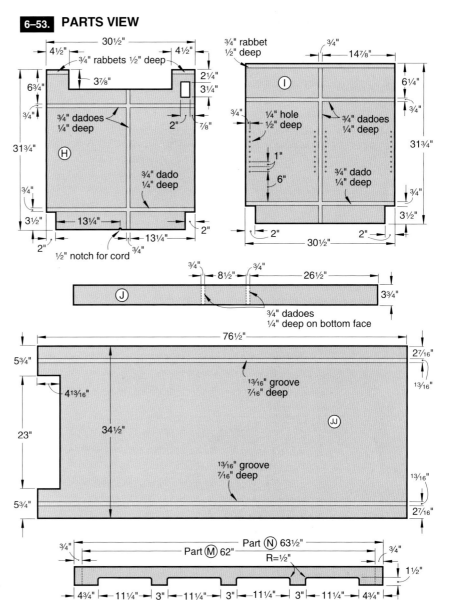

6–53. PARTS VIEW

the same procedure you just used on the bottom, mark and cut the opening in this part, and then set it aside. Repeat the process for the hole in the left back (C), and set it aside.

3 Clamp a scrapwood block along the bottom edge of the left end panel (H), and lay out the location of the ½" notch, where

shown on **6–53**. Position the tip of the 1" bit in the scrapwood/panel seam, and then drill the semicircular notch. Cut the large notch and switch-box hole in the upper end of part H. Complete this panel by marking and cutting the toekick notches in the bottom corners. Cut identical toekick notches in the right end panel (I).

6–53 continued. **PARTS VIEW**

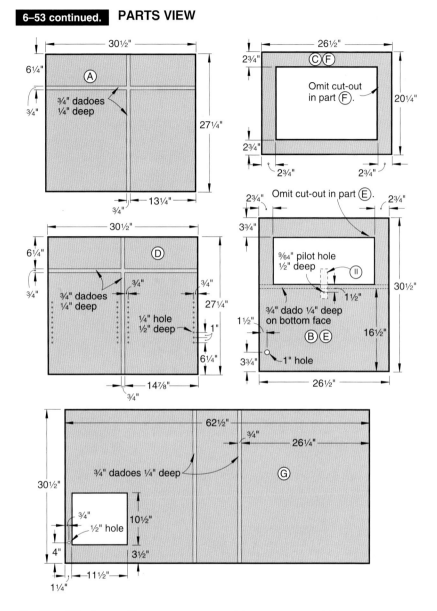

A Jig Makes It Easy to Drill the Shelf-Pin Holes

1 Make a jig for drilling opposing identically located shelf-pin holes by cutting a ¾"-thick piece of scrapwood to 3½ x 18". Referring to **6–53** for the right divider (D), lay out the centerpoints of one vertical column of shelf-pin holes on the scrapwood.

Then use your drill press to create the jig by accurately drilling the holes through the piece of scrap.

2 To use the jig, clamp it to the right center divider (D), making certain that the jig's edge is flush with the edge of the panel, and that the jig's bottom end is flush with the panel's lower end. Use a stop collar on the bit in the your handheld drill so you don't

drill too deep (masking tape works, too). It would be difficult to clamp the jig in position for the holes near the center of this part, but fortunately there's an easy solution. Simply slide a scrap strip of MDF into the vertical dado. Then hold the jig against the strip while you drill the shelf-pin holes.

3 Before you use the jig to drill the holes in the right end panel (I), you need to cut ¼" of length from the jig's bottom end. This cut is needed because the bottom ¼" of part D is housed in a dado. Register the jig's end against an MDF strip in the lower horizontal dado, and hold the jig's edge flush panel's edge to drill the outer columns of holes.

Put the Carcase Together

1 When you work with MDF, you must pre-drill for screws to avoid splitting the panel's edges and ends. The #8 flathead wood screws we used require a 5⁄32" shank hole, a 7⁄64" pilot hole, and a countersink. Although you can drill all of these components with separate bits, you'll speed your work by using a combination bit.

We assembled the panels by driving screws spaced 2" from the edges, and then about every 6" along the joint. For some of the joints, we used screws only, while others required both glue and screws. But attaching the end panels (H, I) was an operation that used glue only. Read along, and we'll tell you the method we used for each joint.

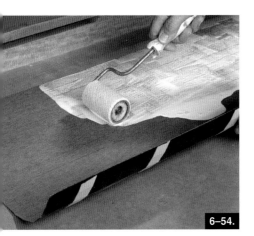

6–54.

Masking tape holds the laminate strips while you roll on contact cement. Brush cement into the carcase's corners.

2 Using #8 × 1½" flathead wood screws, attach the left divider (A) to the left drawer support (B), and then add the left back (C) to the assembly. Make certain that all edges are flush. Assemble the right divider (D), the right drawer support (E), and the right back (F) in the same way.

3 Hold the bottom (G) on edge, and slide the two assemblies (A/B/C) and (D/E/F) against it. Fasten the bottom with #8 x1½" screws.

4 Roll the assembly so the bottom rests on the floor. Then put the assembly up on blocks to create clearance for installing the left and right end panels (H, I). We used two thickness of 2x stock and one thickness of 1x stock, elevating the assembly 3¾". We supported the assembly at six points: near each corner and at the midpoint on each side.

5 Use pipe clamps to dry-assemble (no glue) the end panels (H, I) to the assembly. The pipe clamps don't have to span the entire length of the assembly; you can clamp to the left and right dividers (A, D). Then use a level and a long straightedge (we used a jointed and ripped 2 x 4 on edge) to check that the assembly is level along its length and side to side. Make any necessary adjustments by tapping tapered shims between the bottom of the assembly and the blocks.

6 Glue and clamp the end panels (H, I) in place, and then glue and screw the stretchers (J) into position. Also glue and screw the kickers (K) and outer drawer support (L) into position. Double-check that the assembly is level and square, and let the glue dry, preferably overnight.

Add the Workbench Base

1 Unclamp the carcase, and turn it upside down. This assembly weighs quite a bit, so the easiest way to do this is by placing some old blankets or carpet scraps on the floor and rolling the assembly onto its back, and then upside down. Use shims under the carcase to level it.

2 Double-check the length of the sub-base (M) and the base (N) against your carcase, and then cut these parts to size. Referring to **6–53** (*page 114*), lay out the notch pattern where shown on the base (N). Using glue and #8 × 2"

flathead wood screws, attach the sub-base (M) to the base (N). Be certain that your screws avoid the areas that will be removed. Using a bandsaw or jigsaw, cut the notches, and then sand the edges with 100-grit sandpaper.

3 Fasten these sub-base/base assemblies to the carcase with glue and #8 x 2" wood screws driven through the bottom (G).

Add Edging Strips to the Carcase

1 To conceal the edges of the plywood ends and the MDF carcase, you'll apply hardwood edge banding. Rip enough cherry into ¾ x ¾" strips for the bands (O, P, Q, R, S).

2 To ensure you get a perfect fit, gauge each piece against the carcase, and cut it to length. Work through the pieces in alphabetical order, gluing and clamping them to the carcase.

3 Remove the clamps after the glue dries. Then sand the bands flush, using 100-grit sandpaper and a sanding block.

4 Cut the doorstops (U) and false drawer cleats (V) from ¾"-thick maple. Screw doorstops to the bottom side of the right drawer support (E), as shown on **6–52**. Center the doorstops side to side and ¾" back from the banded face of the carcase. Referring to **6–52**, screw the false drawer cleats (V) to the left divider (A) and the left end panel (H).

2½" wood knob

³⁄₁₆" shank hole

½" rabbet
½" deep

31½"

(X)   (Y)

(Z)

(Y)

31½"

³⁄₁₆" shank hole

½" rabbet
½" deep

5⅞"

(X)

12⁵⁄₁₆"

25⅞"

#8 x 1" F.H.
wood screw

½" rabbets
½" deep

6–55. **DRAWER**

CORNER DETAIL

(Y)

¼"

¼"

¾"   ½"

(X)

Finish Before You're Finished

1 The best time to apply finish to the interior of the carcase is now. (We used two coats of water-based polyurethane.) But before you brush it on, mask the corners that receive the laminate strips (W). These pieces serve as friction-reducing drawer glides, and will be added later.

2 Use your tablesaw to cut a piece of plastic laminate 30½" long. Then slice it into 1"-wide strips. Place the strips face down on your workbench. Butt the strips tightly together and join them with several strips of masking tape. Turn the strip assembly right side up, and roll on contact cement (we used a water-borne variety), as shown in **6–54**.

3 Remove the masking tape from the carcase and brush

on a coat of contact cement. Following label directions, adhere the laminate strips where shown.

Make the Drawer and Pull-Out Next

1 From ¾"-thick maple, rip and crosscut the drawer front/back (X), the drawer sides (Y), the pull-out front/back (AA), the pull-out top/bottom (BB), and the pull-out shelf edges (EE). (Refer to **6–55**.) Double-check the size of both sets of front/backs (X, AA) against the opening. These parts are designed to show a ¹⁄₁₆" reveal around their perimeter, so their overall dimensions are ⅛" smaller than the opening.

2 To make the lock joint for the corners of the drawer and the pull-out, put a ¼" dado blade in your tablesaw. If you removed the 6"-high auxiliary face from your tablesaw's rip fence, replace it now and position the fence ¼" from the inner edge of the blade. Referring to Cut 1 in **6–56** and the Corner Detail drawing in **6–57**, raise the dado blade ¼" high, and make this cut in the drawer and pull-out box pieces you made (X, Y, AA, BB). Set parts Y and BB aside.

3 Referring to Cut 2 of **6–56**, raise the dado blade ¾" high above the table, and make this cut into parts X and AA.

4 Put a ½" dado blade into your tablesaw, and move the tablesaw fence (with the wooden auxiliary face still attached) until the edge of the blade just touches the face. Raise the blade ½" high, and cut the rabbet along one edge of the drawer and pull-out box parts (X, Y, AA, BB).

5 Lower the dado blade to ¼" above the table; then lock the

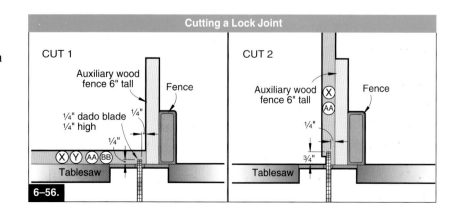

Cutting a Lock Joint

CUT 1

Auxiliary wood fence 6" tall

Fence

¼" dado blade ¼" high

¼"

¼"

(X)(Y)(AA)(BB)

Tablesaw

CUT 2

Auxiliary wood fence 6" tall

(X)
(AA)

Fence

¼"

¾"

Tablesaw

6–56.

6–57. **PULL-OUT**

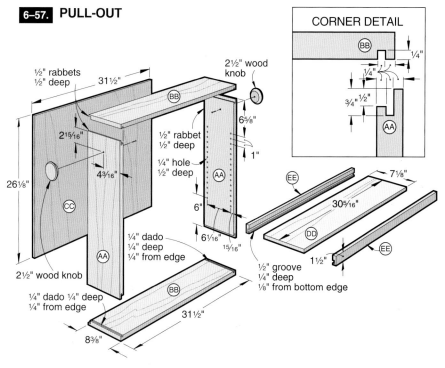

CORNER DETAIL

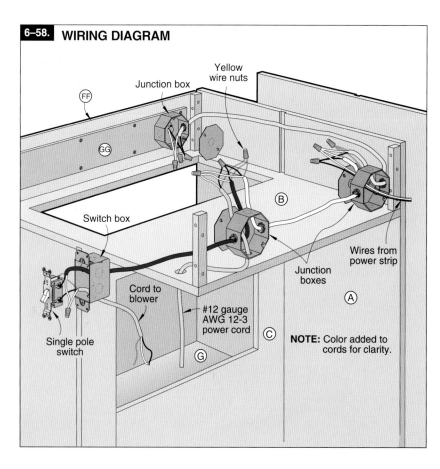

6–58. **WIRING DIAGRAM**

NOTE: Color added to cords for clarity.

fence ⅛" from the inner edge of the blade to cut the grooves in the pull-out shelf edges (EE).

6 Referring to **6–57**, lay out the hole locations in parts AA, and use your drill press to drill them. Mark the centerpoints of the ³⁄₁₆" knob-mounting holes in parts X and AA, and drill them.

7 Assemble the drawer and pull-out by gluing the corners and clamping with a band clamp. Make certain that each assembly is flush, flat, and square.

8 From ½"-thick birch plywood, cut the drawer bottom (Z), the pull-out side (CC), and the pull-out shelves (DD). Attach the drawer bottom (Z) with #8 × 1" flathead wood screws. Attach the pull-out side (CC) with glue only. Glue and clamp the pull-out shelf edges (EE) to the pull-out shelves (DD).

Add the Power Strips

1 Referring to the Materials List, cut the backers (GG) from ½"-thick birch plywood. Cut the drawer false fronts (FF) from ¾"-thick maple. You'll note that these parts are sized for a snug fit into their openings.

You simulate a reveal around their perimeter by cutting a ¹⁄₁₆" rabbet ¼" deep. You can cut this rabbet with your tablesaw's regular blade. To do that, lower the blade below the surface of the table and move the scrapwood face attached to your rip fence above the blade. Turn on the

6–59. **POWER STRIP**

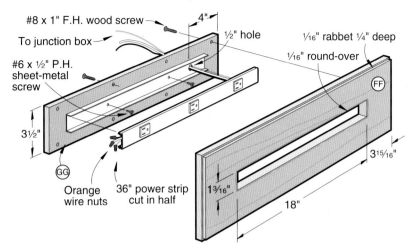

#8 x 1" F.H. wood screw

To junction box

#6 x ½" P.H. sheet-metal screw

4"

½" hole

3½"

GG

Orange wire nuts

36" power strip cut in half

¹⁄₁₆" rabbet ¼" deep

¹⁄₁₆" round-over

FF

3¹⁵⁄₁₆"

1³⁄₁₆"

18"

motor, and slowly raise the blade into the edge of the scrapwood face. Make any adjustments necessary, and cut the rabbets.

2 Take the 36" power strip you purchased, and hacksaw it in half. Screw wire nuts onto the cut wires, and wrap them with electrical tape for extra security. Referring to **6–59**, screw the mounting bracket to the backer (GG); then drill the ½" hole through the backer.

3 Make the cutout in each drawer false front (FF), and rout the round-over along its perimeter. Guiding the wires through the hole, snap the power strip to its mounting bracket. Screw the backer (GG) to the drawer false front (FF).

You're Ready for Wiring

Important Safety Note: *The wiring of this project is not difficult, but if you have even the slightest concern about your ability to safely complete it, call a licensed electrician to do the wiring for you.*

1 Purchase three 4" octagonal junction boxes with plain covers. Remove a knockout from the back of each box, and screw them to the backers (GG), where shown, on **6–58**. Remove knockouts from the sides of the junction boxes where shown, and install the clamp connectors. Screw the backer/drawer false front assemblies (GG/FF) to the carcase.

2 To wire this project, we purchased a 25', 12-gauge power cord with ground. Cut off the plug, and then cut an 8' length of wire from the cord. You'll cut this length into shorter pieces to run between the components in the bench.

3 Attach a length of wire to the blower motor, and then place the blower unit in place. Drill pilot holes into the bottom (G), and screw the unit into place with #12 × 1¼" sheet-metal screws through the rubber grommets in the base. Install a switch box into the hole in the left end panel (H).

4 Make all the connections shown in **6–58**. Secure each connection with a wire nut and electrical tape. Double-check your work, and then install the cover plates. Plug in the cord to make sure everything works properly.

6–60.

FILTER-HOLDING BLOCK

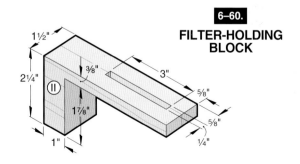

1½"

2¼"

II

1"

⅜"

1⅞"

3"

⅝"

⅝"

¼"

Install the Doors and Filters

1 Edge-glue ¾"-thick solid maple into blanks for the doors (HH). Then, rip and cross-cut these parts to size, keeping in mind that there is a ¹⁄₁₆" clearance gap around the door opening and where two doors meet. Round over the outer perimeter of each door with a ¹⁄₁₆" round-over bit or by hand-sanding.

2 Drill the ³⁄₁₆" holes for the knobs and the 1⅜" holes for the hinges, where shown on **6–52**. The measurements shown are for the exact hinges we used. Other hinges may require different dimensions. Follow the installation instructions supplied with the hinges to mount the doors.

3 Referring to **6–60**, make the filter-holding block (II).

Screw it into position, and install the furnace filters.

4 Use ⅝" brads to install weatherstripping around the perimeter of the doors that conceal the blower motor. Position the weatherstripping ¾" back from the banded face of the carcase.

Add the Top to the Bench

1 The top (JJ) consists of three pieces of MDF. Here's an easy way to make certain that all three parts are identical. Carefully cut one piece, including the notch, to the dimensions shown on **6–52**. This will be your pattern. Then cut two blanks that are ½" oversized in both length and width. Center the pattern on top of one of the blanks and clamp the parts. Use a flush-cutting bit in your router to

clone the pattern. Repeat for the other blank.

2 Set one of the tops aside. Carefully align the other two tops, and then screw them together with #8 x 1¼" flathead wood screws. Space these screws about 2" from the perimeter and 12" apart.

3 Position the two-layer top on the workbench. Be sure the notched end is flush with the outside of the left end panel (H) and centered side-to-side. Then, clamp it in position.

4 Referring to **6–52**, drill four ½" holes through the two layers of the top and through the stretchers (J).

5 Cut ½" dowels 2¼" long and put a slight taper on the first ⅜" of each one to ease insertion. Drive the tapered ends of the dowels into the four holes until

6–61. HOOD

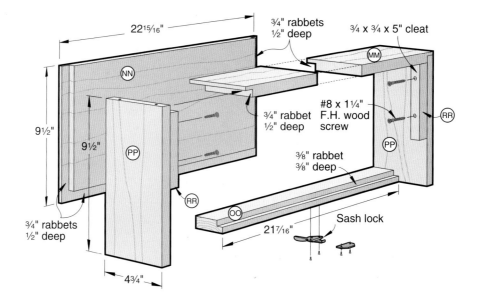

¾" remains above the top. Now apply glue around the dowels, and drive them flush. The glue will secure the dowels, enabling you to remove the top.

6 As an optional step, cut dadoes for the metal track system (refer to the supplies section of the Materials List) for anchoring hold-downs and an accessory router/sanding pad. Secure these tracks by drilling and countersinking holes for #6 × ¾" flathead sheet-metal screws. Drill the first holes 2¼" from each end, and then every 6". Add the final top piece. Carefully align the edges and ends and clamp them. Drill pilot holes from the bottom of the top assembly, then screw the layers together with #8 × 2" flathead wood screws. Space these screws about 1" from the perimeter and 12" apart.

7 Cut and add the edge bands (KK) and one of the end bands (LL). Do not install the end band that spans the notch at the end of the bench. You'll add that piece after installing the dust-hood assembly. We used a biscuit joiner to cut slots for biscuits that helped align these pieces during glue-up and clamping.

Fashion the Hood

1 Begin making the hood by ripping and crosscutting ¾"-thick solid maple to size for the top (MM), back (NN), bottom (OO), sides (PP), bracket (QQ), and stop blocks (RR).

2 Put a dado blade in your tablesaw, and cut the rabbets in these parts, where shown on **6–61**. Glue and clamp together parts (MM) through (PP) to make the hood assembly.

3 Use glue and screws to attach the hood bracket (QQ) to the left end panel (H). Also glue and screw the stop blocks (RR) to the dust hood assembly.

4 Place the dust hood assembly into the notch at the end of the workbench and temporarily clamp it into the raised position. Install the sash lock centered on parts OO and QQ, where shown on **6–52**. Glue and clamp the end band (LL) into position.

Make Some Helpful Accessories

1 To make the router/sanding pad, we cut four sanding-mat strips (SS) of solid cherry. Then we attached a pair of these strips at opposing ends of a rubber mat with polyurethane glue and ½" brads. Placing the strips on a piece of scrapwood, drive the brads through the strips, and then turn the assembly over to clinch the brad tips into the strips.

To use the pad, put one end into one of the metal channels and slightly stretch the mat to put the other end into the other channel. This slight tension keeps the mat in place.

2 If you want to install a vise similar to the one we used, you'll most likely need to cut spacers (TT) to the size listed in the Materials List. We used 1½" brads in the corners to hold the spacers in position; then drilled pilot holes for the lag screws.

Final Steps

1 After removing the hardware, we applied two coats of water-based polyurethane to the exterior of the workbench. We applied two coats of oil finish to the top to prevent glue drippings from adhering. Reinstall the fittings after the finish dries.

2 Move the bench to its shop position, and put shims under the feet to level it. Route the power cord through a floor-mounted cord cover available at a hardware or office-supply store. The cover protects the cord, and securing the cover to the floor with double-faced tape helps reduce a tripping hazard.

❖

WORKBENCH TOOL CRIB

Designed to fit beneath an open workbench, this nifty storage tray (**6–62**) securely tucks out of sight (and away from your legs) when not in use. Store portable power tools, or whatever else you wish, in it.

Simply build the crib as shown to fit between the legs of your workbench. Fasten it in place with two $\frac{3}{8}$" carriage bolts. To put your crib into action, simply pull on the handle cutout and swing it out. A pair of stops screwed to the bottom of your workbench prevent the crib from falling too far forward.

As an added feature, cut a hole in the back piece and install a multiple electric outlet in the tool crib.

❖

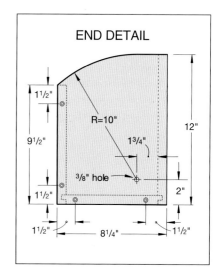

END DETAIL

6–62. **EXPLODED VIEW**

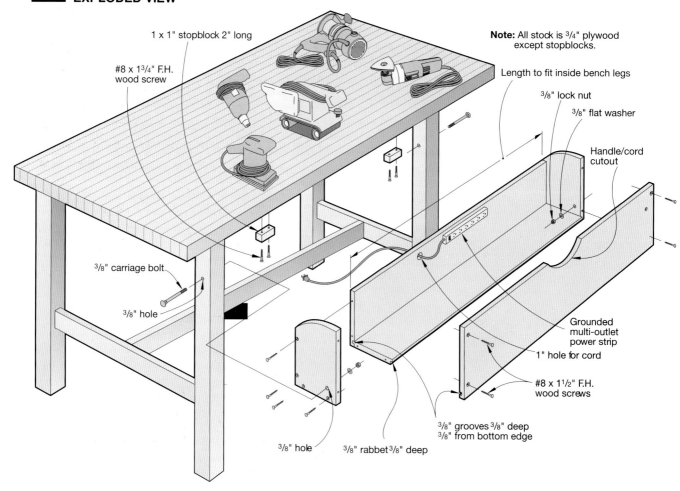

1 x 1" stopblock 2" long

#8 x 1$\frac{3}{4}$" F.H. wood screw

Note: All stock is $\frac{3}{4}$" plywood except stopblocks.

Length to fit inside bench legs

$\frac{3}{8}$" lock nut

$\frac{3}{8}$" flat washer

Handle/cord cutout

$\frac{3}{8}$" carriage bolt

$\frac{3}{8}$" hole

Grounded multi-outlet power strip

1" hole for cord

#8 x 1$\frac{1}{2}$" F.H. wood screws

$\frac{3}{8}$" grooves $\frac{3}{8}$" deep $\frac{3}{8}$" from bottom edge

$\frac{3}{8}$" hole

$\frac{3}{8}$" rabbet $\frac{3}{8}$" deep

FLATTEN YOUR WORKBENCH'S TOP

Professional woodworkers know that a flat workbench top is essential for assembling square boxes, flat panels, or four-legged projects that don't rock. They don't even want to work on a benchtop that's not true. And if it's not, here's the best way to level a troublesome one.

Note: These procedures work for any solid-wood top that's at least 1" thick. We recommend you flatten your benchtop if it's uneven by more than $1\frac{1}{4} \times \frac{11}{32}$". (You can check for flatness by moving a straightedge across your benchtop.) Do not try these flattening steps on laminated tops made of plywood, particleboard, hardboard, or similar materials.

To show you how this technique works, we searched for a bench badly in need of flattening. It took us awhile, but we eventually found a doozy. As you can see in **6–63**, this turn-of-the-century bench had a warped, irregular surface that was out of flat by more than ½". It was a great-looking antique, but it needed its usefulness restored to become a woodworking tool once again. Although the flattening process described here bared new wood on the benchtop, the rest was restored to its antique patina by rubbing in a combination of stains afterwards.

6–63.

Can this benchtop be saved? Absolutely.

6–64.

Prior to flattening the benchtop, make any necessary repairs to the base and the workbench top.

Start by Preparing the Bench

Before you flatten the bench, you need to make any necessary repairs. As shown in **6–64**, our sample bench had delaminated edge boards that required our attention. You may have to remove the top and retighten, reglue, or reinforce the joints in the base to make it solid and rack-free.

Because you will be routing into the top in the following steps (**6–65**), you need to remove any embedded metal fasteners (such as brads and staples). A metal sensor will aid your search, and save you from dulling or destroying a router bit.

Next, place the bench in the spot in your shop where you will be using it. Check the top for level (**6–65**), and add wood shims

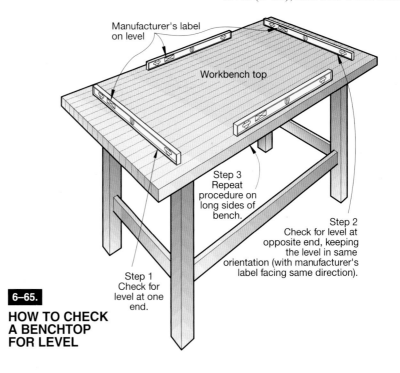

Manufacturer's label on level

Workbench top

Step 3
Repeat procedure on long sides of bench.

Step 2
Check for level at opposite end, keeping the level in same orientation (with manufacturer's label facing same direction).

Step 1
Check for level at one end.

6–65.

HOW TO CHECK A BENCHTOP FOR LEVEL

6–66.

The flattening jig clamps to the bench-top and the router rides back and forth in the router carrier to even the wood.

to the bottom of the base's legs as necessary.

With the top as level as possible, attach the shims with adhesive or fasteners. Mark the position of each leg onto the floor so that you can always return the bench to its level location.

Now, Make and Attach the Flattening Jig

With the jig shown in **6–66**, you can slide a router over your benchtop to remove all the high areas and leave a flat surface. The router slides back and forth across the top in a router carrier that slides along a perfectly level and straight carrier guide on each side of the benchtop. These pieces are mounted to the bench with carrier-guide supports

To mount the carrier guides and carrier-guide supports, follow **6–67**. (You may find it necessary to remove one or more vises.) To make the carrier guides, joint one edge of a 2 × 4 and rip the opposite edge to give you a straight board with parallel edges.

After mounting the carrier guides, use a level to make sure they are level along their length and level with each other. If the benchtop is twisted you may find it necessary to shim the carrier guides with thin pieces of wood or paperboard, as shown in **6–68**. Use only as many shims as necessary.

Now, use another jointed board as long as the carrier guides to check them for straightness. If the carrier guides bowed during mounting, you will need to correct the situation with shims.

These leveling and shimming steps can prove time-consuming, but stick with them—you'll be rewarded with a flat bench in the end. Finally, construct a router carrier according to **6–70**.

Rev Up the Router and Start Flattening

To adjust your router's straight bit for the correct cutting depth, use a straightedge and tape measure, as shown in **6–69**, to find the lowest spot on the bench. Then, place the

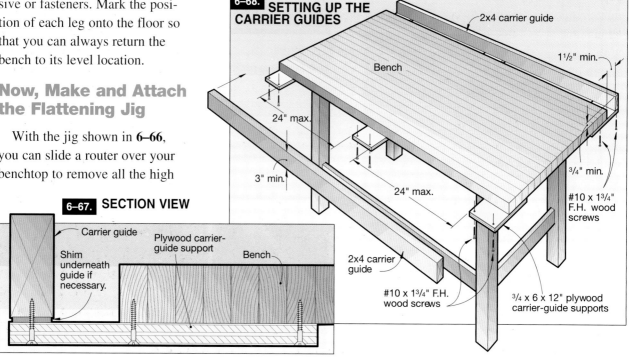

6–68. SETTING UP THE CARRIER GUIDES

2x4 carrier guide

Bench

1½" min.

24" max.

3" min.

24" max.

¾" min.

#10 x 1¾" F.H. wood screws

2x4 carrier guide

#10 x 1¾" F.H. wood screws

¾ x 6 x 12" plywood carrier-guide supports

6–67. SECTION VIEW

Carrier guide

Plywood carrier-guide support

Bench

Shim underneath guide if necessary.

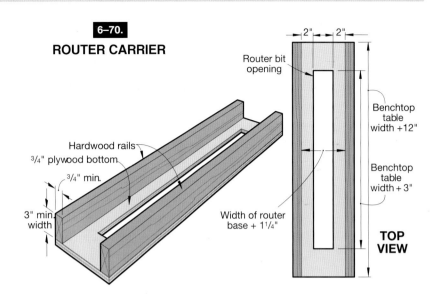

6–70.
ROUTER CARRIER

Router bit opening

Hardwood rails
³/₄" plywood bottom
³/₄" min.
3" min. width

Width of router base + 1¹/₄"

2" 2"

Benchtop table width +12"

Benchtop table width + 3"

TOP VIEW

Slide a straightedge along the carrier guides, and use a tape measure to find the low spot on the benchtop.

router carrier onto the carrier supports so its slot centers over the low spot. Mount a 1"- to 1½"-diameter straight bit in your router, place the router into the router carrier (directly above the low spot), and set the bit for a paper-thin cut. You must securely tighten the router bit into its collet, and the router motor housing to the base, to eliminate any possibility of the cutting depth changing while you complete the following steps.

Next, use four clamps to secure the router carrier to the carrier guides at either end of the bench. Make a full cut across the width of the bench by first running the router away from you and along the left hardwood side of the router carrier. Then, make a return cut toward you along the right side of the router carrier. Without moving the router carrier, repeat these router passes once or twice so the bit cuts to its

full depth. As you move the router, remember to place minimal downward pressure on it to avoid bowing the router carrier.

Repeat this procedure at three evenly spaced locations along the length of the bench, as shown in **6–71**. These cuts confirm that you adjusted the router bit to the correct depth. Adjust the bit for a deeper cut if it passes over any areas without removing stock. If the router bogs down during these cuts, switch to a smaller bit or a more-powerful router.

Now, measure the width of any of your four router cuts. Subtract ¼" from your measurement and use this figure to space out marks along the length of each carrier guide. For instance, our router cuts were 2¼" wide, so we made marks every 2". Use these marks in the following step to guide your placement of the router carrier so your cuts overlap by ¼".

Next, flatten the entire top by starting at either end of the bench

and making router passes in increments along its length. For consistently deep cuts you must clamp the router carrier for each router cut. You can speed things along by having a helper reposition and reclamp one end of the router carrier while you clamp the other end and operate the router.

As you approach each of your initial, spaced-apart cuts, check to see that your cutting depth has not changed. If it changes, the bit or motor housing is slipping up or down, and you will need to fix the problem. In working on the benchtop shown here, we were tripped up by the brand-new router we were using. Factory lubrication on the outside of the motor housing was causing the housing to slip, even though we tightened the base securely. So, we cleaned the motor housing and base with mineral spirits, and then started the routing process over again. Another lesson learned the hard way!

6–71.

Make four evenly spaced cuts across the benchtop to test the cutting depth of the router bit.

6–72.

A cabinet scraper and random-orbit sander help make quick work of the tiny ridges left on the benchtop.

The Final Touches

No matter how careful you were to clamp the router carrier and put minimum downward force on the router, you will still wind up with fine ridges where one router pass meets the other. Fortunately, you can quickly smooth away these ridges while keeping your bench flat.

As shown in **6–72**, we lowered the ridges with a cabinet scraper and followed this with a light sanding using a random-orbit sander. Be careful to remove the ridges and no more.

❖

CHAPTER 7

Shop Cabinet Craftsmanship

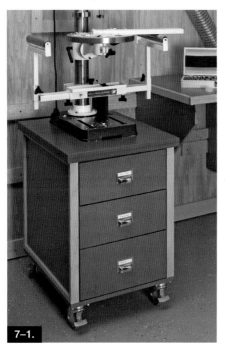

7–1.

This basic three-drawer cabinet can be easily built.

WANT TO SEE A LOT *of cabinets? Visit the kitchen-planning area of a large home center. You'll see dozens and dozens of styles and sizes in both base and wall varieties, right on the sales floor. Then, if you want even more options, peek into the special-order catalog.*

Now your woodworking shop may not be the heart of the home as is your kitchen, but there is a kindred need for cabinets. And depending on your work flow and the amount of tools, supplies, and hardware you need to store, your shop-cabinet requirements may be every bit as special and varied as those for your kitchen.

Sure, you can buy ready-made cabinets for use in your shop. But it's a bet that you'll eventually make a few anyway down the line to meet special needs. So why not start off with custom cabinets in the first place (7–1)? You'll not only save money, you'll gain in quality, too. And, just like a basic box, cabinets aren't really difficult to build.

In this chapter you'll learn how pros tackle cabinetmaking, making it seem like child's play. With the tips and tricks you pick up, you can go on to build your choice of the cabinets presented, from specialized tool versions to the nifty, yet simple, shop reference center on *page 170*. We've done everything possible to help you with the cabinetmaking process, so go ahead, start building!

START WITH A BASE CABINET

You need only basic woodworking skills to build the standard cabinet box, so start with a typical base cabinet, as shown in **7–4**. (It's Euro-style, with no exposed hinges.) Vary the dimensions and the number of drawers to make nearly any cabinet unit you want.

You'll need enough solid wood (hardwood for a natural finish or softwood for painting) to build doors, drawers, and various trim pieces. You can use ¼" plywood, either hardwood or softwood, for door panels. Buy ¾"-thick melamine-coated sheet goods for the bottoms and sides of base cabinets and the sides, top, and bottom of wall cabinets. Use ¼"-thick material for the backs. (For the cabinets shown, we used a special "thermo-fused" melamine that has one easy-to-clean shiny white face and one brown surface

7–2.

Equip your router with a good edge guide before you start making grooves for cabinet backs. Set the guide, and then do all the backs at once.

great for adding veneer. It comes in oversized 49 × 97" sheets to allow for kerfs and cost about $30 each from a builders' supply outlet.)

Build the Cabinet

1 To avoid chipping the melamine, outfit your tablesaw with a sharp, 80-tooth blade. Then, cut a sheet in half lengthwise, and cut one half into three equal pieces, each approximately 24 × 32".

2 One piece at a time, clamp each sheet of melamine to your workbench, with the white face up. This surface will be visible inside the cabinet. Equip your router with an edge guide and a ¼" straight bit, and rout a groove ⅜" deep and ⁹⁄₁₆" from the edge, as shown in **7–2** and **7–4**. This edge is going to serve as the back edge of the cabinet sides and bottom.

3 Now, cut the two sides to 23¼ × 30½". Make the cabinet bottom 23¼" deep, and as wide as the finished width of the cabinet, minus 1½". From the same material, cut two rails for the top, and a drawer divider, each piece 5" wide and as long as the bottom's width. Rout a groove in the back rail, sized and located to match the grooves in the sides.

4 Now, because they will show on the finished cabinet, put veneer or PVC tape on the front edges of the sides, bottom, and rails. We used a hardwood veneer, but PVC tape in several colors is readily available.

Apply veneer and PVC tape with a common household iron, as shown in **7–3**. Secure the workpiece vertically in a vise or in a handscrew clamp that is, in turn, clamped to the top of your workbench. Cut enough tape to protrude beyond both ends of the piece. With your iron set at

7–3.

All you need to apply veneer is an inexpensive household iron. If you plan to do a lot of cabinet work, you can buy veneer tape in 250' rolls.

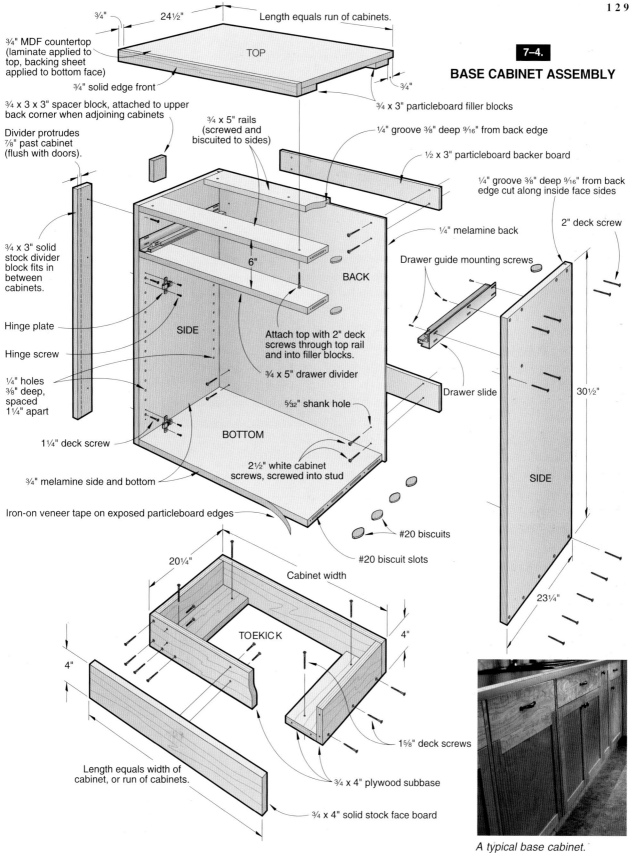

7–4.

BASE CABINET ASSEMBLY

¾" 24½" Length equals run of cabinets.

¾" MDF countertop
(laminate applied to
top, backing sheet
applied to bottom face)

TOP

¾" solid edge front

¾" ¾ x 3" particleboard filler blocks

¾ x 3 x 3" spacer block, attached to upper
back corner when adjoining cabinets.

¾ x 5" rails
(screwed and
biscuited to sides)

¼" groove ⅜" deep ⁹⁄₁₆" from back edge

Divider protrudes
⅞" past cabinet
(flush with doors).

½ x 3" particleboard backer board

¼" groove ⅜" deep ⁹⁄₁₆" from back
edge cut along inside face sides

¼" melamine back

2" deck screw

¾ x 3" solid
stock divider
block fits in
between
cabinets.

BACK

Drawer guide mounting screws

6"

Hinge plate

SIDE

Attach top with 2" deck
screws through top rail
and into filler blocks.

Hinge screw

¾ x 5" drawer divider

Drawer slide

¼" holes
⅜" deep,
spaced
1¼" apart

⁵⁄₃₂" shank hole

30½"

1¼" deck screw

BOTTOM

SIDE

¾" melamine side and bottom

2½" white cabinet
screws, screwed into stud

Iron-on veneer tape on exposed particleboard edges

#20 biscuits

23¼"

#20 biscuit slots

20¼" Cabinet width

4"

TOEKICK

4"

1⅝" deck screws

¾ x 4" plywood subbase

Length equals width of
cabinet, or run of cabinets.

¾ x 4" solid stock face board

A typical base cabinet.

medium heat, run it along the length of the piece. This melts enough of the adhesive backing to hold the tape in place.

Then, go over the tape again, holding the iron in place for three or four seconds in each spot. Move an iron's length ahead, and also press a wood block on the previous spot. The wood absorbs some of the heat, and the pressure helps to set the veneer in place.

Score the tape from beneath at one end of the workpiece, using the wood block as a backing surface. Snap off or cut the veneer, then trim it flush. Repeat at the opposite end.

5 There are several handy little tools out there made to trim both edges of the veneer at the same time, as shown in **7–5**. Or, you can shave the veneer flush with a sharp chisel or craft knife.

6 See Make a Shelf Hole Jig, on *page 135*, for details about making a drilling jig that you use to position shelf holes on the cabinet sides. Use this jig to drill a series of holes in each area where you want to place a shelf, to allow for some adjustment in shelf spacing.

Most factory-built cabinets have holes from the top of the cabinet clear down to the bottom, but there's no need to do that. Your cabinets will look better on the inside with fewer holes. In addition, you won't wind up with a hole, or half a hole, near the bottom, where sawdust or

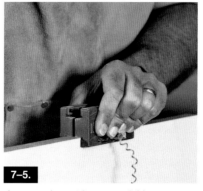

7–5.

A razor-sharp trimmer quickly cuts the lightweight veneer. This spring-loaded model handles both sides at once, and works on stock of various thicknesses.

liquids could wind up in the particleboard core.

7 Attach the sides to the bottom with #20 biscuits and #8 × 2" production screws. Keep it simple, and line up the edge of the biscuit joiner with the edge of the cabinet side, as shown in **7–6**, and then the edge of the bottom. That way, you don't even need to make pencil marks. Use four biscuits per edge.

8 Also, cut biscuit slots for the drawer divider. A shopmade jig will do this. It consists of a square piece of particleboard, 5" on each side, with wider guide boards screwed to two adjoining sides. Fit and clamp the jig against the corner of a cabinet side, line up the biscuit joiner, as shown in **7–7**, and you're ready to cut, again without need for measuring or marking.

9 Attach the sides to the bottom with biscuits and yellow glue. Before that glue has set, install

the drawer divider and front top rail, as shown in **7–8**, with biscuits and glue. Rails provide all the strength you need, and the extra access area makes installation easier than it would be with a full top.

10 Slide the back into its grooves from above, as shown in **7–9**. Spread the corners just enough to fit the back top rail and its biscuits into place. Add five #8 × 2" self-tapping, flathead screws along each bottom edge and two screws into each end of each rail. Drill a 1/8" countersunk starter hole for each screw. Always drill pilot holes into particleboard.

11 Lay the cabinet on its face to install the top and bottom backer boards, made from 1/2" particleboard. Cut them 3" wide and the same length as the rails. You'll drive screws through these pieces when you install the cabinets.

7–6.

Keep joinery simple and avoid measurements completely, when possible. Here, the biscuit joiner is lined up with the edge of the cabinet side.

7–7.

Make a simple jig to guide your biscuit joiner for the drawer divider slots, and then label it. Again, you eliminate the need for measurements.

12 Glue the backer boards in place with yellow glue. Press down on each board as you drive brads through the cabinet sides and into the ends of the board, as shown in **7–10**. The pressure closes any gap on the visible side of the back, and the brads hold the boards in place while the glue dries. The brads, like the screwheads, won't be visible once the cabinets are set in place.

7–9.

Slide the back into place without glue. That flexibility will come in handy if you need to adjust the cabinet slightly for squareness during installation.

7–8.

After you've attached the sides to the bottom, there's still enough "give" to allow the installation of the drawer divider and top front rail.

13 Euro-style cabinets are designed to butt against each other, but we modified that arrangement. We attached a ¾" solid-wood divider between each adjacent pair of cabinets. Each divider protrudes ⅞" from the front edge of the cabinet. That measurement puts the dividers flush with the cabinet doors, which are held ⅛" away from the cabinet by their hinges.

7–10.

Glue the backer boards in place, and then add small fasteners to hold them while the glue dries. Use a pneumatic gun, or just tap in some brads with a hammer.

NOW, TRY A WALL CABINET

The wall cabinets are a shallower, simpler version of the base cabinets. All you need is the basic box and shelves. You don't have to deal with drawers. See the details in **7–11**, on *page 132*.

Make the Box and Shelves

1 For the sides, cut ¾" melamine 12" wide and to the length needed. Cut the top and bottom 12" wide and to a length equal to the finished width of the cabinet minus 1½". Rout a groove for the back, as in the base cabinets. Drill the shelf support holes, again assuming two shelves per cabinet.

2 Make two biscuit slots per joint. Assemble the bottom and sides with biscuits and yellow glue, slide the back into its grooves, and add the top. Add four screws per joint, following the procedure described previously.

3 To make each shelf, cut ¾" melamine the same length as the width of the cabinet opening, and ¼" narrower than the inside depth of the cabinet. Cover the exposed edge on the front of each shelf by ironing on PVC tape.

Make the Doors

We built frame-and-panel cabinet doors with exposed, stub tenons joining rail to stile,

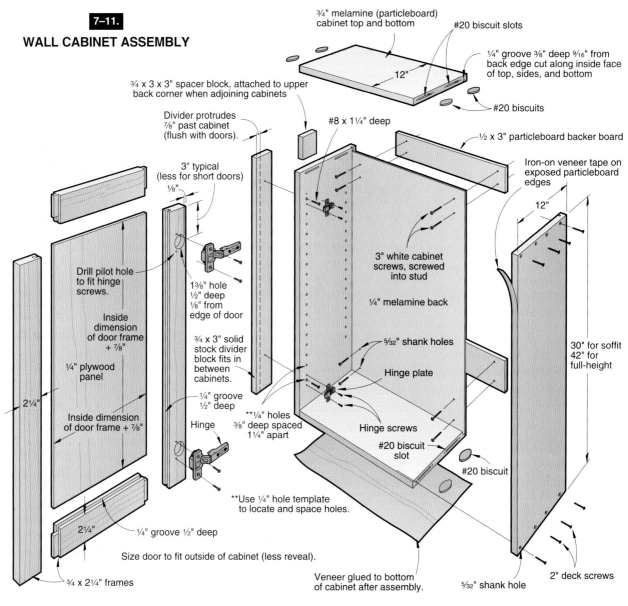

7-11.

WALL CABINET ASSEMBLY

¾" melamine (particleboard) cabinet top and bottom

#20 biscuit slots

¼" groove ⅜" deep ⁹⁄₁₆" from back edge cut along inside face of top, sides, and bottom

#20 biscuits

¾ x 3 x 3" spacer block, attached to upper back corner when adjoining cabinets

Divider protrudes ⅞" past cabinet (flush with doors).

#8 x 1¼" deep

½ x 3" particleboard backer board

3" typical (less for short doors)

⅛"

Iron-on veneer tape on exposed particleboard edges

12"

Drill pilot hole to fit hinge screws.

1⅜" hole ½" deep ⅛" from edge of door

3" white cabinet screws, screwed into stud

Inside dimension of door frame + ⅞"

¼" plywood panel

¾ x 3" solid stock divider block fits in between cabinets.

¼" melamine back

¼" groove ½" deep

5⁄32" shank holes

Hinge plate

Inside dimension of door frame + ⅞"

Hinge

30" for soffit 42" for full-height

2¼"

**¼" holes ⅜" deep spaced 1¼" apart

Hinge screws

#20 biscuit slot

Inside dimension of door frame + ⅞"

#20 biscuit

**Use ¼" hole template to locate and space holes.

2¼"

¼" groove ½" deep

5⁄32" shank hole

Size door to fit outside of cabinet (less reveal).

¾ x 2¼" frames

Veneer glued to bottom of cabinet after assembly.

5⁄32" shank hole

2" deck screws

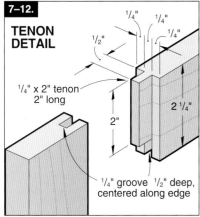

7-12.

TENON DETAIL

¼"   ¼"

½"   ¼"

¼" x 2" tenon 2" long

2¼"

2"

¼" groove ½" deep, centered along edge

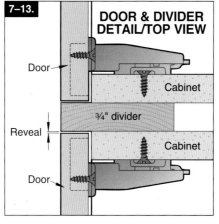

7-13.

DOOR & DIVIDER DETAIL/TOP VIEW

Door

Cabinet

¾" divider

Reveal

Cabinet

Door

and flat panels. We used solid hardwood for the frames and ¼" hardwood plywood for the panels. We kept a 90° profile around the outside edges, to match the cabinet dividers. See **7-11** to **7-13** for the basic door dimensions and construction details. Do the following:

1 For a single-door cabinet, make the door width ¼" less than the width of the cabinet itself. Plan on double doors for any cabinet that's more than 22" wide. In that case, subtract ⅜" from the cabinet width and divide the result by 2, to get the width of each door.

2 Rip enough solid stock 2¼" wide to make the rails and stiles for your doors. Then, take some time to choose the straightest pieces for the stiles. Straight stiles produce flat, good-fitting doors. Cut the longest stiles to length first, if you're making cabinets of various heights, then cut the remaining stiles, and finally the rails. To determine the length of the rails, subtract 4½" (the combined width of two stiles) from the finished door width; then add 1" to account for the tenons.

3 With a standard blade on your tablesaw, cut a groove on the inside edges of the stiles and rails, as shown in **7–14**. Make it ½" deep and just wide enough to accept the plywood. Usually, ¼" plywood won't measure exactly ¼", so run some test cuts on scrap pieces first. Cut the length of the piece, and then flip the other face against the fence and cut in the opposite direction. This step centers the groove on the workpiece. Adjust the fence until the width is perfect.

4 Equip your tablesaw with a dado set and cut ½"-long tenons on both ends of each rail, sized to fit the grooves. We used

7–14.

A feather board adds a lot to your accuracy when you're cutting grooves on the edges of the door rails and stiles.

a sliding carriage that rides in the miter gauge slots and a hold-down clamp, as shown in **7–15**.

5 Ease the inside edges of the grooves with 150-grit sandpaper on a block. The door will look nicer. Besides, a sharp edge doesn't take finish well.

6 Dry-fit the doors, measure for the panels, and then cut the panels with a plywood blade mounted on the tablesaw. Allow for a ¹⁄₁₆" gap at each side and each end to make for a smooth assembly. Ease the edges of this panel, too, with 150-grit sandpaper on a block. Rounded edges slip easily into the grooves without catching and possibly damaging the veneer. When everything fits perfectly, glue the tenons in place with yellow glue and clamp each door with pipe clamps or bar clamps. Again, use 150-grit sandpaper to ease the exposed edges of the rails and stiles next to the panel. Check the door for square by measuring the diagonals, as shown in **7–16**.

7 Now drive two ⅝" brads into each tenon from the back side

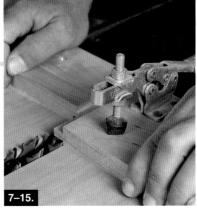

7–15.

This shopmade tenoning jig features a hold-down toggle clamp. Make your own carriage-style jig, or use a miter gauge.

of the door and remove the clamps. Lay each door on a flat surface and weight it down with more doors, or boards, while the glue dries. Crisscross the doors as you stack them. Otherwise, squeeze-out at the joints could glue them together.

8 We recommend cup-style hinges, which are unseen when the doors are closed; and they allow adjustments. Home centers and hardware stores carry various brands, or check a woodworking-supply catalog, which also offers the accessories that help you position the hinges accurately. You'll need a 1⅜" Forstner bit to make the cup holes in the doors.

Design the Drawers

The typical cabinet drawer is 22" deep. Its width depends not only on the opening, but also on the kind of slides you choose. The height must be at least ¼" less than the height of the opening but might have to be reduced, again depending on the slides.

See **7–18**, on *page 134*, for the dimensions and details of one of our typical drawers. Maple is always a good choice for the

7–16.

Before the glue sets, measure across the diagonals of each door. Equal measurements mean the corners are square.

7–17.

Buy an affordable dovetail jig and practice with it, and you'll be able to make the drawer joints quickly.

SHOP TIP

Skid Control

As you complete each base cabinet, cut small pieces of particleboard scrap for skids, and fasten them underneath with a brad nailer. Now you can slide the heavy units around your shop without damaging them while you work on the doors and drawers.

7–19.

sides, front, and back of the drawer box. (Poplar also works.)

Any of the popular dovetail jigs, along with a router and dovetail bit, will produce the half-blind dovetails you want in a quality drawer. Our setup is shown in **7–17**. Of course, you might choose another type of joint that's easier to make.

For a clean look, we installed undermount drawer slides, which don't show on the sides of the drawers. Following the manufacturer's directions, mount the guides on the ¾" melamine divider rail that separates the drawer compartment from the rest of the cabinet. These under-

mount slides require just ³⁄₁₆" clearance on each side. A good hardware catalog will specify the side and height clearances for each style of slide.

It's Toekick Time

The 3" setback at the bottom of a base cabinet lets you get right up close without stubbing your toes. Furniture factories build this "toekick" into their cabinets.

Some pros prefer a separate toekick assembly that's easy to

adjust on the site, just in case the floor isn't perfectly level. Once you adjust the toekick, you can place the cabinets on top and know that they're sitting level, too.

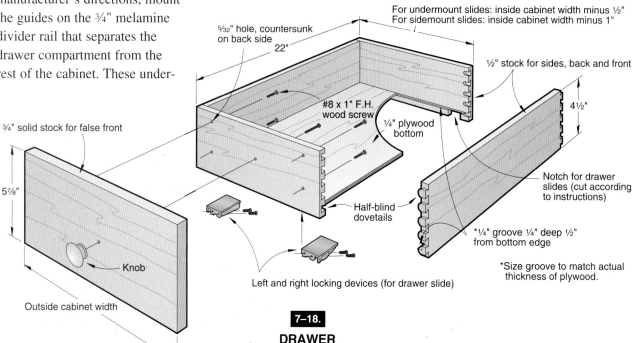

⁵⁄₃₂" hole, countersunk on back side

22"

For undermount slides: inside cabinet width minus ½"
For sidemount slides: inside cabinet width minus 1"

½" stock for sides, back and front

#8 x 1" F.H. wood screw

¼" plywood bottom

4½"

¾" solid stock for false front

5⅞"

Knob

Outside cabinet width

Half-blind dovetails

Left and right locking devices (for drawer slide)

Notch for drawer slides (cut according to instructions)

*¼" groove ¼" deep ½" from bottom edge

*Size groove to match actual thickness of plywood.

7–18.

DRAWER

Make a Shelf Hole Jig

To make sure your shelf holes line up accurately, you need a drilling jig. Make the simple one shown in 7–20 out of scrap material, and you'll be all set to construct any standard cabinetry.

To use the jig, fit its guide boards against one top corner of a cabinet side. Mark the guide holes you want to use, set a depth stop on your ¼" brad-point bit, and then drill the shelf holes ⅜" deep. Flip the jig over to fit against the opposite top corner, mark the correct holes, and drill again, being careful not to enlarge the guide holes.

Register with top edge of side panels.

¾ x ¾" stock

¼" groove ¼" deep, centered

5½"

1¼"

1¼"

30"

*¼" holes spaced 1¼" apart

¼" hardboard

*Drill ¼" holes ⅜" deep into cabinet with a ¼" brad-point drill bit.

1½"

7–20.

To make the toekick, do the following:

1 Cut scraps of ¾" particleboard or plywood to a 4" width. For each run of cabinets, you'll need two pieces the length of that run, and enough 18¾" pieces to make the vertical and horizontal cross members set at 16" intervals. Join the cross members into L-shaped structures with screws, as shown in **7–4** on *page 129*, and then screw on the sides, making a support 20¼" wide.

2 Apply a durable finish to all surfaces. For solid-wood surfaces, polyurethane varnish stands up to use.

❖

CLAMP A LARGE BOX WITH CONFIDENCE

To demonstrate the techniques involved in carcass clamping, we assembled a typical base cabinet. (This cabinet will later get a face frame, but our demonstration doesn't include that step.) You can adapt many of the techniques you see here for clamping boxes of all sizes: jewelry boxes, wall cabinets, and drawers, for example.

1 When you build a plywood cabinet, be certain to gauge the width of the rabbets and dadoes from the actual thickness of the stock, not its "nominal" or advertised size. Strive for dadoes that let you assemble the cabinet

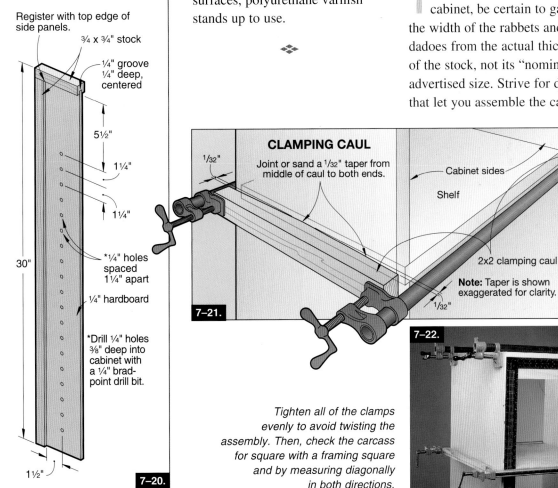

CLAMPING CAUL

1/32"

Joint or sand a 1/32" taper from middle of caul to both ends.

Cabinet sides

Shelf

2x2 clamping caul

Note: Taper is shown exaggerated for clarity.

1/32"

7–21.

Tighten all of the clamps evenly to avoid twisting the assembly. Then, check the carcass for square with a framing square and by measuring diagonally in both directions.

7–22.

Coat the edges of a glue block with glue and rub it back and forth until the glue grabs. No clamping is necessary.

without hammering on the plywood or particleboard.

2 Dry-clamp the assembly to check the fit of all the pieces and to prepare your clamps. Make a pair of clamping cauls, as shown in **7–21**, for each fixed shelf in the cabinet. These cauls will help distribute clamping pressure evenly over the length of the dadoes holding these shelves.

3 Apply glue to mating surfaces and clamp up the assembly, as shown in **7–22**. Check for square with a framing square, and double-check by measuring diagonally in both directions across the front of the carcass. If the diagonals aren't equal, place a clamp on the corners of the longer diagonal and apply pressure to equalize the diagonals. Leave all clamps in place for at least one hour, and then remove them. Install the back immediately to avoid straining the joints.

4 You can reinforce the joints with glue blocks positioned inconspicuously, as shown above, in **7–23**. These blocks are easy to make, install quickly, and add strength to the cabinet's joints.

❖

BUILD A FLOOR-CABINET SYSTEM: THE BASIC CABINET

If you've decided to build the cabinet-based workbench shown in **7–24** and described in Chapter 6 on *pages 93* to *95*, you'll first have to build the cabinets. But guess what? Once you've got that process down pat, you can go on to build other cabinets of the same basic type that adapt to other uses, as you'll see on the following pages.

Inexpensive materials, easy joinery, and flexibility describe this sturdy workshop floor cabinet. You can build it with legs and adjustable levelers, as shown in **7–24**, or without legs so it will ride a mobile base (see *page 250*). If you're interested in possibly making several of these cabinets, check out the production tricks on *pages 138* and *139*. You'll find several ways to save time.

Start with a Pair of Side Assemblies

1 To make the cabinet with legs, cut the legs (A) and leg cleats (B) to the sizes listed in the Materials List, or for the cabinet without legs, cut the optional stiles (C) to the size listed. Then cut the side rails (D), front rail (E), and the back rail (F) to size.

2 With a dado blade adjusted to match the width of your ½" plywood, cut the ¼"-deep rabbet in part E, where shown on **7–25**. Then cut centered grooves in parts A and B (or C), D, and F, where shown on **7–25** and **7–26**.

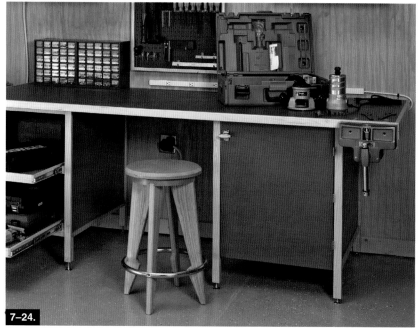

Fit two cabinets like this with a top (see Chapter 6) and, presto, you've got a sturdy workbench.

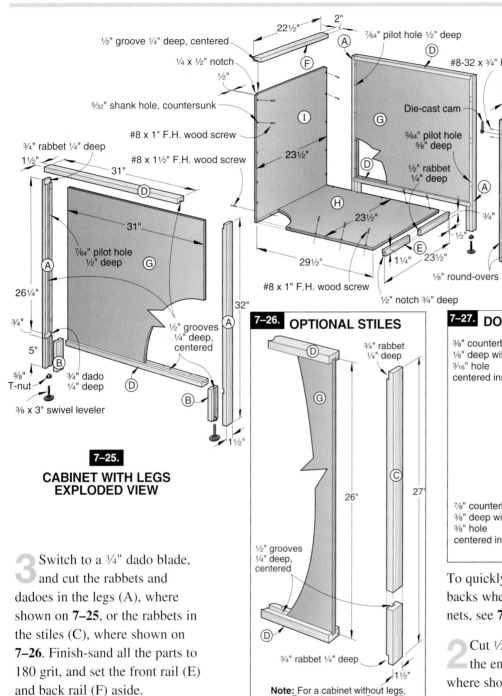

7–25.

CABINET WITH LEGS EXPLODED VIEW

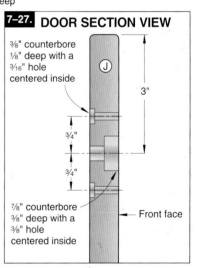

7–26. **OPTIONAL STILES**

Note: For a cabinet without legs, part Ⓒ replaces parts Ⓐ and Ⓑ.

7–27. **DOOR SECTION VIEW**

To quickly and cleanly notch the backs when making multiple cabinets, see **7–29**.

Cut ½" notches ¾" deep in the ends of the front rail (E), where shown on **7–25**. Glue and clamp the rail to the bottom (H), keeping the ends flush. Glue and clamp the back rail (F), centered, to the back (I). Clamp the back to the bottom and drill pilot and countersunk shank holes through the back and into the bottom. Drive the screws.

Switch to a ¾" dado blade, and cut the rabbets and dadoes in the legs (A), where shown on **7–25**, or the rabbets in the stiles (C), where shown on **7–26**. Finish-sand all the parts to 180 grit, and set the front rail (E) and back rail (F) aside.

Cut the side panels (G) to the size listed. Glue and clamp the legs and rails or the stiles and rails to the panels. For the cabinet with legs, glue and clamp the leg cleats (B) to the legs (A), where shown on **7–25**.

A Bottom and Back Complete the Cabinet

Cut the bottom (H) and back (I) to the sizes listed. Cut the ¼ x ½" notches on the back's top corners, where shown on **7–25**.

3 Place the first side assembly flat on a horizontal surface. Squeeze a bead of glue on the side panel along the rear leg or stile and the lower side rail. Place the back/bottom assembly (E/F/H/I) on the side assembly, and clamp it in place. Drill countersunk pilot holes through the bottom and back and into the rear leg or stile and the lower side rail, and drive the screws. To quickly and accurately drill these angled pilot holes, see **7–30**.

4 Apply glue, and position the second side assembly on the upturned edges of the bottom and back, as shown in **7–28**, and clamp it in place. Turn the cabinet over. As before, drill countersunk pilot holes, and drive the screws.

Now, Add a Door

1 Cut the door (J) to the size listed. Drill pilot holes and

Spread glue on top edges

7–28.

Production Tips for Speedy Cabinet Construction

When you're mounting just one pair of hinges or one handle or pull, it makes sense to lay out their locations by simple measurement. But when making several copies of the same project, this can be time consuming and lead to errors as well. The floor cabinets shown in this section require a number of repetitive operations. Here's how to save time on these tasks.

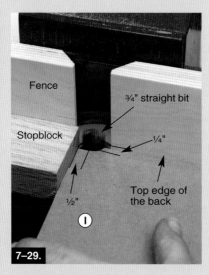

Fence
¾" straight bit
Stopblock
¼"
Top edge of the back
½"
I

7–29.

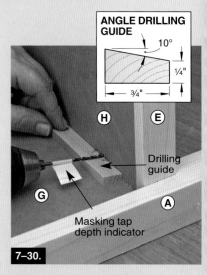

ANGLE DRILLING GUIDE
10°
¼"
¾"
H   **E**
Drilling guide
G   **A**
Masking tap depth indicator

7–30.

PRODUCTION TIP #1:

Notch the Backs with a Router Bit

Sure, you can cut the ¼ x ½" notches in the top corners of the back (I) with a handsaw. But when making several cabinets, a ¾" straight bit in your table-mounted router and a stopblock clamped to its fence make quick work of this task (7–29).

PRODUCTION TIP #2:

Drill Angled Pilot Holes Using a Guide

To drill pilot holes for the screws that join the bottom (H) and back (I) to the side assemblies, bevel-rip an 8"-long drill guide to the profile shown on the inset in 7–30. Resting the bit on the guide, drill the pilot holes. Then remove the guide and countersink the holes.

screw the L-shaped halves of the hinges to the door, where shown on **7–25**. The door is shown hinged on the right; you may hinge it on the left if you wish. Position the door with its top edge ¼" down from the top of the leg (A) or

Apply glue to the upward-facing edges of the bottom and back panels. Lower the second side assembly into place.

optional stile (C), and transfer the hinge locations to the cabinet. Remove the hinges from the door, position them on the leg or stile, and mark and drill the pilot holes. For speedy and accurate hinge installation, see **7–31** and **7–32**.

2 To install the T-handle, mark the center of the ⅞" counterbore on the front of the door at the

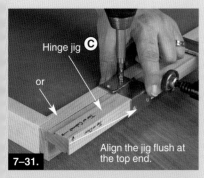

Hinge jig **C**

or

Align the jig flush at the top end.

7–31.

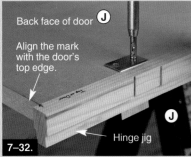

Back face of door **J**

Align the mark with the door's top edge.

J

Hinge jig

7–32.

PRODUCTION TIP #3:

Position the Hinges Perfectly

Make the jig shown on **7–34**. Mark "Top of Cabinet" and "Top of Door," where shown on **7–35**. Align the end marked "Top of Cabinet" with the top of the leg (A) or the optional stile (C), and clamp it in place. Position the hinges in the jig's dadoes, and drill pilot holes, as shown in **7–31**. Now, place the door inside face up. Capturing the hinges in the jig's dadoes, align its "Top of Door" mark with the door's top edge. Clamp the jig to the door, and drill the pilot holes, as shown in **7–32**.

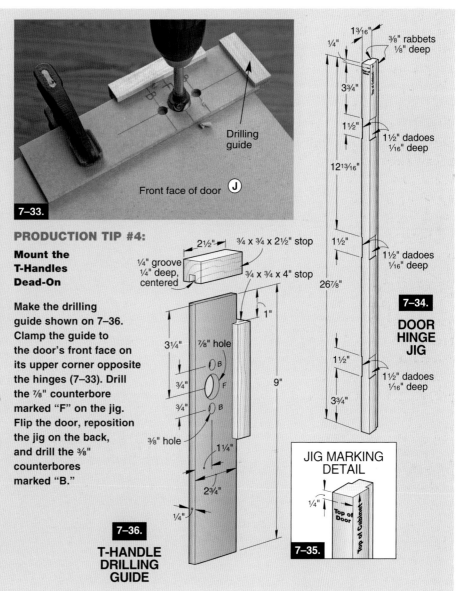

Drilling guide

Front face of door **J**

7–33.

PRODUCTION TIP #4:

Mount the T-Handles Dead-On

Make the drilling guide shown on **7–36**. Clamp the guide to the door's front face on its upper corner opposite the hinges (**7–33**). Drill the 7/8" counterbore marked "F" on the jig. Flip the door, reposition the jig on the back, and drill the 3/8" counterbores marked "B."

2½" — ¾ x ¾ x 2½" stop

¼" groove ¼" deep, centered

¾ x ¾ x 4" stop

1"

3¼"

7/8" hole

9"

¾" F B

3/8" hole

1¼"

2¾"

¼"

7–36.

T-HANDLE DRILLING GUIDE

1 3/16"

¼"

3/8" rabbets 1/8" deep

3¾"

1½"

1½" dadoes 1/16" deep

12 13/16"

1½"

1½" dadoes 1/16" deep

26 7/8"

7–34.

DOOR HINGE JIG

1½"

3¾"

1½" dadoes 1/16" deep

JIG MARKING DETAIL

¼"

Top of Door

Top of Cabinet

7–35.

upper corner opposite the hinged edge, and the 3/8" counterbores on the back, where dimensioned on **7–25** and **7–27**. Using Forstner bits, drill the counterbores. Then using your drill press, drill the holes centered in the counterbores. To save repetitive layout time and increase accuracy when installing several handles, see **7–33**.

3 Rout 1/8" round-overs along all the door's edges, except the inside edge on the hinged side. Finish-sand the door.

Install The Hardware

1 For a cabinet with legs, lay the cabinet on its back, center the T-nuts in the channel formed by the mating grooves in the legs and leg cleats, and hammer them in place. Screw in the levelers.

2 Secure the T-handle to the door with #8–32 x ¾" round-head machine screws. Slide the cam over the handle's shaft, position it snug against the door, and tighten the setscrew. Trim the protruding shaft with a hacksaw.

Cutting Diagram

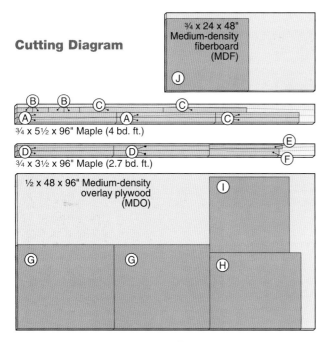

¾ x 24 x 48" Medium-density fiberboard (MDF)

Ⓙ

Ⓑ Ⓑ Ⓒ Ⓒ
Ⓐ Ⓐ Ⓒ
¾ x 5½ x 96" Maple (4 bd. ft.)

Ⓓ Ⓓ Ⓔ
¾ x 3½ x 96" Maple (2.7 bd. ft.) Ⓕ

½ x 48 x 96" Medium-density overlay plywood (MDO)

Ⓘ

Ⓖ Ⓖ Ⓗ

Materials List for Basic Cabinet

| PART | FINISHED SIZE | | | | |
	T	W	L	Mtl.	Qty.
A legs	¾"	1½"	32"	M	4
B leg cleats	¾"	1½"	5"	M	4
C optional stiles	¾"	1½"	27"	M	4
D side rails	¾"	1½"	31"	M	4
E front rail	¾"	1¼"	23½"	M	1
F back rail	¾"	2"	22½"	M	1
G side panels	½"	26"	31"	MDO	2
H bottom	½"	23½"	29½"	MDO	1
I back	½"	23½"	25¾"	MDO	1
J door	¾"	22¼"	26⅝"	MDF	1

Materials Key: M = maple; MDO = medium-density overlay plywood; MDF = medium-density fiberboard.
Supplies: #8 x 1", #8 x 1½", and #4 x ½" flathead wood screws; #8-32 x ¾" roundhead machine screws; solid stock and ¼" hardboard for the T-handle and hinge jigs.
Blades and bits: Stack dado set; ½" round-over router bit; ⅜" and ½" Forstner bits.
Buying Guide: *Cabinet without feet.* 1½" reversing hinges (2 pairs); 2¾" T-handle; die-cast cam. *Cabinet with feet.* Add the following items : ⅜" 6-prong T-nuts (10); ⅜ x 3" swivel levelers (4).

3 Fasten the hinges to the door, and then to the cabinet's leg (A) or optional stile (C), with #4 x ⅝" flathead wood screws.

4 To outfit the basic cabinet for its desired use, see the following pages.

THREE-DRAWER UTILITY CABINET

Follow the basic cabinet-building technique you saw on the preceding pages to create this hard-working shop cabinet (**7–37**). Even the drawers won't slow you down. With their side/slide hardware, they practically build themselves.

Start with the Case

1 Cut the stiles (A), side rails (B), front rails (C), and back rail (D) to the sizes listed in the Materials List. With a dado blade adjusted to match the thickness of the ½" plywood side panels (E), cut centered grooves in parts A, B, and D, where shown on **7–38**. Then cut the ¼"-deep rabbet along the top edge of the lower front rail (C) and the ½" notches ¾" deep in both front rails. Now cut the rabbets in the stiles (A). Finish-sand all the parts to 180 grit.

2 Cut the side panels (E) to the size listed. Glue and clamp the stiles and rails (A/B) to the side panels, checking the assemblies for square.

3 Cut the bottom (F) and back (G) to the sizes listed. Cut the ¼ x ½" notches in the back's top corners.

4 Retrieve the lower front rail (C), and glue and clamp it to the bottom (F), keeping the ends flush. Retrieve the back rail (D), and glue and clamp it, centered, on the back (G). Now, clamp the back to the bottom as shown, and drill screw holes through the back and into the bottom. Drive the screws.

5 Place the first side assembly (A/B/E) flat on a horizontal surface. Squeeze a bead of glue on the side panel along the rear stile and the lower side rail. Place the bottom/back assembly (C/F/D/G) on the side assembly, and clamp it in place. Now drill angled countersunk screw holes through the bottom and back and into the rear stile and the lower side rail. Drive the screws.

6 Apply glue to the upward-facing edges of the bottom and back panels. Position the second side

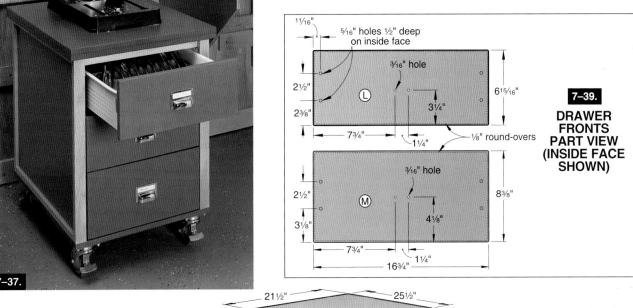

Beef Up the Mobile Base to Carry a Load

To take the strain out of moving tool-bearing cabinets or the cabinet shown combined with another, add a sturdy frame to the mobile base. Use ¾" thick hardwood to make a supporting frame consisting of ends, sides, and two parallel center supports. Size the frame to fit inside the casters.

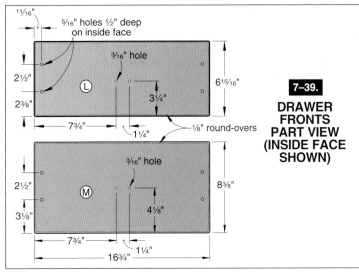

7–39.

DRAWER FRONTS PART VIEW (INSIDE FACE SHOWN)

- 11/16"
- 5/16" holes ½" deep on inside face
- 3/16" hole
- 6¹⁵/₁₆"
- 2½"
- 2⅜"
- 3¼"
- 7¾"
- 1¼"
- ⅛" round-overs
- L
- 3/16" hole
- 2½"
- 8⅜"
- 3⅛"
- 4⅛"
- 7¾"
- 1¼"
- 16¾"
- M

7–37.

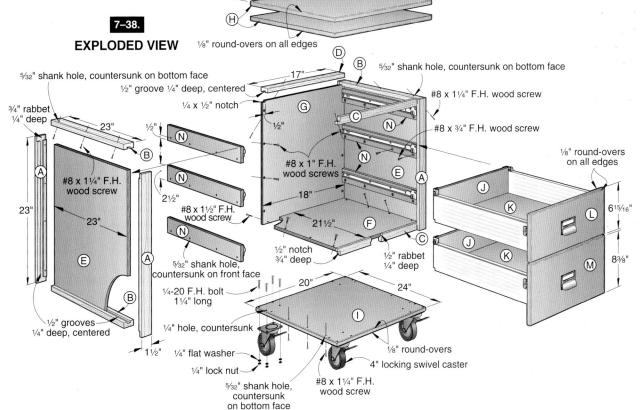

7–38.

EXPLODED VIEW

- 21½"
- 25½"
- H
- ⅛" round-overs on all edges
- 5/32" shank hole, countersunk on bottom face
- 17"
- D
- B
- 5/32" shank hole, countersunk on bottom face
- ½" groove ¼" deep, centered
- ¼ x ½" notch
- #8 x 1¼" F.H. wood screw
- ¾" rabbet ¼" deep
- 23"
- ½"
- G
- N
- C
- N
- #8 x ¾" F.H. wood screw
- ½"
- A
- B
- #8 x 1¼" F.H. wood screw
- 23"
- N
- #8 x 1" F.H. wood screws
- N
- E
- A
- ⅛" round-overs on all edges
- 23"
- A
- 2½"
- 18"
- J
- K
- L
- 6¹⁵/₁₆"
- #8 x 1½" F.H. wood screw
- N
- 21½"
- F
- C
- 5/32" shank hole, countersunk on front face
- ½" notch ¾" deep
- ½" rabbet ¼" deep
- J
- K
- M
- 8⅜"
- B
- ½" grooves ¼" deep, centered
- ¼-20 F.H. bolt 1¼" long
- 20"
- 24"
- I
- 1½"
- ¼" hole, countersunk
- ¼" flat washer
- ⅛" round-overs
- 4" locking swivel caster
- ¼" lock nut
- #8 x 1¼" F.H. wood screw
- 5/32" shank hole, countersunk on bottom face

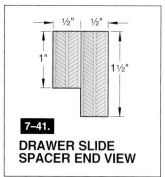

7-41.

DRAWER SLIDE SPACER END VIEW

Use a spacer to position the drawer slide ½" down from the cleat's top edge. With the slide's front end ⅛" from the cleat's front edge, drill the pilot holes.

assembly, and clamp it in place. Turn the cabinet over. Then, as before, drill angled countersunk screw holes, and drive the screws.

7 Clamp the upper front rail (C) in place. Drill angled countersunk screw holes through the front rail and into the side rails (B). Drive the screws.

8 Cut two ¾ x 21½ x 25½" pieces of medium-density fiberboard (MDF) for the top (H). Glue and clamp them together, keeping the ends and edges flush.

With their ends against the bottom, use a pair of 18"-long spacers to position the top cleats. Then glue and screw the cleats in place.

7-42.

Spacers

Ⓝ

Sand the edges smooth, and rout ⅛" round-overs along all the top's edges.

9 Cut the mobile base (I) to size, and rout ⅛" round-overs along the top and bottom edges.

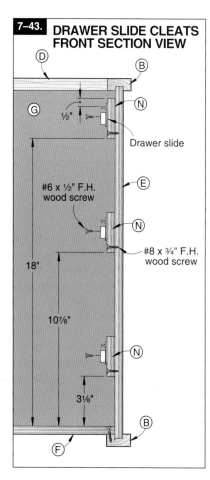

7-43. DRAWER SLIDE CLEATS FRONT SECTION VIEW

Ⓓ   Ⓑ

Ⓖ   Ⓝ

½"

Drawer slide

#6 x ½" F.H. wood screw   Ⓔ

Ⓝ

18"   #8 x ¾" F.H. wood screw

10⅞"

Ⓝ

3⅛"

Ⓑ

Ⓕ

Using a caster's mounting plate as a guide, mark mounting-hole locations at all four corners. Drill ¼" holes and countersink them on the panel's top face.

Add the Drawers

1 Cut the drawer backs (J), drawer bottoms (K), and drawer fronts (L, M) to size. Ease the edges of the bottoms and backs with a sanding block. Rout ⅛" round-overs along the front and back edges of the drawer fronts. Drill the screw holes for the pulls, where shown on **7-39**.

2 Finish-sand the parts. Apply two coats of satin polyurethane to the bottoms and backs and three coats to the fronts, sanding between coats. Now, to assemble the drawers using the metal side/slides, see the section Make Drawers Quickly on the *next page.*

3 With the drawers made, cut the cleats (N) to size. Mark the bottom front corner of each cleat, making sure you have three mirror-image pairs. Drill countersunk screw holes through the cleats ½" up from their bottom edges, where shown on **7-38**.

4 Make the drawer slide spacer shown on **7-40** from 12"-long scraps of ½" plywood. Using the holes in the drawer slides' cabinet members as guides, drill pilot holes in the cleats, as shown in **7-40**.

Make Drawers Quickly

Initially developed for the commercial kitchen-cabinet trade, the all-in-one metal drawer side/slide hardware used in this project allows you to build sturdy drawer boxes in record time. But that's not its only advantage. The side/slide hardware costs about the same as regular drawer slides alone while eliminating the wood for drawer box sides and fronts. And, adjustable brackets attach the finished drawer fronts to the sides, allowing you to fine-tune the gaps between drawers during final cabinet assembly. Here's how to build a drawer using this time- and money-saving hardware.

Note: The metal sides have a pair of raised tabs (see drawing below) used for positioning in mass-production applications. Clip off these tabs before assembling your drawers.

STEP 1

Drill screw holes and fasten the drawer back (J) to the drawer bottom (K), where shown on 7–44. Position the metal sides on the back/bottom assembly, drill pilot holes, and screw them in place, as shown in 7–45. For easy assembly, a 12 x 18" scrapwood assembly frame 6" high holds the back and sides clear of your worktop.

7–44.

DRAWER
(Viewed from the back)

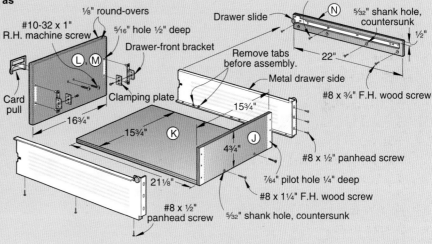

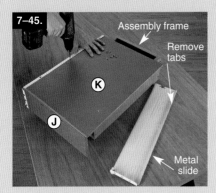

7–45.

STEP 2

To accurately place the drawer-front brackets on both the small drawer fronts (L) and the large drawer front (M), make the drilling guide according to 7–46. To avoid confusion, draw brackets on both of the jig's faces to connect the pairs of holes for parts L and M, as shown. Mark the door front's centerline on its back face. Place the guide's cleat against the front's bottom edge, align the guide's edge with the marked centerline, and clamp it in place. Drill ⁵⁄₁₆" holes ½" deep, as shown in 7–47. Flip the jig, and repeat at the drawer front's other end.

7–46.

DRAWER FRONT
DRILLING GUIDE

7–47.

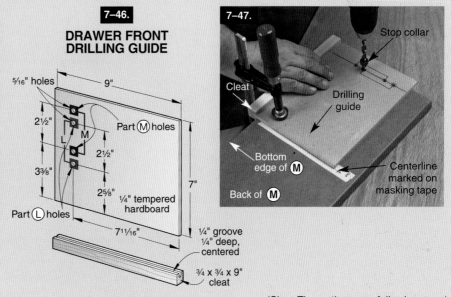

(Shop Tip continues on following page)

Make Drawers Quickly (continued)

STEP 3

With the bracket's drawer-side flange to the outside, drive its plastic inserts into their holes, as shown in 7–48. Use a scrap of wood to evenly distribute the force. Loosely fasten the clamping plates to the brackets with the provided machine screw. Slide the drawer sides between the brackets and the clamping plates, as shown in 7–49, and tighten the screws.

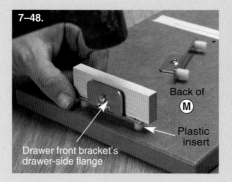

7–48.

Back of **M**

Plastic insert

Drawer front bracket's drawer-side flange

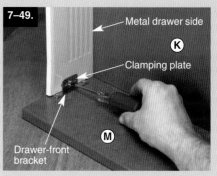

7–49.

Metal drawer side

K

Clamping plate

M

Drawer-front bracket

Cutting Diagram

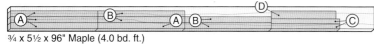

Ⓐ Ⓑ Ⓐ Ⓑ Ⓓ Ⓒ

¾ x 5½ x 96" Maple (4.0 bd. ft.)

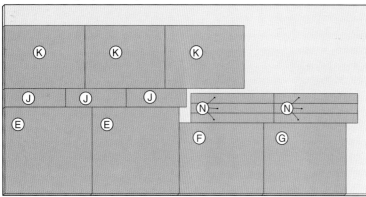

Ⓚ Ⓚ Ⓚ
Ⓙ Ⓙ Ⓙ Ⓝ Ⓝ
Ⓔ Ⓔ
Ⓕ Ⓖ

½ x 48 x 96" Medium-density overlay plywood (MDO)

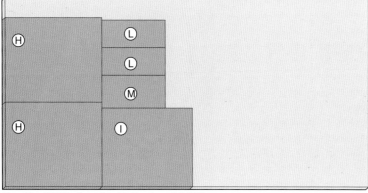

Ⓗ Ⓛ
Ⓛ
Ⓜ
Ⓗ Ⓘ

¾ x 48 x 96" Medium-density fiberboard (MDF)

Materials List for Three-Drawer Utility Cabinet

PART	FINISHED SIZE			Mtl.	Qty.
	T	W	L		
CABINET					
A stiles	¾"	1½"	23"	M	4
B side rails	¾"	1½"	23"	M	4
C front rails	¾"	1¼"	18"	M	2
D back rail	¾"	2"	17"	M	1
E side panels	½"	22"	23"	MDO	2
F bottom	½"	18"	21½"	MDO	1
G back	½"	18"	21¾"	MDO	1
H top	1½"	21½"	25½"	LMDF	1
I mobile base	¾"	20"	24"	MDF	1
DRAWERS					
J drawer backs	½"	4¾"	15¾"	MDO	3
K drawer bottoms	½"	15¾"	21⅛"	MDO	3
L small drawer fronts	¾"	6¹⁵⁄₁₆"	16¾"	MDF	2
M large drawer front	¾"	8⅜"	16¾"	MDF	1
N cleats	½"	2½"	22"	MDO	6

Materials Key: M = maple; MDO = medium-density overlay plywood; LMDF = laminated medium-density fiberboard; MDF = medium-density fiberboard.
Supplies: #8 x ¾", #8 x 1", #8 x 1¼", #8 x 1½", and #6 x ½" flathead wood screws; #8 x ½" panhead screws; ¼-20 x 1¼" flathead bolts, ¼" flat washers, and ¼" lock nuts (16 of each); #10-32 x 1" roundhead machine screws (6); ¼" hardboard and solid stock for the drawer front drilling guide.
Blades and bits: Stack dado set, ⅛" round-over router bit.
Hardware: 4⅝"-deep, 22"-long drawer side/slides (3 pairs); card pulls (3); 4" locking swivel casters (4).

5 Cut two 2 x 18" pieces of scrap for cleat positioning spacers. Use them to position the top pair of cleats (N), as shown in **7–42**. Now trim the spacers to 10⅞", and then 3⅛" to mount the middle and bottom pairs of cleats, where shown on **7–43**.

Finish and Assemble the Cabinet

1 Finish-sand the case and the top. Apply two coats of satin polyurethane to the case and three coats to the top, sanding between coats.

2 Place the top (H) upside down on a pair of sawhorses. Turn the cabinet upside down, center it on the top, and clamp it in place. From the outside of the cabinet, drill angled countersunk screw holes through the side rails (B) and back rail (D) and into the top. Drive the screws.

3 With the hardware shown on **7–39**, bolt the casters to the mobile base (I). Place the base on the cabinet, drill countersunk screw holes through the base and into the lower side rails (B). Drive the screws, and place the cabinet right side up on the floor.

4 Attach the drawer slides' cabinet members to the cleats (N), driving the screws into the previously drilled holes. Fasten the pulls to the drawer fronts, and slide the drawers into the cabinet. Adjust the drawer fronts to leave ¼" between the top (H) and the top drawer front, and ³⁄₁₆" between drawer fronts, leaving ⅛" between the bottom drawer front and the mobile base.

5 Center your benchtop tool on the top, and mark the locations of the mounting holes. Drill the holes, and bolt the tool to the top. Roll the cabinet out of the way, set the caster brakes, and round up some stuff to fill its drawers.

❖

*To get the flip-top work center up and running, just swivel the turn buttons (**7–50**), rotate the flip panel (**7–51**) and get to work (**7–52**).*

FLIP-TOP WORK CENTER

Although smaller than their full-size counterparts, bench-top tools still take up valuable space when not in use. Those occupied, horizontal surfaces could be used for project assembly, finishing, or just tinkering. And with this cabinet (**7–50** to **7–52**), they'll be available.

Note: *The following instructions show you how to build a single flip-top cabinet. To build the twin cabinets shown, simply double the number of parts on the Materials List and see "Create a dynamic duo" on page 149.*

Start with a Basic Cabinet

1 To construct this mobile-base cabinet, start by building a basic cabinet without legs, using the instructions on *pages 136 to 140*.

2 From ½" MDO plywood, cut the gusset (A) to the shape shown on **7–54**. Clamp it to the inside edges of the cabinet's front stiles, where shown on **7–55**. Drill countersunk screw holes through the gusset and into the stiles. Drive the screws.

What Fits on the Work Center?

With all its parts retracted for storage, the tool's footprint must be slightly smaller than the 21¹³⁄₁₆ x 29⅜" flip panel and fit within the radius shown on 7–53. We recommend that the tool weigh a maximum of 125 pounds and that it be centered on the panel.

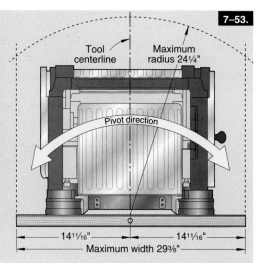

7–53.

Tool centerline

Maximum radius 24¼"

Pivot direction

14¹¹⁄₁₆"   14¹¹⁄₁₆"

Maximum width 29⅜"

Next, the Rail Assembly

1 Cut the lower rear rail (B), upper rear rail (C), lower side rails (D), and upper side rails (E) to the sizes listed on the Materials List. Lay out the rails in the U-shaped configuration shown on **7–55**. Mark the front ends of the four side rails with masking tape, and draw arrows to indicate the mating faces.

2 Cut two 21³⁄₁₆ x 29⅜" pieces of ¾"-thick MDF for the flip panel (F). Position your tablesaw fence to center a ½" dado blade in the length of the flip-panel halves, and cut a ¼"-deep groove in each half. Hacksaw a 27" length of ½" steel rod for the flip-panel pivot, and use it to align the mating grooves. Clamp the halves together. The rod should fit without slop, but still turn freely.

3 To cut mating rod grooves in the side rails (D, E), leave the tablesaw fence in the same posi-

tion used to groove the flip panel. Place the miter gauge in the right-hand slot. Supporting the rails with the miter gauge, and with the ends that are marked with tape against the fence, cut grooves in the parts' mating faces.

4 Glue and clamp parts B, C, D, and E together to form the rail assembly, with the bottom of the assembly facing up. To align the side rails (D, E), lay the ½" steel rod in its mating groove. Check the assembly for square. Drill countersunk screw holes through the lower rails (B, D) and into the upper rails (C, E), where shown on **7–55**. Drive the screws, and remove the rod.

5 Rout ⅛" round-overs on all edges of the rail assembly (B/C/D/E), except for the rear bottom edge of the lower rear rail (B). Sand the assembly smooth with 180-grit sandpaper.

Build the Flip Panel

1 Retrieve the flip-panel (F) halves, and rub their grooves with paraffin. Using a foam roller to spread an even coat, apply glue to the grooved face of one flip-panel half. Clamp the panel halves together with their edges flush. Drill pilot and countersunk shank holes, keeping them 1½" from the panel's edges, and drive the screws.

2 Cut 1⅛" grooves ¾" deep, centered in the flip panel's edges, as shown in **7–56**.

3 Cut the edge inserts (G) to size. Drill ½" holes centered in the inserts' length and width, as shown on **7–55**. Now, using the ½" rod to align the inserts, glue and clamp them in the grooves. Remove the rod.

4 To anchor the shop-made turn buttons to the flip panel, drill ½" counterbores ⅜" deep in the panel's corners, and then drill centered ¼" holes in the counterbores, where shown on **7–55**. Epoxy lock nuts in the counterbores. Rout ⅛" round-overs along the ends and edges of the flip panel, and finish-sand it. To make the turn buttons, cut four 2"-long pieces of ⅛ x ¾" aluminum bar. Drill holes, where shown on **7–57** on *page 148*.

5 Remove the basic cabinet's hardware. Apply three coats of satin polyurethane to the cabinet door, flip panel (F), and the

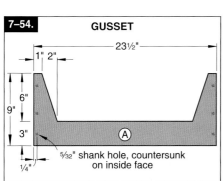

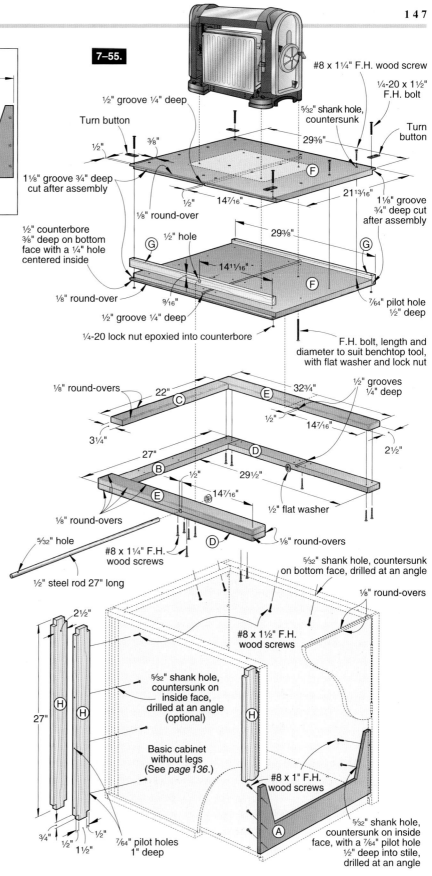

rail assembly (B/C/D/E) and two coats to the cabinet and gusset (A), sanding between coats. To seal the edges of the door, flip panel, and rail assembly, double-coat them as you apply each coat of finish.

Assemble the Top

1 Attach the flip panel to the rail assembly by pushing the ½" steel rod through the ½ x ½" channels in the side rails and the panel, inserting ½" flat washers between the panel and the rails.

2 To pin the rod ends to the rail assembly, drill countersunk shank holes through the lower side rails (D) and the rod, and a pilot hole into the upper side rails (E), where shown on **7–58**. Drive the screws.

3 With the rail assembly positioned screw head side down and the flip panel lock nut side down, fasten the turn buttons, as shown on **7–59**. Orient the front pair of turn buttons to bear on the side rails, and the rear pair to bear on the back rail. Tighten the bolts so the turn buttons are snug, but still able to be turned by hand.

7-56.

Supporting the flip panel (F) with a tall auxiliary fence, cut grooves in its edges for the edge inserts (G). To center the grooves, make one pass with each of the flip panel's faces against the fence.

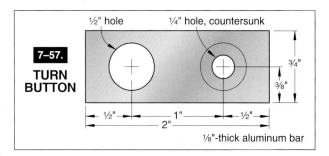

7-57.

TURN BUTTON

½" hole   ¼" hole, countersunk

¾"

⅜"

½"   1"   ½"

2"

⅛"-thick aluminum bar

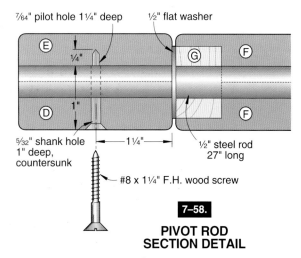

7/64" pilot hole 1¼" deep   ½" flat washer

E   G   F

¼"

D   1"   F

5/32" shank hole 1" deep, countersunk   1¼"   ½" steel rod 27" long

#8 x 1¼" F.H. wood screw

7-58.

PIVOT ROD SECTION DETAIL

Mount the Top, and Then the Tool

1 Clamp the rail/flip panel assembly on the cabinet, flush at the rear with the back edge of the cabinet's upper rear rail, and overhanging the cabinet's upper side rails by ¾".

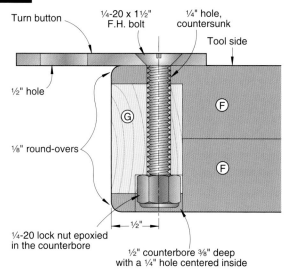

7-59. **TURN BUTTON SECTION DETAIL**

Turn button   ¼-20 x 1½" F.H. bolt   ¼" hole, countersunk

Tool side

½" hole

⅛" round-overs

G   F

F

¼-20 lock nut epoxied in the counterbore

½"

½" counterbore ⅜" deep with a ¼" hole centered inside

2 Drill angled countersunk holes through the cabinet's upper side and rear rails and into the top's rail assembly, where shown on **7–55**. Drive the screws.

3 Center your benchtop tool on the flip panel. Mark and drill mounting holes through the panel. To keep the worktop flush when the tool is in its stored position, countersink the holes, and use flat-head bolts to fasten the tool to the panel.

4 To stow the tool, steady it with one hand while you turn the front turn buttons so they point straight forward. Now, with both hands on the tool, rotate it to the front. When the flip panel is horizontal again, the turn buttons that were in the front now contact the underside of the lower rear rail (B), and the turn buttons that were in the rear now point straight forward. Rotate these turn buttons to the side so they tuck underneath the lower side rails (D).

Caution: *When rotating the flip panel, be sure to keep your fingers away from the pinch zones between the flip panel and the rail assembly.*

5 To attach two or more cabinets together, see Create a Dynamic Duo, *at right*. Build a mobile base for this cabinet, as shown on *page 150*.

¾ x 5½ x 96" Maple (4 bd. ft.)

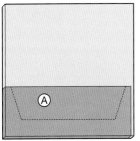

½ x 24 x 24" Medium-density overlay plywood (MDO)

Cutting Diagram

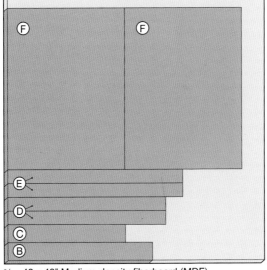

¾ x 48 x 48" Medium-density fiberboard (MDF)

DUAL-PURPOSE SANDING CENTER

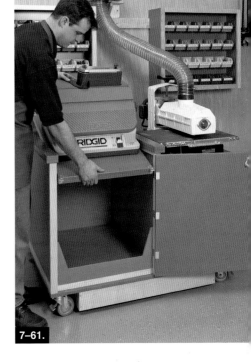

7–61.

Materials List for Dual-Purpose Sanding Center

PART	FINISHED SIZE			Mtl.	Qty.
	T	W	L		
A gusset	½"	9"	23½"	MDO	1
B lower rear rail	¾"	3¼"	27"	MDF	1
C upper rear rail	¾"	3¼"	22"	MDF	1
D lower side rails	¾"	2½"	29½"	MDF	2
E upper side rails	¾"	2½"	32¾"	MDF	2
F flip panel	1½"	21¹³⁄₁₆"	29⅜"	LMDF	1
G edge inserts	¾"	1⅛"	29⅜"	M	2
H optional fillers	¾"	2½"	27"	M	3

Materials Key: MDO = medium-density overlay plywood; MDF = medium-density fiberboard; LMDF = laminated medium-density fiberboard; M = maple.
Supplies: #8 x 1", #8 x 1¼", and #8 x 1½" flathead wood screws; ½" steel rod 27" long; ½" flat washers (2); ¼-20 lock nuts (4); ¼-20 x 1½" flathead bolts (4); ⅛ x ¾" aluminum bar 10" long; quick-set epoxy; flathead bolts, flat washers, and lock nuts for fastening your benchtop tool to the flip panel.
Blades and bits: Stack dado set, ⅛" round-over router bit, ½" Forstner bit.

SHOP TIP

Create a Dynamic Duo

To join the cabinets together, as shown on *page 145,* cut three optional fillers (H) to the dimensions in the Materials List. Glue and clamp two of them together, keeping their ends and edges flush. Then cut pairs of ¾ x ½" notches in the ends, where shown on 7-55. Finish the fillers with satin polyurethane.

Clamp the fillers to the first cabinet, where shown on 7-60. Drill screw holes through the cabinet's side panels into the fillers, angled to the front and rear, as shown. Drive the screws. Clamp the second cabinet in place, drill the screw holes, and drive the screws.

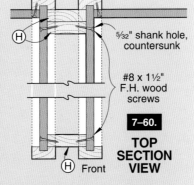

⁵⁄₃₂" shank hole, countersunk

#8 x 1½" F.H. wood screws

7–60.

TOP SECTION VIEW

Front

Combine the flip-top machine cabinet shown on the last few pages with the utility cabinet from *pages 140* to *145,* add a sander platform and a mobile base, and you'll have sturdy support for a couple of tools, ample storage, plus mobility (**7–61**).

We set up our center to partner two related tools—a 16" drum sander and an oscillating spindle/belt sander—in one compact package. If your needs differ, though, that's no problem. You can alter the design to suit your needs. Any tool with a footprint no larger than 20 x 32" can rest

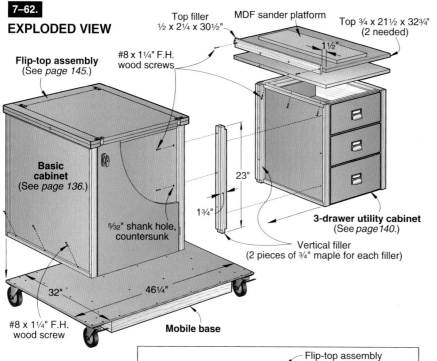

7-62.

EXPLODED VIEW

Top filler ½ x 2¼ x 30½"

MDF sander platform

Top ¾ x 21½ x 32¾" (2 needed)

Flip-top assembly (See *page 145*.)

#8 x 1¼" F.H. wood screws

1½"

Basic cabinet (See *page 136*.)

⁵⁄₃₂" shank hole, countersunk

23"

1¾"

3-drawer utility cabinet (See *page 140*.)

Vertical filler (2 pieces of ¾" maple for each filler)

32"

46¼"

#8 x 1¼" F.H. wood screw

Mobile base

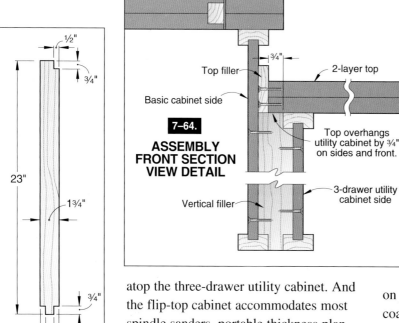

Flip-top assembly

Top filler

¾"

Basic cabinet side

2-layer top

7-64.

ASSEMBLY FRONT SECTION VIEW DETAIL

Top overhangs utility cabinet by ¾" on sides and front.

Vertical filler

3-drawer utility cabinet side

½"

¾"

23"

1¾"

¾"

½" ¾" ½"

7-63.

VERTICAL FILLER DETAIL

---

atop the three-drawer utility cabinet. And the flip-top cabinet accommodates most spindle sanders, portable thickness planers, and other benchtop tools that measure less than 29" wide and 24" tall.

Create the Components

1 Start by building the basic cabinet and flip-top assembly. Next, construct a three-drawer utility cabinet, omitting the top (H) and mobile base (I). Build the top, shown in **7-62**, by gluing and screwing together two pieces of ¾" medium-density fiberboard (MDF). Don't attach the top to the cabinet yet.

2 Now make the two vertical fillers by face-gluing two pieces of ¾ x 1¾ x 23" maple for each, and then rabbeting the ends, as dimensioned on **7-63**. Screw the fillers to the left side of the three-drawer utility cabinet, where shown on **7-62**.

3 Screw the basic cabinet to the fillers to temporarily join the two units. Now set the top on the utility cabinet and mark its position. Measure for the top filler, and then cut it to size and screw it to the top, where shown on **7-62** and **7-64**. Detach the two cabinets from each other, and screw the top on the utility cabinet, as shown, using your marks as guides.

4 Next, build the mobile base, using the platform dimensions shown in **7-62** and the instructions in "Beef up a Mobile base to Carry a Load" on *page 141*. When done, apply two coats of polyurethane finish to it and the completed cabinets.

5 After the finish dries, set the cabinets on the mobile base and reattach them to one another. Then, secure the assembly to the mobile base platform by driving #8 x 1¼" screws at an angle through the

cabinets' lower side and lower back rails, where shown on **7–62**.

Mount Your Sanders

1 To position your drum sander, first measure its base dimensions. Now measure its height from the base to the top of the drive belt. Compare this measurement to the height difference between the two

cabinet tops (4"). For best performance, the top of the drive belt should sit 1/16" or slightly more above the top surface of the flip-top assembly. In our case, the sander needed to be raised just over 3/4". To do this, we made a sander platform from 3/4" MDF, sized just larger than the dimensions of the drum sander's base.

2 Techniques for mounting your drum sander to the sander platform may vary. We drilled four holes to correspond with holes in the sander's metal base, and then counterbored them on the underside to receive flathead machine screws. Next, we bolted the sander to its platform, shimming it with a couple of flat washers on each bolt to get the exact height we needed.

3 Now position the sander platform so there's about a 1/4" gap between the outfeed end of the drive belt and the edge of the flip-top. Secure the sander platform in one of two ways. If space exists at each end, drill countersunk pilot holes and drive 1 1/2" flathead screws through the platform and into the utility-cabinet top. Or, you can mark the outline of the sander platform, lift the sander off, and then drill four shank holes through the utility-cabinet top. Now countersink the holes from below, set the sander back in place, drill pilot holes into the platform, and drive 2 1/2" screws from underneath.

4 Mount your spindle sander to the flip-top assembly by first positioning it centered on the flip-top. Mark and drill holes where needed, and bolt the spindle sander in place.

With the spindle sander stowed, the flip-top provides outfeed support and a work surface. The drawers offer ample storage.

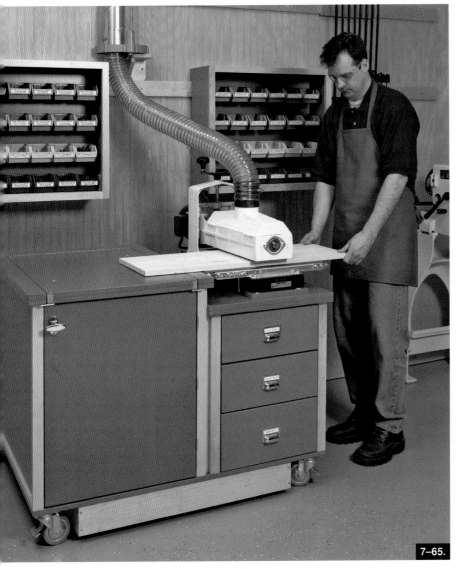

7–65.

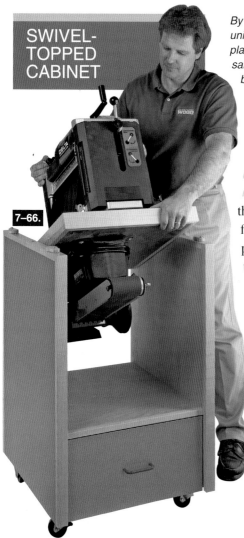

SWIVEL-TOPPED CABINET

7–66.

By simply flipping over the top, this unit switches from being a portable planer stand to a stationary belt/disc sander stand. The drawer in its bottom provides handy storage of sanding supplies.

sizes listed in the Materials List from ¾" plywood.

2 Mark the locations of the dado and rabbet on the inside face of two of the plywood side panels (A). Cut or rout them to the sizes listed on **7–67** and **7–68**. (We fitted our tablesaw with a dado blade and wooden auxiliary rip fence. Then we made test-cuts in scrap pieces of plywood first to verify a

snug fit of the mating plywood pieces in their respective rabbets and dadoes.)

3 Band the front and back edges of the middle shelf (B) with banding strips (D), where shown on **7–68**.

4 To match the reveal formed later between the drawer front (O) and carcass, cut or rout a ⅛" rabbet ⅛" deep along the top and side edges along the outside face of the back (C). See **7–67** and **7–68** for reference.

5 Glue and clamp the shelves (B) and back (C) between the inside (machined) pieces (A),

With this storage-conscious cabinet (**7–66**), you can have two tools, plus a spacious drawer in the amount of space usually dedicated to one tool. The swivel rotates 180° and locks in place, allowing you to switch tools in just seconds.

Begin with the Carcass Assembly

1 Rip and crosscut the cabinet sides (A), bottom and middle shelves (B), and back (C) to the

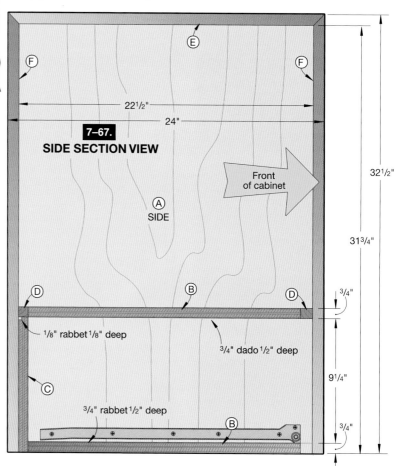

7–67.
SIDE SECTION VIEW

22½"
24"

Front of cabinet

(A) SIDE

32½"

31¾"

¾"

⅛" rabbet ⅛" deep

¾" dado ½" deep

9¼"

¾" rabbet ½" deep

¾"

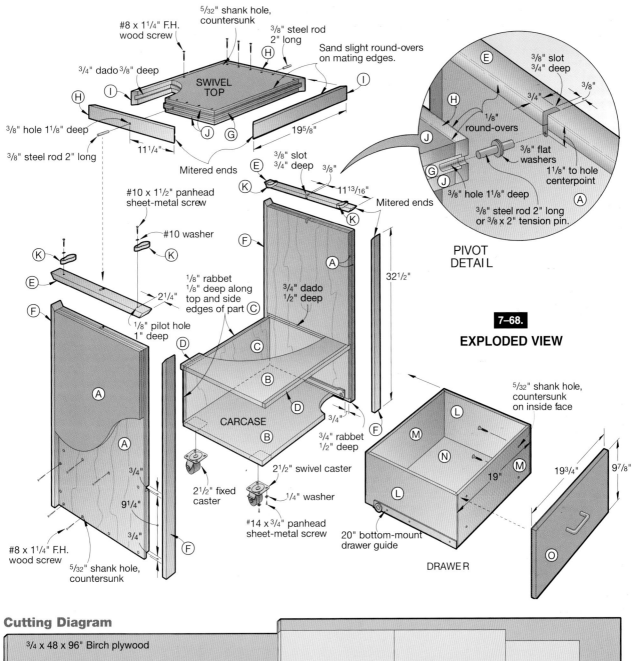

#8 x 1¼" F.H.
wood screw

⁵/₃₂" shank hole,
countersunk

³/₈" steel rod
2" long

Sand slight round-overs
on mating edges.

³/₄" dado ³/₈" deep

SWIVEL
TOP

³/₈" hole 1⅛" deep

³/₈" steel rod 2" long

11¼"

Mitered ends

19⅝"

³/₈" slot
³/₄" deep

³/₈" slot
³/₄" deep

³/₈"

³/₄"

³/₈"

1/8"
round-overs

³/₈" flat
washers

1⅛" to hole
centerpoint

³/₈" hole 1⅛" deep

³/₈" steel rod 2" long
or ³/₈ x 2" tension pin.

PIVOT
DETAIL

11¹³/₁₆"

Mitered ends

#10 x 1½" panhead
sheet-metal screw

#10 washer

1/8" rabbet
1/8" deep along
top and side
edges of part

³/₄" dado
½" deep

32½"

1/8" pilot hole
1" deep

2¼"

CARCASE

³/₄"

³/₄" rabbet
½" deep

2½" swivel caster

1/4" washer

#14 x ³/₄" panhead
sheet-metal screw

20" bottom-mount
drawer guide

2½" fixed
caster

³/₄"

9¼"

³/₄"

#8 x 1¼" F.H.
wood screw

⁵/₃₂" shank hole,
countersunk

7–68.

EXPLODED VIEW

⁵/₃₂" shank hole,
countersunk
on inside face

19"

DRAWER

19¾"

9⅞"

Cutting Diagram

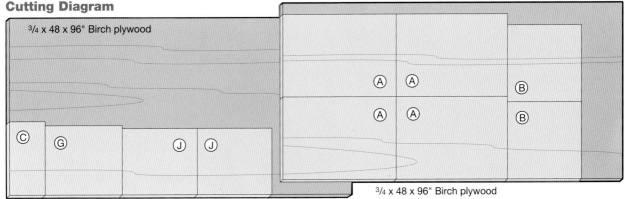

³/₄ x 48 x 96" Birch plywood

Ⓐ Ⓐ Ⓑ

Ⓐ Ⓐ Ⓑ

Ⓒ Ⓖ Ⓙ Ⓙ

³/₄ x 48 x 96" Birch plywood

7–69.

Check for tight miter joints, and then glue and clamp the solid maple banding pieces around the plywood center section for the swivel top.

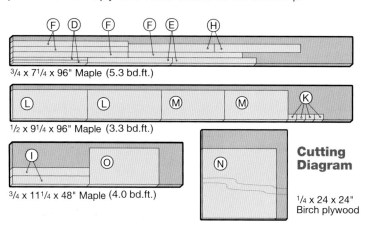

3/4 x 7 1/4 x 96" Maple (5.3 bd.ft.)

1/2 x 9 1/4 x 96" Maple (3.3 bd.ft.)

3/4 x 11 1/4 x 48" Maple (4.0 bd.ft.)

Cutting Diagram

1/4 x 24 x 24" Birch plywood

Materials List for Swivel-Topped Cabinet

PART	FINISHED SIZE			MTL.	QTY.
	T	W	L		
CARCASS					
A sides	3/4"	22 1/2"	31 3/4"	BP	4
B shelves	3/4"	21"	21"	BP	2
C back	3/4"	20"	10"	BP	1
D banding	3/4"	3/4"	21"	M	2
E* banding	3/4"	1 1/2"	24"	M	2
F* banding	3/4"	1 1/2"	32 1/2"	M	4
SWIVEL TOP					
G center	3/4"	18 7/8"	21 3/4"	BP	1
H* banding	3/4"	2 1/4"	22 1/2"	M	2
I* banding	3/4"	2 1/4"	19 5/8"	M	2
J top & bottom	3/4"	18 1/16"	20 15/16"	BP	2
K turn tabs	1/2"	1 1/2"	2 1/2"	M	4
DRAWER					
L sides	1/2"	8"	21"	C	2
M front & back	1/2"	8"	18 1/4"	C	2
N bottom	1/4"	19"	21"	BP	1
O front	3/4"	9 7/8"	19 3/4"	M	1

*Cut parts oversized. Trim to finished size according to the instructions.

Materials Key: BP = birch plywood; M = maple; C = choice of poplar or maple.
Supplies: #8 x 1", #8 x 1 1/4", #8 x 1 1/2" flathead wood screws; 2 1/2" locking swivel casters (2); 2 1/2" fixed casters (2); #14 x 3/4" panhead sheet-metal screws with flat washers (16); one pair of 20" bottom-mount drawer guides (we used Knape Vogt #1300 guides); 3/8" steel rods 2" long (or 1/8 x 2" tension pins) with 3/8" flat washers (4); #10 x 1 1/2" panhead sheet-metal screws with four #10 washers (4); 3/4" wire pull; paint; clear finish.

checking for square. Drill and countersink the mounting holes, and strengthen the assembly with screws.

6 With the edges and ends perfectly flush, glue the outside panels (A) to the carcass assembly.

7 From solid stock (we used maple), cut and miter-cut the banding pieces (E, F) to size, and glue and clamp them in place. Later, sand the surfaces flush.

8 Drill the mounting holes, and screw the swivel and fixed casters to the bottom corners of the bottom shelf (B).

Construct the Swivel Top

1 Cut the 3/4" plywood panel (G) for the top.

2 Measure the combined thickness of three pieces of plywood you'll be using for the top (G, J). Cut the width of the top banding (H, I) to the measured thickness. (Because plywood is not exactly 3/4" thick, you'll need to measure the thickness of the three pieces of plywood to determine the exact width necessary for the banding.) Crosscut the banding to the listed length plus 2".

3 Cut a dado 3/8" deep, centered along one face of each piece of banding (H, I), for the center top piece (G) to fit into snugly. (We made test-cuts in scrap stock first.)

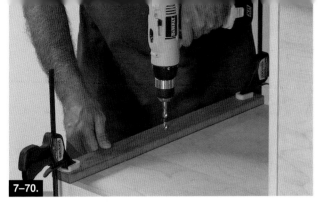

For accurately placed holes in the swivel top and inside surface of the side panels, use a hole guide to locate the holes.

Using a ⅜" straight bit and an L-shaped template, rout a ⅜"-wide slot on the inside face of each side panel.

4 Miter-cut the banding pieces (H, I) to wrap around part G.

5 To ensure matching holes in the swivel top to those on the inside face of the cabinet, form a hole guide. To do this, cut a piece of ¾" stock to 2¼" wide by 22½" long. Then, measure from end to end to find the exact center of the strip, and mark a centerpoint. Use your drill press to drill a ⅜" hole through the strip, and use this as a guide to drill a ⅜" hole through each sidepiece H, where shown on **7–68**. (Later, as shown in **7–70**, we'll use the template to drill the mating holes in the cabinet sides.)

6 Glue the banding pieces (H, I) in place around part G, as shown in **7–69**. Later, measure the openings, and cut the top and bottom pieces (J) to fit snugly inside the banding. Sand a slight round-over along the mating edges of the banding (H, I) and the top and bottom pieces (J), where shown on the Pivot Detail accompanying **7–68**.

7 Drill countersunk mounting holes, and screw the top and bottom pieces (J) in place.

8 Using the ⅜" holes in both H banding pieces as guides, drill ⅜" holes into each E banding pieces for a total depth of 1⅛", where shown on the Pivot Detail in **7–68**.

9 Lay the cabinet on its side on your workbench. As shown in **7–70**, clamp the hole template in place, and drill a ⅜" hole ¾" deep into the inside face of the cabinet sides, making sure that it is centered from end to end.

10 Using a square, carefully mark the location for the ⅜" slot ¾" deep centered over the hole

drilled in the previous step. For reference, see the layout lines in **7–71**. Fit your router with a ⅜" straight bit, and use an L-shaped template, as shown in **7–71**, to rout the slot to shape. Flip the cabinet over, and form the opposite slot.

11 Cut two ⅜" smooth steel dowel rods to 2" long each. Epoxy one into each ⅜" hole in the top. See the Pivot Detail in **7–68** for reference. Place two ⅜" flat washers on each steel dowel rod, and place the swivel top in place.

12 Transfer the full-size Turn Tab Pattern and hole centerpoint in **7–73** (on *page 156*) to ½" solid stock four times. Drill the 3/16" holes where marked. Then, bandsaw or scrollsaw the turn tabs (K) to shape. Sand or carefully rout a ⅛" round-over along the top and bottom edges of each tab. Screw the turn tabs to the top of the top banding (E), where dimensioned on **7–68**. Tighten the screws just enough so it takes just a small amount of pressure for you to rotate the tabs.

Add a Drawer for Handy Storage

1 From ½" stock (we prefer maple or poplar), cut the drawer sides (L) and front and back (M) to size.

2 Drill four countersunk mounting holes through the back surface of the front piece (M) for attaching the solid drawer face (O) later.

3 Cut a ½" rabbet ⅛" deep along the ends of each sidepiece, where shown on **7–72**.

4 Glue and clamp the drawer parts (L, M) together, checking for square. Drill countersunk holes, and then reinforce the joints with #8 × 1½" wood screws.

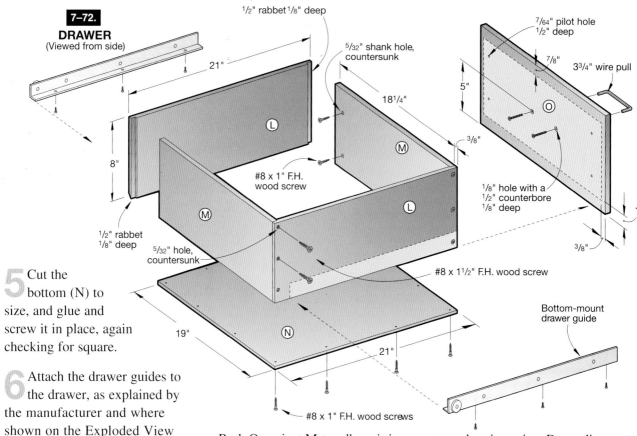

7–72.
DRAWER
(Viewed from side)

21"

8"

1/2" rabbet 1/8" deep

5/32" shank hole, countersunk

18 1/4"

7/64" pilot hole 1/2" deep

7/8"

3 3/4" wire pull

5"

3/8"

1/8" hole with a 1/2" counterbore 1/8" deep

1"

3/8"

#8 x 1" F.H. wood screw

1/2" rabbet 1/8" deep

5/32" hole, countersunk

#8 x 1 1/2" F.H. wood screw

Bottom-mount drawer guide

19"

21"

#8 x 1" F.H. wood screws

5 Cut the bottom (N) to size, and glue and screw it in place, again checking for square.

6 Attach the drawer guides to the drawer, as explained by the manufacturer and where shown on the Exploded View (**7–68**). Then, attach the mating guide pieces to the inside of the drawer opening.

7 Cut the solid-stock drawer front (O) to size.

8 Cut two pieces of double-faced tape to 1" long. Place the tape on opposite corners of the drawer front (M). Slide the drawer into the opening in the carcass. Position the drawer front face (O) against the drawer front (M), centered in the opening.

Push O against M to adhere it in place. Remove the drawer from the opening; then screw the face (O) to the front (M).

9 Locate the centerpoints, and then drill the holes for the drawer pull.

Finish Up the Cabinet and Add a Pair of Tools

1 Remove all the hardware and touch-up sand if necessary. Apply the paint and finish. (We used a textured industrial enamel for the painted surfaces and satin polyurethane on the remainder.)

2 Reattach all the hardware.

3 Mount your tools by centering them so the weight of each tool rests

over the pivot pins. Depending on the weight of the tools, you may want to secure them with bolts instead of screws. Also, when mounting the second tool, position it so the weight of the opposing tool is evenly distributed. This will prevent the top with attached tools from whipping around when you release the turn tabs. (We found it useful to have a helper on hand when locating and securing the tools to the swivel top.)

Note: *When releasing the turn tabs and rotating the top with the mounted tools, grab the ends (not the edges) of the swivel top to prevent accidentally pinching your fingers between the swivel top and carcass sides. A firm grasp also helps prevent the top from swiveling too quickly.*

1/8" round-over

3/16" hole

K

TURN TAB
FULL-SIZE PATTERN
(4 needed)

7–73.

7-76.

The bevel-ripped hanger strip on the cabinet fits onto the mating angled piece on the tool board, allowing you to position the cabinet wherever you want to along the tool board.

7-75. **EXPLODED VIEW**

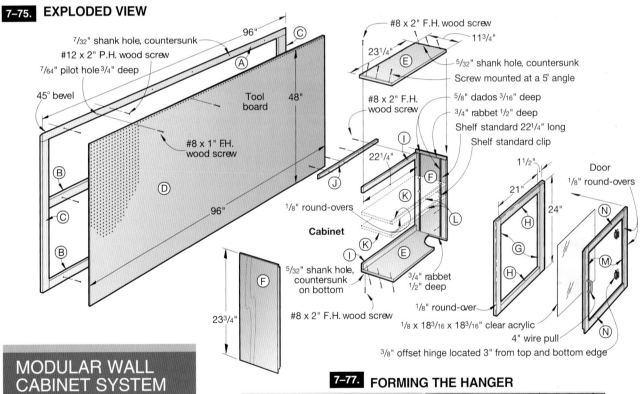

96"

$^{7}/_{32}$" shank hole, countersunk
#12 x 2" P.H. wood screw
$^{7}/_{64}$" pilot hole $^{3}/_{4}$" deep

Ⓒ
Ⓐ

45° bevel

Tool board
48"

#8 x 1" F.H. wood screw

Ⓑ
Ⓒ
Ⓑ
Ⓓ

96"

Ⓕ

$23^{3}/_{4}$"

#8 x 2" F.H. wood screw

$^{5}/_{32}$" shank hole, countersunk on bottom

#8 x 2" F.H. wood screw
$11^{3}/_{4}$"
$23^{1}/_{4}$"
Ⓔ
$^{5}/_{32}$" shank hole, countersunk
Screw mounted at a 5° angle

#8 x 2" F.H. wood screw

$^{5}/_{8}$" dados $^{3}/_{16}$" deep
$^{3}/_{4}$" rabbet $^{1}/_{2}$" deep
Shelf standard $22^{1}/_{4}$" long
Shelf standard clip

$22^{1}/_{4}$"
Ⓘ
Ⓙ
Ⓕ
Ⓚ
Ⓛ

$^{1}/_{8}$" round-overs

Cabinet

Ⓚ
Ⓘ
Ⓔ

$^{3}/_{4}$" rabbet $^{1}/_{2}$" deep

$1^{1}/_{2}$"
21"
Ⓗ
Ⓖ
Ⓗ
24"

Door
$^{1}/_{8}$" round-overs
Ⓝ
Ⓜ
Ⓝ

$^{1}/_{8}$" round-over
$^{1}/_{8}$ x $18^{3}/_{16}$ x $18^{3}/_{16}$" clear acrylic
4" wire pull
$^{3}/_{8}$" offset hinge located 3" from top and bottom edge

MODULAR WALL CABINET SYSTEM

This cabinet package (**7-74**) can be anything you want it to be. We'll even tell you how to customize these cabinets for your shop.

7-77. **FORMING THE HANGER**

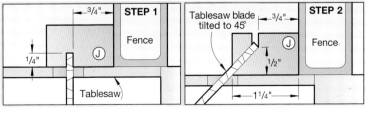

$^{3}/_{4}$" **STEP 1**
Fence
Ⓙ
$^{1}/_{4}$"
Tablesaw

Tablesaw blade tilted to 45° $^{3}/_{4}$" **STEP 2**
Fence
Ⓙ
$^{1}/_{2}$"
$1^{1}/_{4}$"

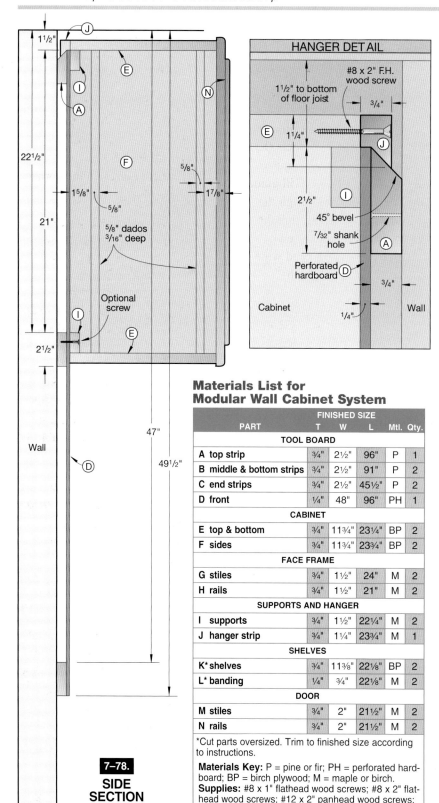

HANGER DETAIL

#8 x 2" F.H. wood screw

1½" to bottom of floor joist

3/4"

E 1¼"

J

2½"

I

45° bevel

7/32" shank hole

A

Perforated hardboard D

3/4"

Cabinet   Wall

1/4"

7–78.
SIDE SECTION VIEW

Materials List for Modular Wall Cabinet System

PART	FINISHED SIZE			Mtl.	Qty.
	T	W	L		
TOOL BOARD					
A top strip	3/4"	2½"	96"	P	1
B middle & bottom strips	3/4"	2½"	91"	P	2
C end strips	3/4"	2½"	45½"	P	2
D front	1/4"	48"	96"	PH	1
CABINET					
E top & bottom	3/4"	11¾"	23¼"	BP	2
F sides	3/4"	11¾"	23¾"	BP	2
FACE FRAME					
G stiles	3/4"	1½"	24"	M	2
H rails	3/4"	1½"	21"	M	2
SUPPORTS AND HANGER					
I supports	3/4"	1½"	22¼"	M	2
J hanger strip	3/4"	1¼"	23¾"	M	1
SHELVES					
K* shelves	3/4"	11⅜"	22⅛"	BP	2
L* banding	1/4"	3/4"	22⅛"	M	2
DOOR					
M stiles	3/4"	2"	21½"	M	2
N rails	3/4"	2"	21½"	M	2

*Cut parts oversized. Trim to finished size according to instructions.

Materials Key: P = pine or fir; PH = perforated hardboard; BP = birch plywood; M = maple or birch.
Supplies: #8 x 1" flathead wood screws; #8 x 2" flathead wood screws; #12 x 2" panhead wood screws; one pair of 3/8" offset self-closing hinges; 4" wire pull, shelf standards and clips; 1/8 x 3/8" diameter cushions; acrylic; clear finish.

Begin with the Perforated Hardboard Tool Board

1 Select the straightest 1 x 3s available for the mounting strips (A, B, C). When purchasing these, make sure they're 3/4" thick. We ran across some furring strips that measured only 5/8" thick. Use the straightest of the bunch for the top strip. Now, bevel-rip the top mounting strip (A) at 45°, where shown on **7–75** and on the Hanger Detail in **7–78**. Crosscut the mounting strips to length.

2 To position the top mounting strip A (the one with the beveled edge), temporarily screw or nail a 2 x 4 to the bottom side of the floor joists, assuming that you have 7' ceilings in your basement. For other situations with taller ceilings, simply snap a chalk line to establish where you want the top of the pegboard. Using the 2 x 4 as a spacer creates the necessary 1½" gap needed for hanging the cabinet(s) later. Make sure you locate the screws in the 2 x 4 where they can be easily removed after the perforated hardboard has been secured to the mounting strips.

3 Lay out and drill countersunk mounting holes through each mounting strip. Using the 2 x 4 as a spacer, screw the top mounting strip (A) in place, where shown on **7–78**. The type of wall you're screwing into will determine the size and type of fasteners required. Screw the other mounting strips (B, C) in place.

4 Drill mounting holes, and screw the perforated hard-board front (D) to the mounting strips. Remove the 2 × 4 spacer.

Now, Construct the Wall Cabinet

1 Cut the cabinet top and bottom (E) and sides (F) to size.

2 Using a dado blade in your tablesaw, cut ⅝" dadoes ³⁄₁₆" deep for the shelf standards on the inside face of the sides (F). Then, cut ¾" rabbets ½" deep on each end of the sides to house the top and bottom (E) later.

3 Mark the screw-hole center-points on the top and bottom

(E). We located ours 1½" from each end and centered the middle screw. The screws are located ½" in from the outside edge. Now, cut a scrap block with a 5° angle across one end. Glue and clamp the plywood panels (E, F) together, checking for square. Using the block as a drill guide, drill angled holes through the corner joints. Countersink the holes, and drive the screws. Remove the clamps, and wipe off any excess glue with damp cloth.

Add the Face Frames and Supports Next

1 Cut the face frame stiles and rails (G, H) to size, so the

assembled face frame will overhang the plywood cabinet ⅛" on all edges.

2 Glue and clamp the face-frame pieces directly to the front of the plywood cabinet. If you own a biscuit joiner, you can further strengthen the corner joints by adding biscuits.

3 Sand the face frame. Then, rout a ⅛" round-over along the outside edges of the frame, where shown on **7–75**.

4 Measure the cabinet opening, and cut the supports (I) to size. Drill mounting holes, and glue and screw the supports to the top and bottom (E), where shown on **7–75**. The back edges of the supports should be flush with the back edge of the cabinet.

5 Cut the hanger (J) to size. Follow **7–77** to form the angled bottom edge. Now, screw the hanger to the back of the cabinet, where shown on the Hanger Detail in **7–78**.

Next, Add the Shelves

1 Cut the plywood shelves (K) and banding strips (L) to size plus 1" in length. Glue and clamp the banding strips to the front edge of the plywood shelves. After the glue dries, remove the clamps, and cut the shelves/banding to final length.

2 Rout ⅛" round-overs along the front, banded edge of each shelf.

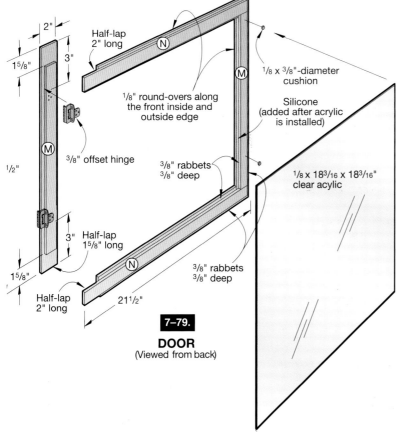

2"

1⁵⁄₈"

3"

½"

1⁵⁄₈"

Half-lap 2" long

Half-lap 2" long

⅜" offset hinge

3" Half-lap 1⁵⁄₈" long

21½"

⅛" round-overs along the front inside and outside edge

⅜" rabbets ⅜" deep

⅜" rabbets ⅜" deep

⅛ x ⅜"-diameter cushion

Silicone (added after acrylic is installed)

⅛ x 18³⁄₁₆ x 18³⁄₁₆" clear acylic

7–79.

DOOR
(Viewed from back)

Half-Lapped Door for the Cabinet

1 Cut the door stiles (M) and rails (N) to size. Using **7–79**, on *page 159*, for reference, cut half-lap joints on the ends of each.

2 Cut or rout a ⅜" rabbet ⅜" deep along the inside edge of each door member.

3 Glue and clamp the door frame together, checking for square and flatness. After the glue dries, remove the clamps, and sand the door frame.

4 Rout a ⅛" round-over along the front inside and outside edge of the door frame. Switch bits or use a dado blade in your tablesaw to cut a ⅜" rabbet ⅜" deep along the back outside edge of the frame.

5 Mark the hinge locations on the doors. Drill pilot holes and screw the hinges in place.

6 Position the door centered in the opening, and mark the hinge-hole locations on the face frame. Position the door to open on the left or right, depending on what's most convenient. Drill the pilot holes, and screw the door/hinges to the face frame. Drill the mounting holes for the wire pull.

Finish It All Up

1 Remove the hardware from the cabinet. Finish-sand, and apply a clear finish to the cabinet, door, and shelves. Leave the pegboard unfinished.

2 Cut the shelf standards to length, and then screw in place.

3 Measure the opening, and cut the acrylic for the door. Run a fine bead of clear silicone to secure the acrylic in place. Later, reattach the hinges, and secure the door to the cabinet.

4 Hang the cabinet on the tool board. Leave the cabinet moveable, or drive a couple of screws through the bottom support (I), the perforated hardboard, and into the bottom mounting strip (B), where shown on **7–78**.

ANOTHER TAKE ON A WALL-CABINET SYSTEM

This wall-cabinet system (**7–80**) has a slightly different look than the one just presented. Although these cabinets are hung uniquely, too, they go together quickly and are just as great for organizing hand tools, safety equipment, power-tool accessories, or anything you want.

Start with the Back and Mounting Strips

1 From ½" plywood, cut the back (A) to the size listed in the Materials List.

2 Using **7–81**, bevel-rip the 29 mounting strips (B) to size from ½"-thick stock. Crosscut the strips to length.

7–80.

3 Mark the screw-hole center-points, where dimensioned on **7–82**, and then drill and countersink a trio of shank holes in each strip.

4 To ensure consistent spacing between the strips and smooth-sliding components, build a spacing jig, as shown in **7–83**.

5 Clamp the back (A) to your workbench. Cut a piece of scrap measuring 2¼ x 22". With the top edges flush, clamp the scrap piece to the top of the back.

6 Starting flush with the bottom edge of the scrap strip (2¼" from the top edge of the back), glue and screw the first mounting strip (B) to the back, where shown in **7–84**. See the Mounting Detail in **7–82** for reference. Check that the ends of the mounting strip are flush with the outside edges of the back. Use only a small amount of glue to avoid squeeze-out. Immediately wipe off excess glue with a damp cloth.

Caution: Glue left between the mounting strips can prevent the tool holders from sliding easily in the grooves later.

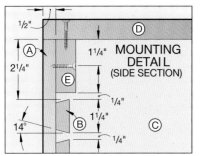

MOUNTING DETAIL (SIDE SECTION)

1/2"
A
2 1/4"
1 1/4"
E
14°
B
1 1/4"
1/4"
1/4"
C
D

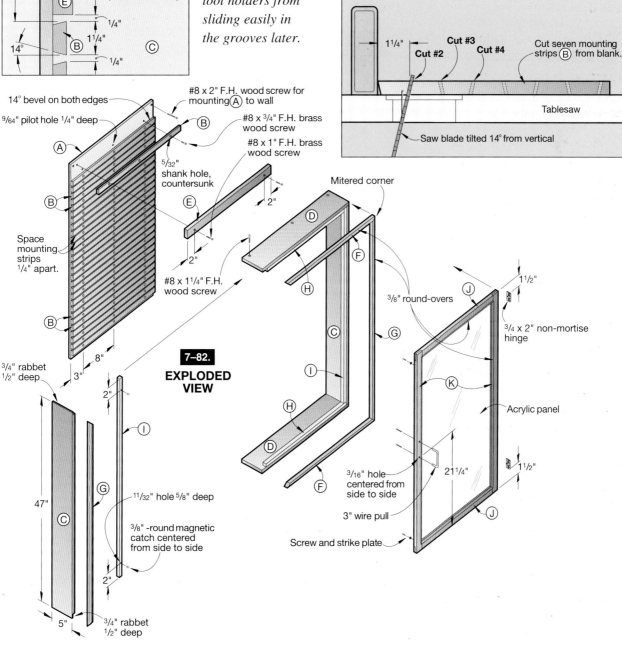

Fence
1/2 x 9 1/4 x 48" maple blank for mounting strips (B)
Cut #1
Waste
Tablesaw
Saw blade tilted 14° from vertical

1 1/4"
Cut #2
Cut #3
Cut #4
Cut seven mounting strips (B) from blank.
Tablesaw
Saw blade tilted 14° from vertical

14° bevel on both edges
9/64" pilot hole 1/4" deep
A
B
#8 x 2" F.H. wood screw for mounting (A) to wall
#8 x 3/4" F.H. brass wood screw
#8 x 1" F.H. brass wood screw
5/32" shank hole, countersunk
E
2"
2"
B
Space mounting strips 1/4" apart.
#8 x 1 1/4" F.H. wood screw
D
Mitered corner
F
H
1 1/2"
J
3/8" round-overs
3/4 x 2" non-mortise hinge
C
G
I
K
Acrylic panel
B
H
D
F
8"
3"
7–82.
EXPLODED VIEW
21 1/4"
1 1/2"
3/16" hole centered from side to side
3" wire pull
Screw and strike plate
J
3/4" rabbet 1/2" deep
2"
I
G
47"
11/32" hole 5/8" deep
3/8"-round magnetic catch centered from side to side
C
2"
5"
3/4" rabbet 1/2" deep

Materials List for Wall Cabinet System

PART	FINISHED SIZE			MTL.	QTL.
	T	W	L		
BACK AND STRIPS					
A back	½"	22"	45½"	BP	1
B mounting strips	½"	1¼"	22"	M	29
CABINET FRAME					
C sides	¾"	5"	47"	BP	2
D top/bottom	¾"	5"	23"	BP	2
E hanger strip	¾"	2"	22"	M	1
F face strips	¾"	¾"	23½"	M	2
G face strips	¾"	¾"	47"	M	2
H door stops	¾"	¾"	22"	M	2
I door stops	¾"	¾"	44"	M	2
DOOR					
J rails	¾"	1½"	21⅞"	M	2
K stiles	¾"	1½"	45⅜"	M	2
L stops	¼"	¼"	19⅜"	M	2
M stops	¼"	¼"	42⅞"	M	2

Materials Key: BP = birch plywood; M = maple.
Supplies: #8 x 1¼" flathead wood screws; #8 x 2" flathead wood screws; #8 x ¾" brass flathead wood screws; #8 x 1" brass flathead wood screws; ⅛" clear acrylic; ½" x #18 brads; paint; clear finish.
Hardware: 3" wire pull; magnetic catches with strike plates (4); pair ¾ x 2½" no-mortise hinges.

7 To ensure consistent gaps between the mounting strips, use the spacing jig, as shown in **7–83**. Working from the top down, glue and screw all the mounting strips to the back.

7–83.
SPACING JIG

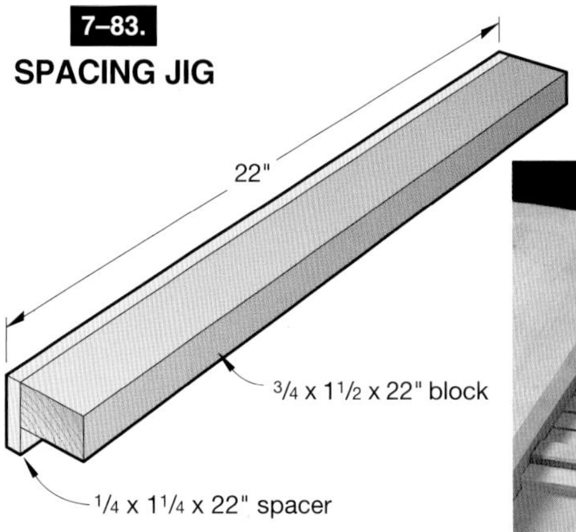

22"

¾ x 1½ x 22" block

¼ x 1¼ x 22" spacer

Build the Basic Cabinet Assembly

1 Cut the cabinet sides (C) and top and bottom (D) to the sizes listed in the Materials List from ¾" birch plywood.

2 Cut a ¾" rabbet ½" deep across both ends of each side piece. Glue and clamp the pieces (C, D). Check for square, and wipe off excess glue. So the cabinet will fit easily onto the back (A) later, the opening is 1/16" larger in length and width than the back.

(Below). Glue and screw the maple mounting strips to the plywood back, using the spacing jig for consistent gaps.

7–84.

3 Cut the hanger strip (E) to size. For mounting the strip to the back later, mark the locations and drill and countersink a pair of mounting holes through the front face of the strip (E), where shown on **7–82**.

4 Glue and clamp the hanger strip to the bottom of the cabinet top (D), ½" in from the back edge. See the Mounting Detail in **7–82** for reference. Drill three mounting holes through the cabinet top, centered into the top edge of the hanger strip (E). Drive a #8 x 1¼" wood screw through each hole just drilled.

5 Rip and miter-cut the face-frame strips (F, G) to size. Glue and clamp them to the front of the cabinet. Sand the strips flush with the cabinet frame.

6 Rout a ⅜" round-over along the outside front edge of the cabinet face strips (F, G).

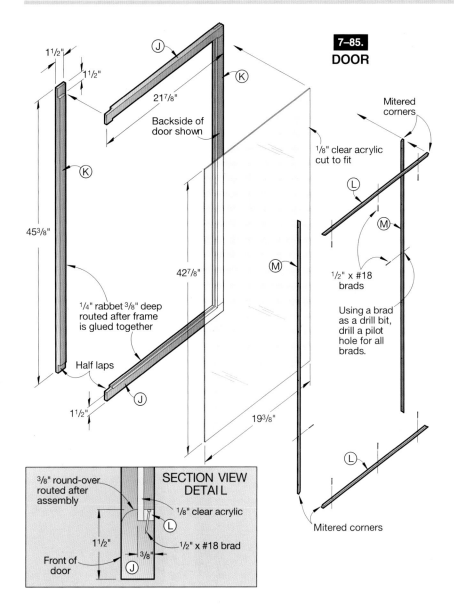

1½"

1½"

21⅞"

Backside of
door shown

J

K

K

45⅜"

42⅞"

¼" rabbet ⅜" deep
routed after frame
is glued together

Half laps

1½"

J

7–85.
DOOR

Mitered
corners

⅛" clear acrylic
cut to fit

L

M

M

½" x #18
brads

Using a brad
as a drill bit,
drill a pilot
hole for all
brads.

L

19⅜"

Mitered corners

SECTION VIEW
DETAIL

⅜" round-over
routed after
assembly

⅛" clear acrylic

L

1½"

½" x #18 brad

⅜"

Front of
door

J

4 Rout a ¼" rabbet ⅜" deep along the back inside edge of the door frame for the acrylic panel and stops. Using a chisel, square the curved corners left by the router.

5 Fit your router with a ⅜" round-over bit, and rout along the front inside edge of the door frame. See the Section View Detail in **7–85** for reference.

6 Using **7–86** for dimensions, drill the mounting holes in the left-hand stile for the pull. (For a flush-closing door, we drilled ⅜" holes 1¹⁄₁₆" deep on the back of the door for the screw heads. Then, we used a combination bolt cutter/wire stripper to snip ⅛" off the end of each screw so the wire pull would draw tight to the door front.) See the Materials List for our hardware needs.

7 Drill the pilot holes, and fasten a pair of no-mortise hinges to the right-hand door stile. Center the door top to bottom in the opening, and mark the mating hinge locations on the inside face of the cabinet side (on the face strip [F]). Next, drill mounting holes, and attach the hinges and door to the cabinet.

8 Cut the acrylic-panel stops (L, M) to size. Snip the head off a ½" x #18 brad, chuck the headless brad into your portable drill, and drill pilot holes through the stops. Do not install the piece of acrylic at this point.

7 Cut the door stops (H, I) to size.

8 For mounting the magnetic catches later, drill a pair of ¹¹⁄₃₂" holes ⅝" deep in one door stop (H), where shown on **7–82** and **7–87**. Hold off installing the door stops until final assembly.

Add the Door

1 Cut the maple door rails (J) and stiles (K) to size.

2 Cut 1½"-long half-lap joints on the ends of each rail and stile.

3 Glue and clamp the door frame together, checking for square and making sure the frame is flat. Later, remove the clamps and sand the door smooth.

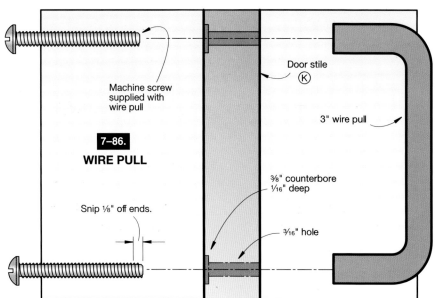

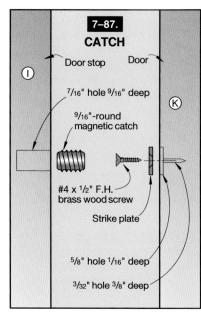

Add the Finish, and Hang the Cabinet

1 Inset a pair of magnetic catches into the holes now provided in the left-hand door stop.

2 Fasten the door stops (H, I) to the cabinet so when swung shut, the front of the door is flush with the front of the cabinet.

3 Close the door tightly against the catches to mark their mating position on the back edge of the door stile (K). Next, using a brad-point bit, drill a ½" hole ¹⁄₁₆" in the door stile, where indented, for each strike plate. Using the centered depression left by the brad-point bit when drilling the ⅝" hole, drill a ³⁄₃₂" pilot hole ⅜" deep centered inside the ⅝"-diameter counterbore. Screw the strike plates in place.

4 Remove the hardware (except for the magnetic catches) from the cabinet and door. Finish-sand the cabinet assembly, back, door, and acrylic panel stops.

5 Mask the surrounding areas and catches, and apply a clear finish to the face strips, door stops, mounting strips, and door.

6 Mask the maple face strips (F, G), then paint the cabinet the color of your choice.

7 Measure the openings, and have an acrylic panel cut to fit. Secure the panel with the stops (L, M).

8 Reattach the wire pull and hinges to the door. Reattach the door to the cabinet.

9 To mount the back (A) to the wall, locate the stud(s), and position the back. Drill mounting holes through the top and bottom of the back, centered over the stud(s). Check for plumb and level, and secure the back to the wall. Fit the cabinet assembly onto the back, and secure it to the back by driving a set of screws through the hanger strip (E) and into the back. After you've built your organizers, remove the cabinet from the back, slide the organizers in place, and reattach the cabinet to the back.

MAKE CUSTOMIZED TOOL HOLDERS

Now that you've built a cabinet or two using our wall-cabinet design, it's time to add some customized holders. But before you begin, familiarize yourself with a holder's parts.

Anatomy of a Tool Holder

As shown in **7–88**, the horizontal part that supports the tool is called the shelf. For heavier items, the shelf fits into a groove in the support. The dovetail strip attaches to the back of the shelf or support and slides between the mounting strips. The banding strips protect the tools from falling off the front or ends of the shelf.

7–88. TOOL HOLDER ANATOMY

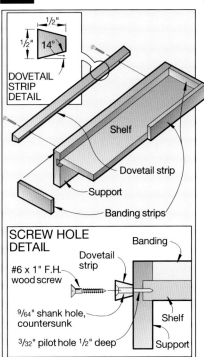

7–89. SCREWDRIVER HOLDER

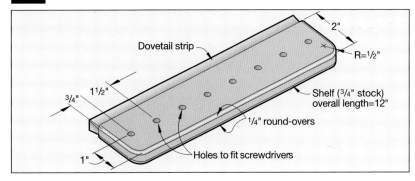

7–90. SQUARE HOLDER

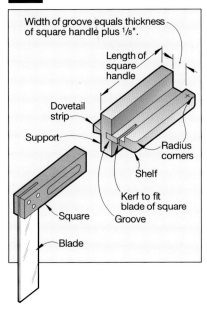

Make Your Holders to Fit

To make the shelf, start by laying the item you want to store on a piece of stock. (We used ½"- and ¾"-thick maple for most shelves.) If the bottom of the tool is square or rectangular, cut the shelf about ⅛" oversize. Or, for screwdrivers, router bits, and other shanked items, cut the shelf to size, and drill holes for the tool shanks, as shown on **7–89**. The distance between

the holes depends on the items you intend to store.

Add the Banding Strips

These should extend high enough above the shelf (usually about ¼") to keep the tool from being bumped off. We used banded shelves for planes, sharpening stones, drill bit index boxes, and other flat-bottomed items. If the tool's outline is irregular like that of the caliper holder on **7–91**, mark a portion of the tool's outline on ¾" stock. Then, cut the outline to shape.

7–91. CALIPER HOLDER

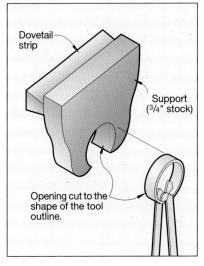

Add the Dovetail Strips

Bevel-rip long lengths of dovetail-strip stock at 14°. See the Dovetail Strip Detail in **7–88** for reference. Crosscut the dovetail strips to length. Drill and countersink mounting holes in the back edge of the strip. Glue and screw the dovetail strip to the back edge of the shelf, support, or holder.

Slide the dovetail strip of the holder between the mounting strips in the cabinet back. If the dovetail strip fits too tightly between the mounting strips, sand it slightly for a smooth sliding action. Remove and finish the holders.

Once dry, slide your holders between the mounting strips in the cabinet back (attached to the wall at this point). Arrange the holders as needed, then secure the cabinet back to the wall.

❖

7–92.

HANDY CABINETS

The perforated hardboard and hooks used for these cabinets (**7–92**) will help you get organized. Not finding the right tool when it's needed most is every woodworker's biggest complaint. Build the double-door cabinet or the single-door cabinet, or all, to really get your shop organized.

Begin with the Double-Door Cabinet and Door Frames

1 Using the Cutting Diagram for reference, lay out and cut the cabinet sides (A), cabinet top and bottom (B), door sides (C), and door top and bottom (D) to the sizes listed in the Materials List. We used a stop when cutting the pieces to length to ensure that the height of the assembled doors would match that of the cabinet.

Materials List for Double-Door Cabinet

PART	FINISHED SIZE			Mtl.	Qty.
	T	W	L		
A cabinet sides	¾"	6¾"	48"	BP	2
B cabinet top & bottom	¾"	6¾"	47¼"	BP	2
C door sides	¾"	2¼"	48"	BP	4
D door tops & bottoms	¾"	2¼"	23¼"	BP	4
E* cabinet top & bottom edging	¼"	¾"	47¼"	M	2
F* cabinet side edging	¼"	¾"	48"	M	2
G* door top & bottom edging	¼"	¾"	23¼"	M	4
H* door side edging	¼"	¾"	48"	M	4
I side cleats	¾"	1½"	43½"	P	6
J cabinet top & bottom cleats	¾"	1½"	46½"	P	2
K door top & bottom cleats	¾"	1½"	22½"	P	4
L cabinet perforated hardboard	¼"	46½"	46½"	PB	1
M door perforated hardboard	¼"	22½"	46½"	PB	2
N door panels	½"	22"	46"	BP	2
O* sides	½"	1"	48"	M	4
P* top & bottom bands	½"	44"	24"	M	4

*Initially cut parts oversized. Then, trim each to finished size according to the instructions.

Materials Key: BP = birch plywood; M = maple; P = pine; PB = perforated hardboard.
Supplies: 4d finish nails; #12 x ¾" panhead sheet-metal screws; 3" deck screws; 1½ x 48" continuous hinges (2); 3" wire pulls; 2½" hasp (2); 2" barrel bolt; padlock; enamel paint; satin polyurethane.

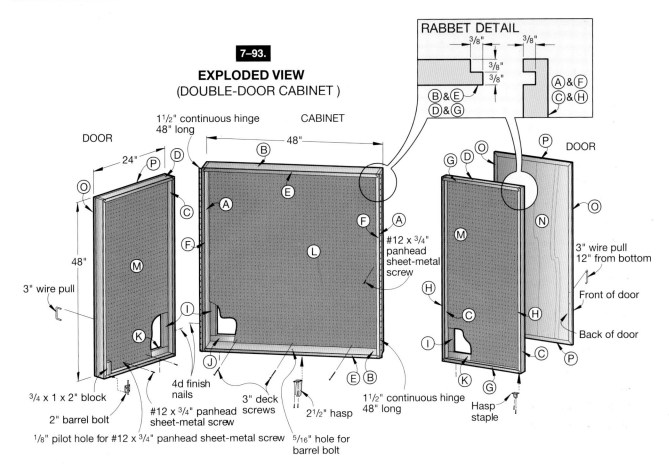

7–93.
EXPLODED VIEW
(DOUBLE-DOOR CABINET)

RABBET DETAIL

Set your tablesaw rip fence ¼" from the inside edge of the blade, and rip the edging strips (E, F, G, H) from the edge of ¾"-thick maple. Crosscut the edging strips to length plus 1". (We used the edging to hide the exposed plies of the birch plywood pieces.)

Glue and clamp the edging strips to the plywood pieces (A, B, C, D), and then later trim the ends of the edging flush with that of the plywood.

As dimensioned on the Rabbet Detail accompanying the Exploded View drawing in **7–93**, cut a ⅜" dado ⅜" deep ⅜"

from the top and bottom ends of the cabinet and door sides (A/F and C/H). Next, cut a ⅜" rabbet ⅜" deep along the ends of the top and bottom pieces (B/E and D/G). (We test-cut scrap pieces of stock first to verify the settings.) Check the fit of the mating cabinet and door pieces.

Glue and clamp the cabinet assembly together, checking for square. Repeat the process to assemble the two doors.

Add the Cleats and Perforated Hardboard

Now cut the cabinet and door cleats (I, J, K) to size from ¾" pine boards.

Glue and nail the cleats to the inside of the cabinet and doors flush with the back edge of the plywood panels, where shown on **7–93** and **7–94**.

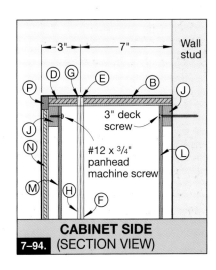

CABINET SIDE
7–94. (SECTION VIEW)

1/2 x 48 x 48" Birch plywood

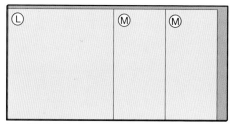

1/4 x 48 x 96" Perforated hardboard

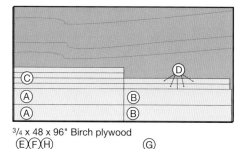

**Cutting Diagram
for Double Door
Cabinet**

3/4 x 48 x 96" Birch plywood

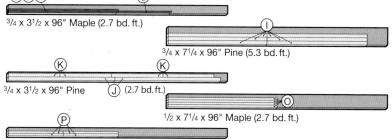

3/4 x 3 1/2 x 96" Maple (2.7 bd. ft.)

3/4 x 7 1/4 x 96" Pine (5.3 bd. ft.)

3/4 x 3 1/2 x 96" Pine   (2.7 bd. ft.)

1/2 x 7 1/4 x 96" Maple (2.7 bd. ft.)

1/2 x 3 1/2 x 96" Maple (1.3 bd. ft.)

7–95.

Double door cabinet

3 Measure the openings and cut the cabinet perforated hardboard (L) and door Pegboard (M) panels to size.

Cut the Door Panels and Banding

1 Cut the two plywood door panels (N) to size from a piece of 1/2" plywood.

2 Cut the door banding pieces (O, P) to size plus 1" in length from maple stock.

3 Rout a 1/16" chamfer along the front mating edges of the door panel (N) and banding (O, P). (See **7–97**.) When joined together in the next step, the chamfers form a decorative V-groove.

4 Miter-cut the banding pieces (O, P) to length. Glue and clamp the banding to the door panels (N).

5 Lay the cabinet, front side facing up, on your workbench. Position the doors, front side (or paneled side) up, on the cabinet with the edges and ends flush. To keep the hinges from binding later, lay a piece of folded writing paper between the cabinet and doors to act as a spacer to create the correct gap.

6 Use masking tape to secure the 1 1/2 x 48" continuous (piano) hinges in place. Using the existing holes in the hinges as guides, drill 1/16" pilot holes into the cabinet and doors. Attach the hinges, set the cabinet assembly upright, and open and close the doors to check the fit.

7 Locate and drill the holes, and attach a 3" wire pull to each door.

8 If locking the cabinet is a consideration, apply the barrel bolt, hasp, and hasp staple to the cabinet and doors.

Finishing Touches

1 Remove the hardware from the assembled cabinet. Sand all the parts smooth.

2 Apply a clear finish (we used satin polyurethane) to the door banding (O, P). (See **7–92**.)

3 After the finish dries, mask off the edging, and apply

Materials List for Single-Door Cabinet

PART	FINISHED SIZE			Mtl.	Qty.
	T	W	L		
A cabinet sides	¾"	6¾"	48"	BP	2
B cabinet top & bottom	¾"	6¾"	29¼"	BP	2
C door sides	¾"	2¼"	48"	BP	4
D door tops & bottoms	¾"	2¼"	29¼"	BP	4
E* cabinet top & bottom edging	¼"	¾"	29¼"	M	2
F* cabinet side edging	¼"	¾"	48"	M	2
G* door top & bottom cleats	¾"	1½"	28½"	P	4
H* door side cleats	¾"	1½"	43½"	M	4
I door panel	½"	28"	46"	BP	6
J side bands	½"	1"	48"	M	2
K door top & bottom cleats	½"	1"	30"	P	2
L door and cab perforated hardboard	¼"	28½"	46½"	PB	2

*Initially cut parts oversized. Then, trim each to finished size according to the instructions.

Materials Key: BP = birch plywood; P = pine; M = maple; PB = perforated hardboard.
Supplies: 4d finish nails; 3" deck screws; #12 x ¾" panhead sheet-metal screws; 1½ x 48" continuous (piano) hinge; 3" wire pull; 2½" hasp; lock; enamel paint; satin polyurethane.

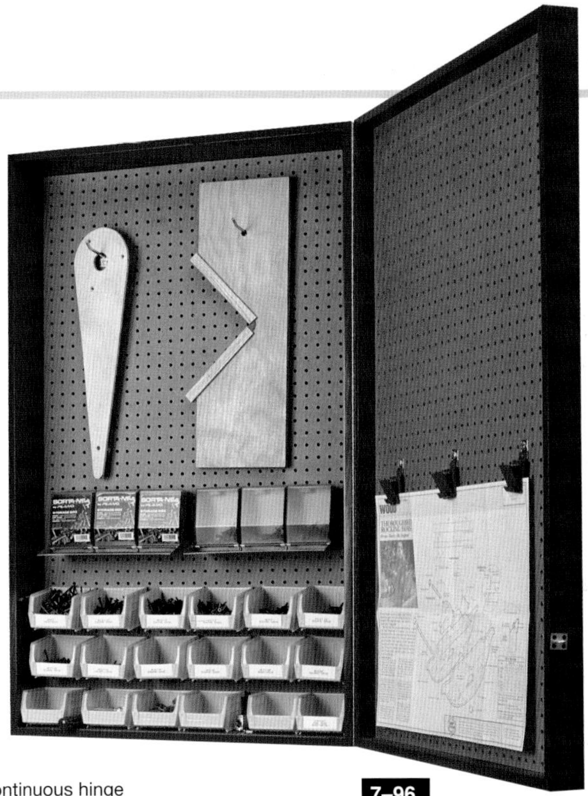

7–96.

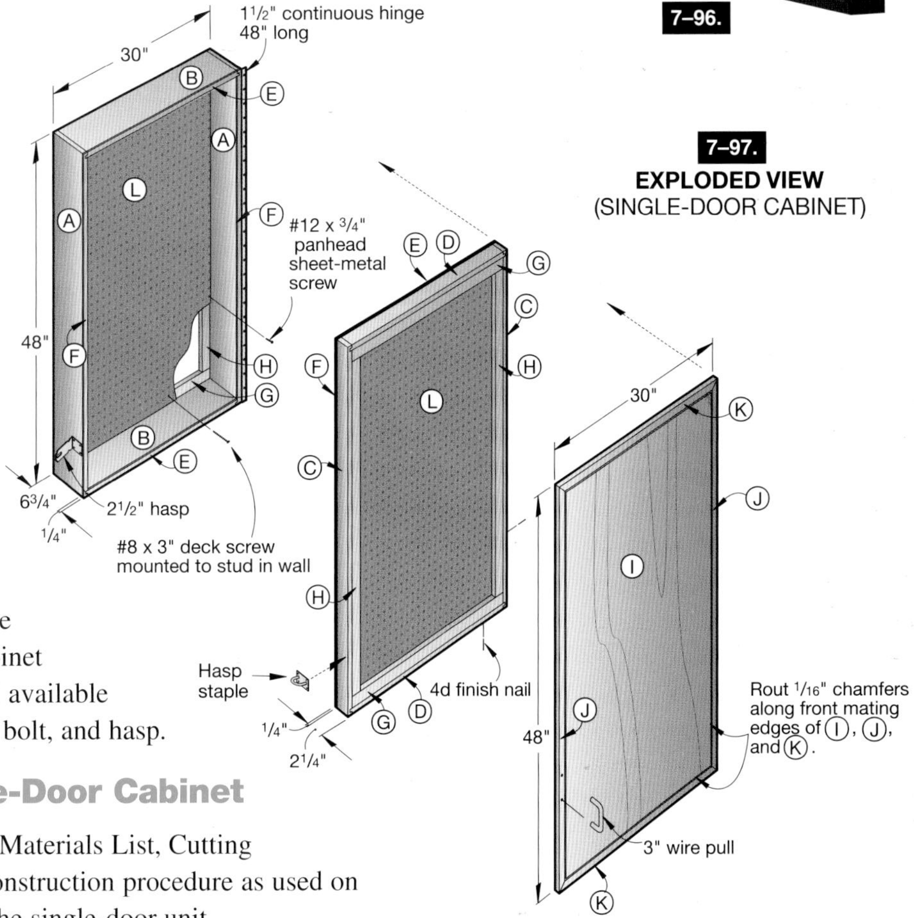

7–97.
EXPLODED VIEW
(SINGLE-DOOR CABINET)

primer to the rest of the surfaces. Later, paint the primed surfaces.

4 Fit the perforated hardboard panels (L, M) in their respective openings. Use panhead sheet-metal screws to secure the panels to the cabinet and doors.

5 Reattach the doors to the cabinet. Finally, with the aid of an assistant, hang the cabinet on the wall, being sure to hit all available studs. Reattach the pulls, barrel bolt, and hasp.

How to Build a Single-Door Cabinet

Using the Single-Door Cabinet Materials List, Cutting Diagram, **7–97**, and the same construction procedure as used on the double-door cabinet, build the single-door unit.

--- (exploded view labels) ---

1½" continuous hinge 48" long

30"

#12 x ¾" panhead sheet-metal screw

48"

6¾"

¼"

2½" hasp

#8 x 3" deck screw mounted to stud in wall

Hasp staple

¼"

2¼"

4d finish nail

30"

48"

Rout 1/16" chamfers along front mating edges of I, J, and K.

3" wire pull

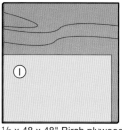

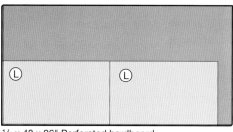

**Cutting Diagram for
Single-Door Cabinet**

½ x 48 x 48" Birch plywood

¼ x 48 x 96" Perforated hardboard

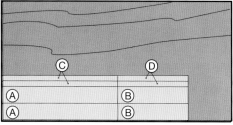

¾ x 48 x 96" Birch plywood

G H

¾ x 7¼ x 96" Pine (5.3 bd. ft.)

F E

¾ x 3½ x 96" Maple (2.7 bd. ft.)

J K

½ x 3½ x 96" Maple (1.3 bd. ft.)

REFERENCE CENTER

Keep reference material close at hand with this two-project setup (**7–98** and **7–99**). Simply swing the table up when you need to take a few notes or consult reference material. When you're not using it, collapse the brackets and the table folds back against the wall. The cabinet, with its drop-down door, provides necessary storage for books, magazines, tool catalogs, and other shop-related papers. Fit the table with an inexpensive stool (see Chapter 11), and you've created a comfortable workstation.

❖

7–98.

7–99.

EXPLODED VIEW

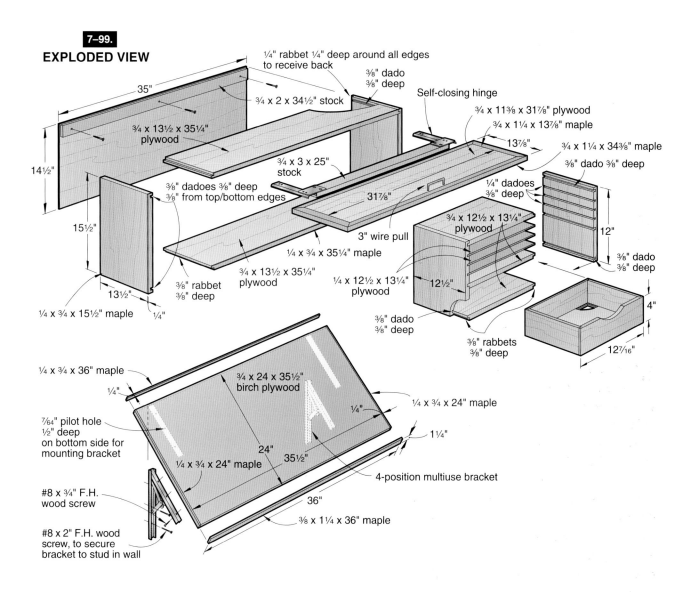

¼" rabbet ¼" deep around all edges to receive back

⅜" dado ⅜" deep

Self-closing hinge

35"

¾ x 2 x 34½" stock

¾ x 13½ x 35¼" plywood

¾ x 11⅜ x 31⅞" plywood

¾ x 1¼ x 13⅞" maple

13⅞"

¾ x 1¼ x 34⅜" maple

⅜" dado ⅜" deep

14½"

¾ x 3 x 25" stock

¼" dadoes ⅜" deep

15½"

⅜" dadoes ⅜" deep ⅜" from top/bottom edges

31⅞"

¾ x 12½ x 13¼" plywood

12"

3" wire pull

⅜" dado ⅜" deep

¼ x ¾ x 35¼" maple

¼ x 12½ x 13¼" plywood

12½"

⅜" rabbet ⅜" deep

¾ x 13½ x 35¼" plywood

13½"

¼ x ¾ x 15½" maple

¼"

⅜" dado ⅜" deep

⅜" rabbets ⅜" deep

4"

12⁷⁄₁₆"

¼ x ¾ x 36" maple

¼"

¾ x 24 x 35½" birch plywood

¼"

¼ x ¾ x 24" maple

⁷⁄₆₄" pilot hole ½" deep on bottom side for mounting bracket

1¼"

24"

35½"

¼ x ¾ x 24" maple

4-position multiuse bracket

#8 x ¾" F.H. wood screw

36"

#8 x 2" F.H. wood screw, to secure bracket to stud in wall

⅜ x 1¼ x 36" maple

CHAPTER 8

Solving Special Storage Needs

IF YOU'VE BEEN WOODWORKING *for any length of time at all, you know how quickly a pile of tools and materials can build up on a bench top. It seems that the best organization quickly gets disorganized during the process of project building. You end up spending as much time searching for the right drill bit as it would take to drill the hole!*

Well, some disarray is always bound to occur, but with the specialty storage you'll see in this chapter, you can make sure less of it happens. As you'll find out, the old adage "a place for everything, everything in its place" can really ring true in your shop.

Start with the flexible storage system that begins on this page. With its interlocking cleats, you can shift the components around on your walls as the need or whim arises.

Then, you'll find not one, but two, accessory storage cabinets designed for tablesaws. There's also storage for hardware, scroll-saw blades, sanders and sanding supplies, specialty drill bits, tools, and a whole lot more. If you build them, you'll be way ahead in the organization department.

Sure, you may still misplace something once in awhile (who doesn't?), but when you do find it, you have a dedicated spot to return it to. So, dig in now to our assortment of specialty storage projects. And try not to lose anything in the process.

❖❖

FLEXIBLE WALL STORAGE SYSTEM

Adding new tools or supplies to your workshop is always a good thing. But sometimes, finding a place to set up a tool or store more supplies means juggling the existing layout of wall cabinets or tool racks. With its interlocking hanging cleats, shown in **8–2**, this system trans-forms a major hassle into a minor task accomplished in minutes.

Here, you'll learn how to build racks for bar and pipe clamps, a perforated hardboard panel for

Clamp racks · Perforated hardboard panel · Clamp rack · General storage cabinet · Wall cleat · Tool cabinet · Wall cleat · Hardware cabinet · Interlocking cleats

8–1.

Build the three cabinets, the perforated hardboard panel for tools, and the clamp racks you see here with instructions on the following pages.

(Right). All of the wall system's components hang by hooking the downfacing bevels of their cleats over the upfacing bevel of a cleat fastened to the wall, as demonstrated here with the perforated hardboard panel.

8–2.

hanging hand tools, and wall cabinets with either clear acrylic or hardboard doors for see-through or covered-up storage. Featured in three handy sizes—12¼" deep for general storage, 8¾" deep for hardware, and 7½" deep for tools—all the cabinets share identical construction details.

Start with the Wall Cleats

1 Measure your shop for the total linear feet of wall cleat (A) you'll need. Our shop has

cleats at the levels shown on **8–3**, although not necessarily on every wall. Plan your wall-cleat mounting heights according to your needs. Joint one edge of your boards straight, rip them to 3" wide, and then bevel one edge, where shown on **8–4**. Sand the cleats, and apply two coats of satin polyurethane.

2 Locate the wall studs, and drill countersunk shank holes in the cleats at these locations. Leveling the cleats, drill pilot holes into the studs, and drive the screws.

8–3. WALL CLEAT POSITIONING

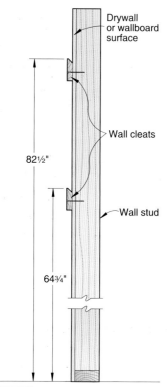

Drywall or wallboard surface

Wall cleats

82½"

Wall stud

64¾"

8–4. CLEAT DETAIL

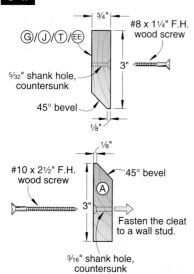

¾"

G/J/T/EE

#8 x 1¼" F.H. wood screw

⁵⁄₃₂" shank hole, countersunk

3"

45° bevel

⅛"

#10 x 2½" F.H. wood screw

⅛"

45° bevel

A

3"

Fasten the cleat to a wall stud.

³⁄₁₆" shank hole, countersunk

Add a Perforated Hardboard Panel

Cut the frame sides (B) and frame top and bottom (C) to size. With a dado blade, cut ¼"

grooves ⅜" deep in the sides, top, and bottom, where shown on **8–5**. Then cut the ¾" rabbets ⅜" deep in the ends of the sides.

Cut the panel (D) to size. Squeeze glue into the frame members' grooves, and clamp them to the panel.

Cut the back rail (E) and spacer (F) to size. Glue and clamp the rail to the back of the panel (D). Clamp the spacer in place, drill screw holes through the spacer and into the framebottom (C), and drive the screws.

Cut the cleat (G) to size, and then bevel one edge, where shown on **8–4**. Glue and clamp it in place, flush with the top of the frame top (C). Drill screw holes, and drive the screws.

Now, Make a Few Clamp Racks

To determine the size of the holder (H), shown on **8–8**, and the size and spacing of the notches for the bar or pipe clamps that you have, see the six steps shown on **8–7**. Cut the holder (H) to size. Then cut the upright (I) and cleat (J) to the widths listed in the Materials List and the same length determined for the holder. Cut 2° bevels on all the parts, where shown on **8–6**. Then cut the 45° bevel on the cleat. To lock the completed rack onto the wall cleat, drill a countersunk screw hole centered in the length of the upright, where shown on **8–6**.

Lay out the notches on the holder (H). With a dado blade in your tablesaw, and an auxiliary extension attached to your miter

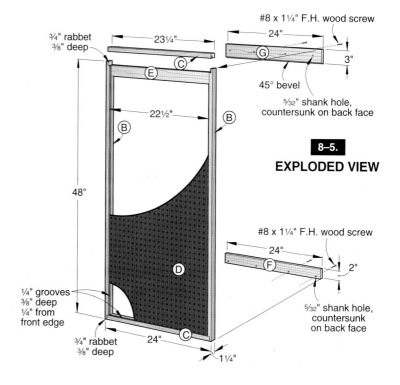

¾" rabbet ⅜" deep

23¼"

C

E

22½"

B

B

48"

D

¼" grooves ⅜" deep ¼" from front edge

¾" rabbet ⅜" deep

24"

C

1¼"

#8 x 1¼" F.H. wood screw

24"

G

3"

45° bevel

⁵⁄₃₂" shank hole, countersunk on back face

8–5. EXPLODED VIEW

#8 x 1¼" F.H. wood screw

24"

F

2"

⁵⁄₃₂" shank hole, countersunk on back face

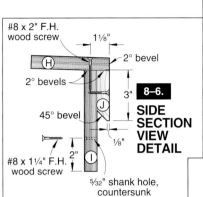

#8 x 2" F.H. wood screw
1⅛"
2° bevel
(H)
2° bevels
3"
45° bevel
(J)
#8 x 1¼" F.H. wood screw
⅛"
2"
(I)
⁵⁄₃₂" shank hole, countersunk

8–6.
SIDE SECTION VIEW DETAIL

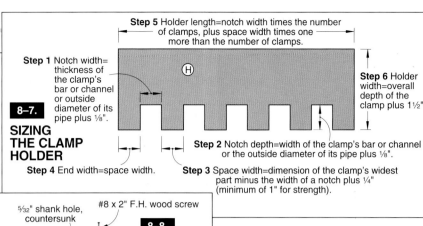

Step 5 Holder length=notch width times the number of clamps, plus space width times one more than the number of clamps.

(H)

Step 1 Notch width= thickness of the clamp's bar or channel or outside diameter of its pipe plus ⅛".

8–7.
SIZING THE CLAMP HOLDER

Step 6 Holder width=overall depth of the clamp plus 1½".

Step 2 Notch depth=width of the clamp's bar or channel or the outside diameter of its pipe plus ⅛".

Step 4 End width=space width.

Step 3 Space width=dimension of the clamp's widest part minus the width of a notch plus ¼" (minimum of 1" for strength).

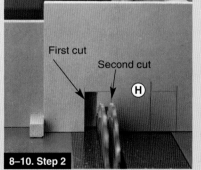

⁵⁄₃₂" shank hole, countersunk
#8 x 2" F.H. wood screw
1⅛"

8–8.
CLAMP RACK

#8 x 1¼" F.H. wood screw
⁵⁄₃₂" shank hole, countersunk on back
(H)
⁷⁄₆₄" pilot hole 1¼" deep
6"
Centered
(J)
3"
(I)
45° bevel
2"
#8 x 1¼" F.H. wood screw
⁵⁄₃₂" shank hole, countersunk

gauge to prevent chip-out, cut the notches. For a method of cutting the notches that avoids tedious layout, see **8–9** to **8–11** in the Shop Tip *below*.

3 Glue and clamp the cleat (J) to the upright (I) with their top edges flush, where shown on **8–6**. Drill screw holes, and drive the screws.

Then glue and clamp the holder (H) to assembly I/J, keeping the back edge of the holder flush with the back face of the cleat. Drill screw holes, and drive the screws.

Next, Build Some Storage Cabinets

*Note: The general storage and tool cabinets differ only in the widths of the sides, top, and bottom, as shown on **8–12**, and*

SHOP TIP

Save Time Notching with a Simple Step-and-Repeat Jig

When making several clamp racks, speed things up by adding an indexing pin the same width as the notches you wish to cut to a miter-gauge auxiliary extension. Then use it just like a box-joint jig to cut the notches in the holder shown in 8–9.

When making racks that call for notches of more than one width, make your jig for the narrowest notch, and cut these holders first. Then use the same jig to cut the wide notches. Illus. 8–9 to 8–11 show you how to cut 1" notches with a jig originally made to cut ½" notches.

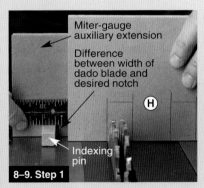

Miter-gauge auxiliary extension
Difference between width of dado blade and desired notch
(H)
Indexing pin

8–9. Step 1

Lay out the holder's first notch, and align it with the dado blade. Adjust the auxiliary extension so the distance between the indexing pin and the end of the holder is the same as the difference between the width of the dado blade and the width of the desired notch. Secure the extension to the miter gauge, and make the first cut.

First cut
Second cut
(H)

8–10. Step 2

Slide the holder over against the indexing pin, and make another cut. In this case, the notch is formed in two cuts because the dado blade is half the width of the desired notch. When the dado blade's width is less than half the desired notch's width, you'll have to make more than two cuts to complete the notch.

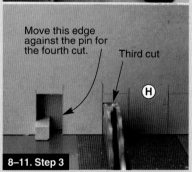

Move this edge against the pin for the fourth cut.
Third cut
(H)

8–11. Step 3

Place the notch just cut over the indexing pin with one side against the pin, and make the third cut. Push the notch's other side against the pin, and make a fourth cut. Once again, if using a narrower dado blade, clean out between the two cuts with additional passes. Now repeat these steps until all the notches are cut.

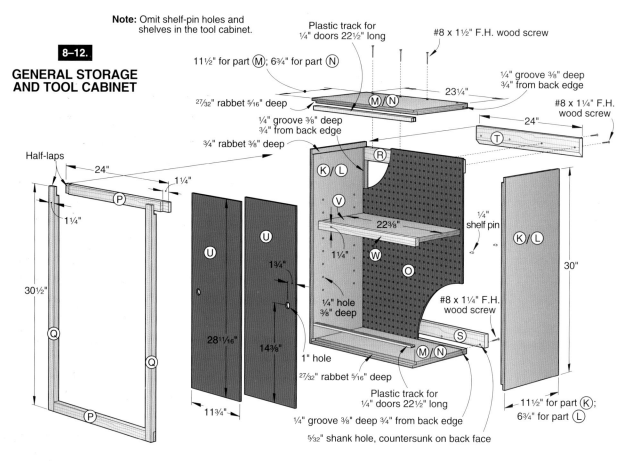

8–12.

GENERAL STORAGE AND TOOL CABINET

Note: Omit shelf-pin holes and shelves in the tool cabinet.

Plastic track for ¼" doors 22½" long

11½" for part (M); 6¾" for part (N)

#8 x 1½" F.H. wood screw

¼" groove ⅜" deep ¾" from back edge

²⁷/₃₂" rabbet ⁵/₁₆" deep

23¼"

#8 x 1¼" F.H. wood screw

24"

¼" groove ⅜" deep ¾" from back edge

¾" rabbet ⅜" deep

Half-laps

24"

1¼"

1¼"

30½"

22⅜"

1¼"

13¾"

¼" shelf pin

30"

28¹¹/₁₆"

14⅜"

¼" hole ⅜" deep

1" hole

²⁷/₃₂" rabbet ⁵/₁₆" deep

#8 x 1¼" F.H. wood screw

11½" for part (K); 6¾" for part (L)

11¾"

Plastic track for ¼" doors 22½" long

¼" groove ⅜" deep ¾" from back edge

⁵/₃₂" shank hole, countersunk on back face

whether or not they have shelves. All the other parts are identical. Although all the parts of the hardware cabinet, shown on **8–13**, differ in size from the other cabinets, the machining operations are the same, except for the backs. The Materials List indicates parts to build one of each cabinet.

1 Cut the sides (K, L, X) and the tops and bottoms (M, N, Y) to size. With a dado blade, cut ¾" rabbets ⅜" deep in the sides, where shown on **8–12** and **8–14**. Then, for the door track, cut the ²⁷/₃₂" rabbets ⁵/₁₆" deep along the front edges of the tops and bottoms. Finally, cut ¼" grooves ⅜" deep in all the parts for the backs (O/Z).

2 Drill shelf-pin holes in the sides (K) of the general storage cabinet, where shown on **8–13**. Do not drill shelf-pin holes in the sides (L, X) of the tool and hardware cabinets.

3 For the general storage and tool cabinets, cut perforated hardboard for the backs (O) to size. Squeeze glue into the top, bottom, and side grooves, and apply it to the sides' rabbets. Capturing the back in the grooves, clamp the case together.

8–13.

GENERAL STORAGE AND TOOL CABINET SECTION VIEW

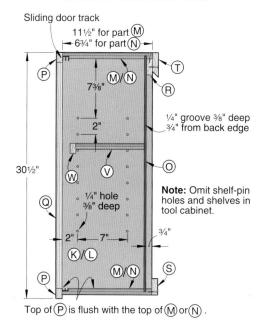

Sliding door track

11½" for part (M)

6¾" for part (N)

7⅜"

2"

¼" groove ⅜" deep ¾" from back edge

30½"

¼" hole ⅜" deep

Note: Omit shelf-pin holes and shelves in tool cabinet.

2"

7"

¾"

Top of (P) is flush with the top of (M) or (N).

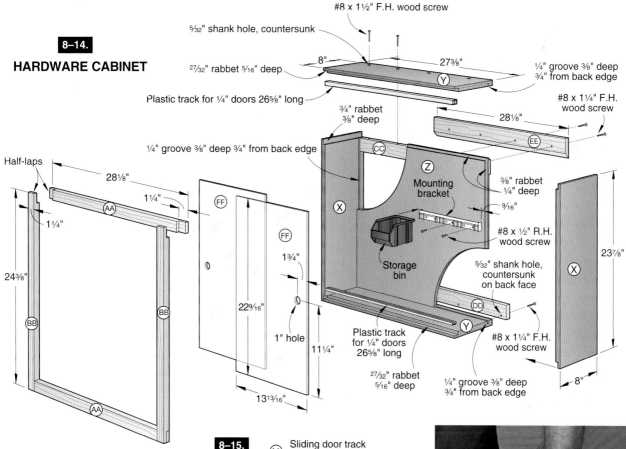

8–14.

HARDWARE CABINET

#8 x 1½" F.H. wood screw

⁵⁄₃₂" shank hole, countersunk

²⁷⁄₃₂" rabbet ⁵⁄₁₆" deep

8"

27⅜"

¼" groove ⅜" deep ¾" from back edge

Plastic track for ¼" doors 26⅝" long

#8 x 1¼" F.H. wood screw

¾" rabbet ⅜" deep

28⅛"

¼" groove ⅜" deep ¾" from back edge

Half-laps

28⅛"

1¼"

1¼"

Mounting bracket

⅜" rabbet ¼" deep

⁹⁄₁₆"

24⅜"

1¾"

#8 x ½" R.H. wood screw

⁵⁄₃₂" shank hole, countersunk on back face

23⅞"

22⁹⁄₁₆"

Storage bin

1" hole

11¼"

8"

Plastic track for ¼" doors 26⅝" long

#8 x 1¼" F.H. wood screw

13¹³⁄₁₆"

²⁷⁄₃₂" rabbet ⁵⁄₁₆" deep

¼" groove ⅜" deep ¾" from back edge

8–15.

HARDWARE CABINET SECTION VIEW

Sliding door track

8"

5½"

#8 x ½" R.H. wood screw

24⅜"

5½"

¾"

Storage bin Mounting bracket

5½"

1½"

Top of (AA) is flush with the top of (Y).

For the hardware cabinet back (Z), cut a piece of ½" plywood to size. Then cut a ⅜" rabbet ¼" deep along its edges, where shown on **8–14**. Orienting the back to leave a ¾" recess at the cabinet's rear, where shown on **8–15**, glue and clamp the hardware cabinet case together.

Checking the cases' exact widths, cut the rails (P, AA) to length, and clamp them in place, where shown on **8–13** and **8–15**. Note that the top edges of both rails (P, AA) are flush with the top faces of the tops and bottoms (M, N, Y). Now measure the exact length of the stiles (Q, BB), and cut them to size. Cut half-lap joints in the ends of the parts, and glue and clamp them in place.

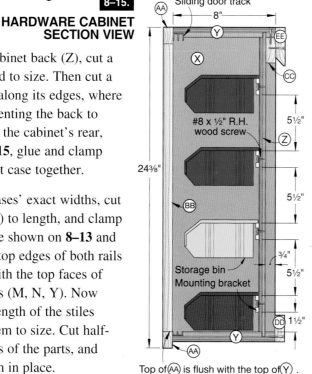

8–16.

Place the 1½" spacer between the cabinet's bottom and the first bracket, and align it with the mark. Drill pilot holes, and drive the screws. Slide the spacer over, and mount the second bracket. Now, using the 5½" spacer, mount the remaining brackets.

Materials List for Flexible Wall Storage System

		T	W	L	Mtl.	Qty.
			FINISHED SIZE			
A	wall cleat	3/4"	3"	†	M	†
	PERFORATED HARDBOARD PANEL					
B	frame sides	3/4"	1 1/4"	48"	M	2
C	frame top/bottom	3/4"	1 1/4"	23 1/4"	M	2
D	panel	1/4"	23 1/4"	47 1/4"	PH	1
E	back rail	3/4"	2"	22 1/2"	M	1
F	spacer	3/4"	2"	24"	M	1
G	cleat	3/4"	3"	24"	M	1
	CLAMP RACK					
H	holder	3/4"	†	†	MDO	1
I	upright	3/4"	6"	†	MDO	1
J	cleat	3/4"	3"	†	M	1
	GENERAL STORAGE AND TOOL CABINETS					
K	storage cabinet sides	3/4"	11 1/2"	30"	MDO	2
L	tool cabinet sides	3/4"	6 3/4"	30"	MDO	2
M	storage cabinet top and bottom	3/4"	11 1/2"	23 1/4"	MDO	2
N	tool cabinet top and bottom	3/4"	6 3/4"	23 1/4"	MDO	2
O	backs	1/4"	23 1/4"	29 1/4"	PH	2
P	rails	3/4"	1 1/4"	24"	M	4
Q	stiles	3/4"	1 1/4"	30 1/2"	M	4
R	rear rails	3/4"	2"	22 1/2"	M	2
S	spacers	3/4"	2"	24"	M	2
T	cleats	3/4"	3"	24"	M	2
U*	doors	1/4"	11 3/4"	28 11/16"	H/A	4
V	shelves	3/4"	8 3/4"	22 3/8"	MDO	2
W	shelf edges	3/4"	1 1/4"	22 3/8"	M	2

		T	W	L	Mtl.	Qty.
			FINISHED SIZE			
	HARDWARE CABINET					
X	sides	3/4"	8"	23 7/8"	MDO	2
Y	top and bottom	3/4"	8"	27 7/8"	MDO	2
Z	back	1/2"	23 1/8"	27 3/8"	MDO	1
AA	rails	3/4"	1 1/4"	28 1/8"	M	2
BB	stiles	3/4"	1 1/4"	24 3/8"	M	2
CC	rear rail	3/4"	2"	26 5/8"	M	1
DD	spacer	3/4"	2"	28 1/8"	M	1
EE	cleat	3/4"	3"	28 1/8"	M	1
FF*	doors	1/4"	13 13/16"	22 9/16"	A	2

†Determined by measurement. See the instructions.
*Parts initially cut oversize. See the instructions.

Materials Key: M = maple; MDO = medium-density overlay plywood; PH = perforated hardboard; H = tempered hardboard; A = acrylic.
Supplies: #8 x 1 1/4", #8 x 1 1/2", #8 x 2", and #10 x 2 1/2" flathead wood screws; #8 x 1/2" roundhead wood screws; construction adhesive.
Blades and bits: Stack dado set, 1" Forstner bit.
Hardware: 48" brown plastic sliding door track, 1/4" shelf pins (20).
Storage bins: Plastic storage bins with mounting bracket (red, yellow, green).

5 Cut the rear rails (R, CC), spacers (S, DD), and cleats (T, EE) to size. Bevel the cleats, where shown on **8–4**. Glue and clamp the rear rails to the backs (O, Z), and the spacers to the bottoms (M, N, Y), where shown on **8–12** to **8–15**. Drill screw holes, and drive the screws. Then glue and clamp the cleats in place, drill screw holes, and drive the screws.

6 Cut the doors (U, FF) 1/16" larger in length and width than the sizes listed. Depending on what you wish to store in the cabinets, choose either tempered hardboard or clear acrylic. (We used both hardboard and acrylic for general storage cabinets, and acrylic for the tool and hardware storage cabinets.) Clean up the edges by jointing off 1/32", and then ease the sharp corners with a sanding block. Chuck a 1" Forstner bit

Cutting Diagram

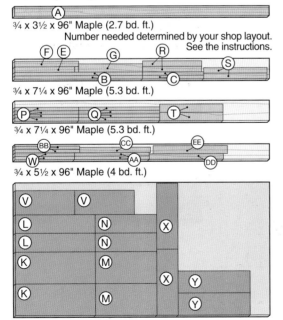

3/4 x 3 1/2 x 96" Maple (2.7 bd. ft.)
Number needed determined by your shop layout.
See the instructions.

3/4 x 7 1/4 x 96" Maple (5.3 bd. ft.)

3/4 x 7 1/4 x 96" Maple (5.3 bd. ft.)

3/4 x 5 1/2 x 96" Maple (4 bd. ft.)

3/4 x 48 x 96" Medium-density overlay plywood (MDO)

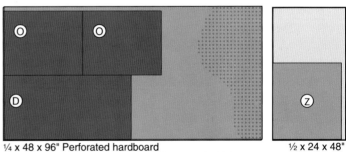

1/4 x 48 x 96" Perforated hardboard

1/2 x 24 x 48" Medium-density overlay plywood (MDO)

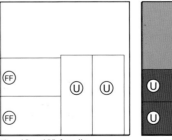

1/4 x 48 x 48" Acrylic

1/4 x 48 x 48" Hardboard

Note: Parts H and I are made of 3/4" medium-density overlay plywood, and part J is made of maple. To determine their dimensions and the quantity of materials needed, see the instructions.

in your drill press, and drill the finger holes. Chamfer their edges with sandpaper. When drilling finger holes in acrylic doors, run your drill press at 250 rpm.

7 For a general storage cabinet, cut the shelves (V) and the shelf edges (W) to size. Glue and clamp the shelf edges to the shelves.

Finish and Hang the System

1 Sand the exposed plywood edges of the clamp racks smooth, and ease all the corners with a sanding block. Sand the perforated hardboard panel frame, cabinet frames, and shelf edges, and ease their corners with a sanding block. Apply two coats of satin polyurethane to all the parts, sanding between coats.

2 Cut sliding door tracks to fit between the cabinet sides, apply a ⅛" bead of construction adhesive, and clamp the tracks in place, as shown in **8–12** and **8–15**.

3 Lay the hardware storage cabinet on its back. Cut 1½"- and 5½"-wide spacers from scrap, and make bracket-alignment marks ⁹⁄₁₆" from one end. Fasten the bin-mounting brackets, where shown on **8–15** and **8–16,** on *page 177.*

4 Lay the perforated hardboard panel on its back, and arrange your tools on it. Once you work out an efficient arrangement, attach perforated hardboard hooks and tool holders.

5 Hang the clamp racks, perforated hardboard panel, and cabinets on the wall cleats. To keep from accidentally dislodging the clamp racks, use the hole drilled in the rack's upright as a guide to drill a pilot hole in the wall cleat. Drive the screw. Install the shelf pins, shelves, and plastic bins in their respective cabinets, and then add the doors. Now, hang up your clamps, refill the perforated hardboard, and arrange your other tools and supplies in the cabinets.

TABLESAW ACCESSORY CABINET

Extension tables for tablesaws make woodworking easier, but they sure eat up a lot of precious floor space. This handsome roll-away cabinet (**8–17** and **8–18**) helps you put that valuable real estate to good use. We store blades, jigs, and accessories in ours.

Start with Carcase Construction

1 Cut the carcase top and bottom (A), side panels (B), and back panel (C) to the sizes shown in the Materials List. Note the grain orientation for each piece, and label the parts.

2 Install a ¾" dado blade on your tablesaw, and set the cutting height to ½". Then, clamp or fasten a wood auxiliary fence to your rip fence. Slide the fence over until the wood face butts against the right side of the dado blade, as shown in **8–19**. Now, cut the rabbets along the edges of the top and bottom panels (A), where shown in **8–20**.

3 Reinstall your standard saw blade and cut the back edging pieces (D) to size and glue them to the back panel (C). (This simple butt joint is fine if you take care to align the pieces. For a self-aligning joint, cut grooves in the edging and the plywood edges and glue splines in between, or use biscuit joints to hold the surfaces flush.)

8–17.

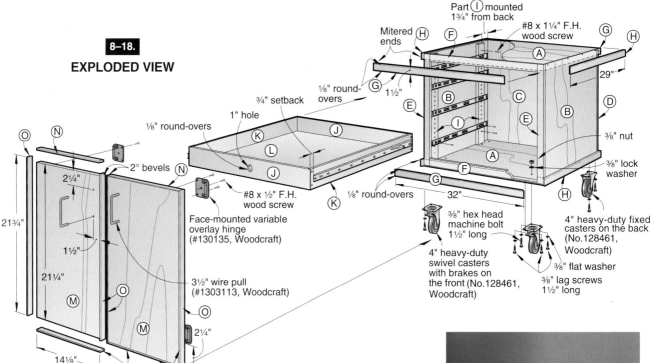

8–18.

EXPLODED VIEW

Part (I) mounted 1¾" from back

#8 x 1¼" F.H. wood screw

Mitered ends (H) (F)

⅛" round-overs

¾" setback

1" hole

⅛" round-overs

2° bevels

2¼"

21¾"

1½"

21¼"

#8 x ½" F.H. wood screw

Face-mounted variable overlay hinge (#130135, Woodcraft)

3½" wire pull (#1303113, Woodcraft)

⅛" round-overs along all outside edges of the doors

14⅛"

2¼"

⅛" round-overs

32"

29"

⅜" nut

⅜" lock washer

4" heavy-duty fixed casters on the back (No.128461, Woodcraft)

⅜" flat washer

⅜" lag screws 1½" long

⅜" hex head machine bolt 1½" long

4" heavy-duty swivel casters with brakes on the front (No.128461, Woodcraft)

4 While the glue dries on those parts, join the other panels. Apply glue to the rabbets on the top and bottom panels (A)—at the sides only, not the front or back— then center and clamp the side panels (B) in place.

5 Drill countersunk holes through the sides (B) for screws to secure the rabbet joints. (To learn about specialized bits for this task, see Triple-Duty Countersink Bits on *page 183*.) After you drive the screws, remove the clamps and place the carcase face down on your bench.

6 Check the fit of the back panel assembly (C/D); then glue and clamp it to the back edges of the carcase. Fasten the

back (C/D) to the top and bottom (A) panels.

Install the Face Frame Pieces

1 You can install the simple face-frame for this cabinet one piece at a time, rather than as an assembly. Start by cutting the two stiles (E) to size; then glue and clamp them to the front edge of the side panels (B).

2 Measure between the stiles to find the exact length for the face frame rails (F). Cut these parts; then glue and clamp them to the rabbeted edges of the top and bottom panels, making sure the rails and stiles are flush where they meet.

3 Next, at each corner of the face frame drill a countersunk hole through the edge of the stile

8–19.

A wood auxiliary fence allows you to use the full width of the ¾" dado blade to cut rabbets in the plywood carcase panels.

and into the rail end. Then drive a #8 × 2¼" screw to connect them.

Edgebanding Guards the Cabinet

We hid the plywood edges around the cabinet top and bottom with maple edgebanding. Doing this also protects these edges from collisions as you move the cabinet around the shop.

Materials List for Tablesaw Accessory Cabinet

PART	FINISHED SIZE T	W	L	Mtl.	Qty
A top/bottom	¾"	27½"	30½"	BP	2
B side panels	¾"	26"	24½"	BP	2
C back panel	¾"	29"	24½"	BP	1
D back edging	¾"	¾"	24½"	M	2
E frame stiles	¾"	1½"	24½"	M	2
F frame rails	¾"	1⅝"	27½"	M	2
G front/back edgeband	¾"	1½"	32"	M	4
H side edgeband	¾"	1½"	29"	M	4
I tray support columns	¾"	2¾"	23½"	M	4
J tray fronts/backs	½"	3"	26½"	M	8
K tray sides	½"	3"	26¼"	M	8
L tray bottoms	½"	26¼"	26¼"	BP	4
M door panels	¾"	14⅛"	21¼"	BP	2
N door edging, top/bottom	¼"	¾"	14⅛"	M	4
O door edging, sides	¼"	¾"	21¾"	M	4

Materials Key: BP = birch plywood, M = maple.
Supplies: #8 x 1¼" flathead wood screws (116); #8 x 1½" flathead wood screws (16); #8 x 2¼" flathead wood screws (4); ¼-20 x ¾" roundhead machine screws (16); ⅜ x 1½" lag screws (12); ⅜ x 1½" hexhead machine bolts (4); ⅜" flat washers (20); ⅜" lock washers (4); ⅜" hex nuts (4); 2 x 8' nonslip rug cushion (1).
Buying Guide: In addition to the lumber and supplies shown above, the following items (or appropriate substitutes) are required for this project: Heavy-duty 4" casters (set of 4); drawer slides, 26" long (4 sets); 3½" wire pulls (2); and face-mount overlay hinges (2 pair).

1 Cut the front and rear edgebanding pieces (G) and side pieces (H) slightly longer than required; then rout a ⅛" round-over along the two outside edges of each piece.

Note: *For best results, mark each piece in place and trim the mitered ends for a tight-fitting joint.*

2 Glue and clamp the pieces to the cabinet. Later, scrape off the excess glue and sand the surfaces flush.

Sliding Trays Make Tool Storage Efficient and Accessible

To make the most of the space inside the cabinet, this design features a set of four shallow drawers, or trays, that are great for storing blades, wrenches, and most other accessories you want close to your tablesaw. Four tray support columns (I), one at each corner of the cabinet, let you adjust the tray positions to suit your storage needs. (Keep in mind that different drawer slide hardware may require changes in the column sizes or in the mounting hole placement.)

Make the Tray Support Columns

1 Cut the columns (I) to length and drill two mounting holes and a series of slide installation holes, as shown in **8–21**.

2 Tap a ¼-20 thread in each slide installation hole.

Cutting Diagram

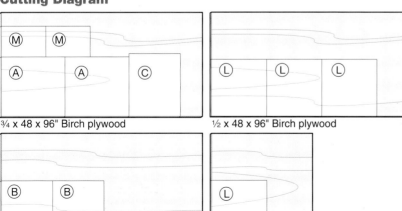

M
M
A
A
C
¾ x 48 x 96" Birch plywood
L
L
L
½ x 48 x 96" Birch plywood
B
B
¾ x 48 x 96" Birch plywood
L
½ x 48 x 48" Birch plywood
H
I
I
¾ x 7¼ x 96" Maple (5.3 bd.ft.)
J
J
J
½ x 7¼ x 96" Maple (2.7 bd.ft.)
K
K
½ x 7¼ x 96" Maple (2.7 bd.ft.)
J
K
K
½ x 7¼ x 96" Maple (2.7 bd.ft.)
N
E
D
O
F
G
¾ x 7¼ x 96" Maple (5.3 bd.ft.)

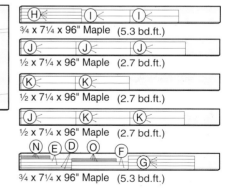

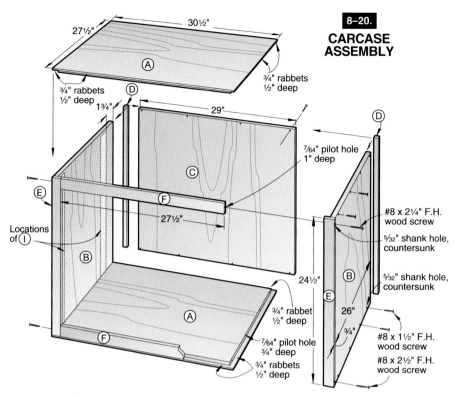

8–20.
CARCASE ASSEMBLY

27½"
30½"
¾" rabbets ½" deep
¾" rabbets ½" deep
1¾"
29"
7/64" pilot hole 1" deep
(A)
(D)
(C)
(D)
(E)
(F)
27½"
Locations of (I)
(B)
24½"
(B)
(E)
(A)
¾" rabbet ½" deep
26"
¾"
(F)
7/64" pilot hole ¾" deep
¾" rabbets ½" deep
#8 x 2¼" F.H. wood screw
5/32" shank hole, countersunk
5/32" shank hole, countersunk
#8 x 1½" F.H. wood screw
#8 x 2½" F.H. wood screw

¼-20 x ¾" roundhead machine screws. *(Note that the front ends of the drawer slides are flush with the front edge of the tray support columns. The other half of each slide mechanism installs on the tray.)* Fasten a second slide to the bottom hole in each support, and then fasten the entire assembly inside the cabinet, as shown in **8–22.**

4 Assemble another pair of supports and slides and fasten them to the other side of the cabinet. Also, you can fasten the additional drawer slides to the supports at this time.

8–21.
TRAY SUPPORT COLUMN

2¾"
2¾"
1½"
2¾"
1"
23½"
5/32" shank hole, countersunk
(I)
13/64" holes Tap holes with ¼-20 tap, countersink 1/16".
5/32" shank hole, countersunk

3 Lay a pair of the tray support columns on your workbench, and set them 18¾" apart (between inside edges). Place the "cabinet" half of a drawer slide on the support columns, and fasten it to the top threaded hole in each using

8–22.

Fasten a pair of drawer slides to the tray support columns for each side; then install the assemblies inside the cabinet.

Next, Make the Trays Themselves

1 Cut the tray fronts and backs (J) and the sides (K) to size. Install a ½" dado blade on your tablesaw; then reinstall the wood auxiliary fence you used earlier to cut the rabbets on the plywood panels. Again, slide the fence until the wood face butts against the dado blade. Adjust the blade height to ⅜" and, guiding your stock with the miter gauge, cut the rabbets at the ends of each tray front and back.

2 Without changing the blade or fence setup, cut rabbets along the lower inside edges of all the tray fronts, backs, and sides by guiding the edge of each piece against the wood auxiliary fence.

3 Drill a centered 1" hole in each tray front. These are finger-pulls, so for comfort rout a ⅛" round-over around the rim of the hole, on both faces.

4 Glue and clamp the tray assemblies together, checking for equal diagonal measurements to make sure each tray is square. Then drill countersunk holes for screws at each corner, as shown in **8–23**.

5 Cut the ½" birch plywood to size for the tray bottoms (L). Place one tray assembly upside down on your workbench, and glue the tray bottom into the rabbets. Drill countersunk holes around the edges of the panel; then fasten it with screws.

6 When all the trays are assembled, rout a ⅛" round-over along all the exposed edges.

7 Install the remaining drawer slide hardware. Allow a ¾" backset from the face of the tray front, so it will end up flush with the face frame. Center the slide on the side of the tray, as shown in **8–23** and **8–24**.

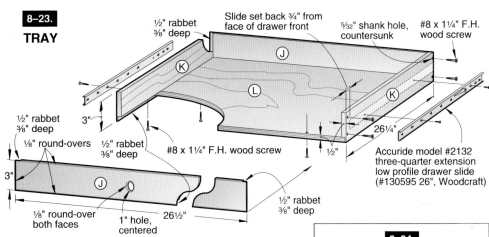

8–23.
TRAY

½" rabbet ⅜" deep
Slide set back ¾" from face of drawer front
⁵⁄₃₂" shank hole, countersunk
#8 x 1¼" F.H. wood screw
K
J
L
K
26¼"
½"
Accuride model #2132 three-quarter extension low profile drawer slide (#130595 26", Woodcraft)

½" rabbet ⅜" deep
3"
⅛" round-overs
½" rabbet ⅜" deep
#8 x 1¼" F.H. wood screw
½" rabbet ⅜" deep

3"
J
1" hole, centered
⅛" round-over both faces
26½"

Make and Install the Doors

1 Cut the door panels (M) from your ¾" birch plywood.

2 Cut the maple edging pieces (N, O) for the door edges. Glue the top and bottom pieces (N) on first, making sure their ends are flush with the edges of the door panels. Then glue the remaining edgebanding (O) on the side edges of each door.

3 Cut a slight back-bevel (2° is plenty) on the inside door edges so they don't bind when you open the cabinet.

4 Sand all the surfaces flush; then rout a ⅛" round-over along the front edges of each assembled door panel.

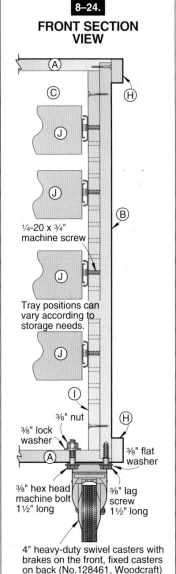

8–24.
FRONT SECTION VIEW

A
C
H
J
J
B
¼-20 x ¾" machine screw
J
Tray positions can vary according to storage needs.
J
I
⅜" nut
⅜" lock washer
A
H
⅜" flat washer
⅜" hex head machine bolt 1½" long
⅜" lag screw 1½" long
4" heavy-duty swivel casters with brakes on the front, fixed casters on back (No.128461, Woodcraft)

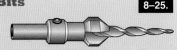

SHOP TIP

Triple-Duty Countersink Bits

8–25.

Drilling for screws often involves a series of sizes—a pilot hole for the threads, a clearance hole for the upper shank, and a countersink or counterbore to recess the screw head. Instead of swapping out bits for each step, use a combination bit, such as the one shown in 8–25. It shapes the hole in just one pass.

5 Make layout marks for positioning the door pulls and hinges; then drill holes for fasteners as required.

6 If you want a durable finish on the cabinet, apply two coats of semigloss polyurethane, sanding between coats. If you're applying finish to the trays as well, remove the slide hardware from the tray sides and the tray support columns and reinstall it when the finish has dried.

7 Set a protective mat (a carpet remnant or even cardboard will do) on your shop floor, and place the cabinet upside down on it. Set the casters in position on the bottom panel; then mark and drill for the bolts and lag screws. Mount the casters and turn the cabinet upright. (The caster set we used features front swivel casters that lock with a foot pedal, so the cabinet won't wander around the shop without permission.)

8 Install the door hinges and pulls, mount the doors, and fit all the trays to check clearance and travel. Make adjustments to the tray positions if necessary. Finally, if you're concerned about tools sliding around in the trays, cut up a nonslip rug cushion to use as liner material. Then fill 'er up!

8–26.

MOBILE TABLESAW BASE

All too often, the space beneath most tablesaws goes to waste, especially benchtop models on a stand. But as you can see here, this needn't be the case. This cabinet (**8–26**) stores a plentiful supply of saw blades, router bits, and other woodworking stuff on the sliding trays. Plus, the entire cabinet is easy to move around in your shop, thanks to a pair of casters. You can just store it against a wall when the cutting is done.

Note: Our cabinet was made to fit a specific saw. You might need to change the overall dimensions to fit your model of benchtop saw.

Materials List for Mobile Tablesaw Base

PART	T	W	L	Mtl.	Qty.
A side	¾"	19½"	19¼"	BP	1
B side	¾"	19½"	20"	BP	1
C back	¾"	23¼"	20"	BP	1
D top & bottom	¾"	23¼"	19⅜"	BP	2
E edging	¼"	¾"	19¼"	M	2
F edging	¼"	¾"	20"	M	2
G edging	¼"	¾"	20"	M	2
H edging	¼"	¾"	23¼"	M	3
I supports	¼"	3¾"	18½"	HB	4
J supports	¼"	5½"	18½"	HB	2
K toekick	¾"	3¼"	22½"	M	1
L strips	¾"	1½"	20"	M	2
M fronts	¾"	4⅜"	22⁷⁄₁₆"	M	2
N front	¾"	6⅛"	22⁷⁄₁₆"	M	1
O bottoms	¾"	18¼"	22⁷⁄₁₆"	BP	3
P cleats	¾"	1"	21⁷⁄₁₆"	M	3
Q holders	1½"	2⁷⁄₁₆"	21⁷⁄₁₆"	F	8
R slides	¼"	1"	18½"	HB	2
S holders	¾"	2¼"	3½"	M	2

Length measured with the grain.

Materials Key: BP = birch plywood; M = maple; HB = hardboard; F = fir.
Supplies: #8 x ¾", #8 x 1¼", #8 x 1½", #8 x 2¼", #8 x 2¾" flathead wood screws; 2¹³⁄₁₆" (overall length) screw eyes (2); ¼ x 1 x 15" aluminum bar stock for locking bar; 2½" rigid casters (2) (3¼" overall height) with #8 x ¾" panhead sheet-metal screws (8); ⅜" dowel stock; ½" dowel stock; primer; red enamel paint; clear finish.

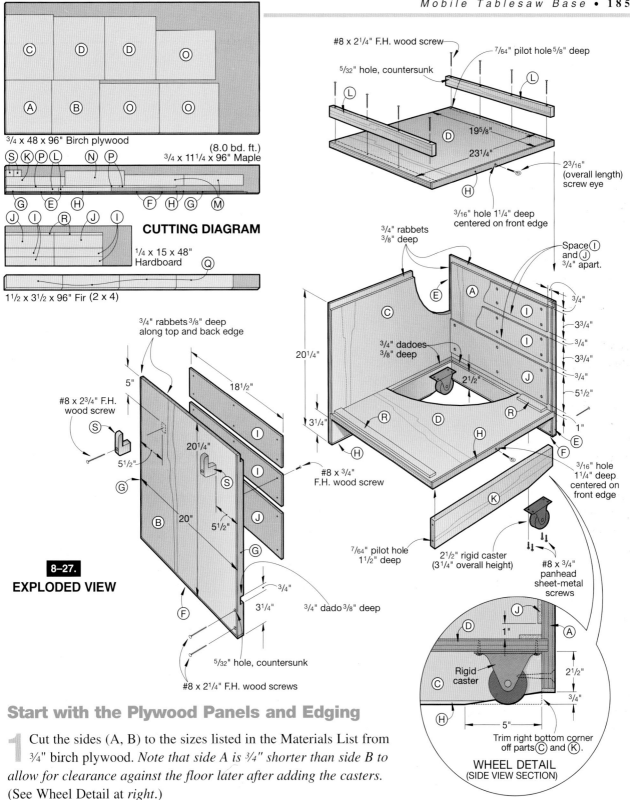

CUTTING DIAGRAM

3/4 x 48 x 96" Birch plywood

(8.0 bd. ft.)
3/4 x 11 1/4 x 96" Maple

1/4 x 15 x 48" Hardboard

1 1/2 x 3 1/2 x 96" Fir (2 x 4)

#8 x 2 1/4" F.H. wood screw
5/32" hole, countersunk
7/64" pilot hole 5/8" deep

19 5/8"
23 1/4"

2 3/16" (overall length) screw eye

3/16" hole 1 1/4" deep centered on front edge

Space (I) and (J) 3/4" apart.

3/4" rabbets 3/8" deep

3/4"
3 3/4"
3/4"
3 3/4"
3/4"
5 1/2"
1"

20 1/4"

3/4" dadoes 3/8" deep

2 1/2"

3 1/4"

3/16" hole 1 1/4" deep centered on front edge

3/4" rabbets 3/8" deep along top and back edge

5"

18 1/2"

#8 x 2 3/4" F.H. wood screw

20 1/4"

20"

5 1/2"

5 1/2"

#8 x 3/4" F.H. wood screw

3/4"

3 1/4"

3/4" dado 3/8" deep

8–27.
EXPLODED VIEW

7/64" pilot hole 1 1/2" deep

2 1/2" rigid caster (3 1/4" overall height)

#8 x 3/4" panhead sheet-metal screws

5/32" hole, countersunk

#8 x 2 1/4" F.H. wood screws

1"

2 1/2"

3/4"

Rigid caster

5"

Trim right bottom corner off parts (C) and (K).

WHEEL DETAIL
(SIDE VIEW SECTION)

Start with the Plywood Panels and Edging

1 Cut the sides (A, B) to the sizes listed in the Materials List from 3/4" birch plywood. *Note that side A is 3/4" shorter than side B to allow for clearance against the floor later after adding the casters.* (See Wheel Detail at *right.*)

2 Cut the back (C) and top and bottom (D) to size.

8–28.

For evenly spaced tray supports, use a piece of ¾"-thick stock to position the pieces before screwing them in place.

3 From the edge of ¾"-thick stock, rip the maple edging (E, F, G, H) to size plus 1" in length. Glue and clamp the edging to the edges of the plywood panels, where shown on **8–27**. Later, trim the ends of the edging flush.

4 Mark the locations of the rabbets, grooves, and dadoes in the plywood panels (A, B, C), and cut or rout them to the sizes listed on **8–27**. (We fit our tablesaw with a ¾" dado blade and wooden auxiliary rip fence to cut the dadoes, rabbets, and grooves. Then, we test-cut scrap pieces first to verify a snug fit of the plywood pieces in the dadoes and grooves.)

Add the Tray Supports

1 From ¼" hardboard, cut the tray supports (I, J) to size.

2 Position and screw the bottom supports (J) 1" above the bottom dado. Then, using a piece of ¾"-thick stock as a spacer, position the remaining supports (I), drill screw-mounting holes, and screw the supports in place, as shown in **8–28**. [Note that we placed a long piece of ¾"-thick walnut in the bottom ¾" dadoes to align the two side panels (A, B) before attaching the supports.]

Begin the Assembly

1 Dry-clamp the plywood panels (A, B, C, D) together to check the fit, and trim if necessary. Then, glue and clamp the pieces, checking for square. (To prevent glue stains on the plywood, we placed masking tape next to the glue joints before gluing. After drying, we peeled the tape, taking the glue squeeze-out with it.)

2 Measure the length of the opening, and cut the toekick (K) to shape, tapering the corner, where dimensioned on the Wheel Detail in **8–27**. Drill the countersunk screw holes, and then glue and screw the toekick into place.

3 Turn the assembled tablesaw base upside down, drill the mounting holes, and screw the 2½" rigid casters in place.

4 Cut the border strips (L) to size. Drill the countersunk screw holes, where shown on **8–27**. Screw the strips to the top of the cabinet.

Add the Three Storage Trays

1 From ¾" maple, rip and crosscut the tray fronts (M, N) to size. From ¾" plywood, cut the tray bottoms (O) to size.

2 Transfer the full-size Handle-Cutout Template in **8–29** to poster board. Cut the poster board template to shape. Center the template on the front face of each tray front. Trace its outline onto each tray front, and cut the radius to shape to create the concave opening (**8–32**). Now, sand the tray fronts smooth.

3 Cut a ¾" rabbet ⅜" deep along the bottom back edge of each tray front (M, N).

4 Cut a cleat (P) to size for each tray. Drill six countersunk holes in each cleat. With the bottom edge of the cleat

FULL-SIZE HANDLE-CUTOUT TEMPLATE

8–29.

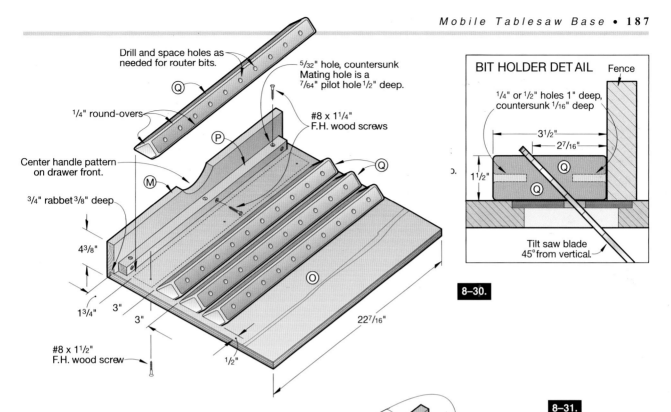

Drill and space holes as needed for router bits.

$^5/_{32}$" hole, countersunk Mating hole is a $^7/_{64}$" pilot hole $^1/_2$" deep.

$^1/_4$" round-overs

#8 x 1$^1/_4$" F.H. wood screws

Center handle pattern on drawer front.

$^3/_4$" rabbet $^3/_8$" deep

4$^3/_8$"

1$^3/_4$" 3"

3"

#8 x 1$^1/_2$" F.H. wood screw

$^1/_2$"

22$^7/_{16}$"

BIT HOLDER DETAIL   Fence

$^1/_4$" or $^1/_2$" holes 1" deep, countersunk $^1/_{16}$" deep

3$^1/_2$"

2$^7/_{16}$"

1$^1/_2$"

Tilt saw blade 45° from vertical.

8–30.

8–31.

BOTTOM TRAY

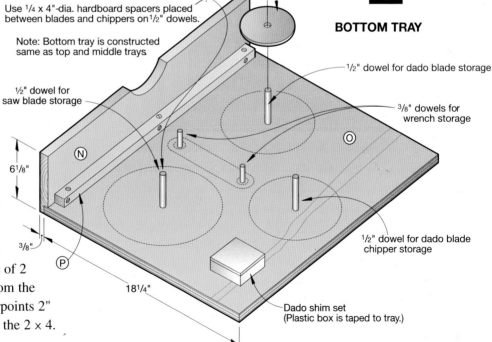

Use $^1/_4$ x 4"-dia. hardboard spacers placed between blades and chippers on $^1/_2$" dowels.

Note: Bottom tray is constructed same as top and middle trays.

$^1/_2$" dowel for saw blade storage

6$^1/_8$"

$^3/_8$"

18$^1/_4$"

$^1/_2$" dowel for dado blade storage

$^3/_8$" dowels for wrench storage

$^1/_2$" dowel for dado blade chipper storage

Dado shim set (Plastic box is taped to tray.)

flush with the top edge of the rabbet, glue and screw the cleats to the tray fronts. Glue and screw a tray front/cleat assembly to each tray bottom (O), checking that the front is square to the tray bottom.

5 To make the router bit holders (Q), cut four pieces of 2 × 4 stock to 21$^7/_{16}$" long. Mark a centerline along both edges (not surfaces) of each piece of 2 × 4 stock. Starting 1$^{23}/_{32}$" from the ends and spacing the centerpoints 2" apart, mark centerpoints on the 2 × 4.

6 Drill $^1/_4$" and $^1/_2$" router-bit shank holes in the 2 × 4 stock. The number of $^1/_4$" holes versus $^1/_2$" holes will be determined by your bit collection. (After drilling into the wood, we wobbled the bit slightly to allow the bit shanks to be removed easily from the holes.)

7 Rout $^1/_4$" round-overs along all edges of each piece of 2 × 4.

8 Using the Bit Holder Detail in **8–30** for reference, angle the blade on your tablesaw, and rip each

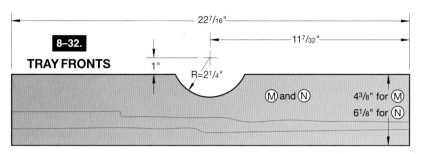

8–32.
TRAY FRONTS

22⁷/₁₆"
11⁷/₃₂"
1"
R=2¹/₄"
Ⓜ and Ⓝ
4³/₈" for Ⓜ
6¹/₈" for Ⓝ

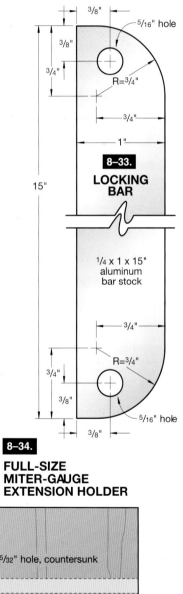

3/8"
⁵/₁₆" hole
3/8"
3/4"
R=³/₄"
3/4"
1"
8–33.
15"
LOCKING BAR
¹/₄ x 1 x 15"
aluminum
bar stock
3/4"
R=³/₄"
3/4"
3/8"
⁵/₁₆" hole
3/8"

8–34.
FULL-SIZE MITER-GAUGE EXTENSION HOLDER

⁵/₃₂" hole, countersunk
Ⓢ

2 x 4 section in two. Sand each of the holders (Q) smooth.

9 Locate and drill countersunk screw holes on the bottom side of each tray (O). Screw the bit holders to the top of each tray, as shown on **8–30**.

10 Using **8–31** as a reference, assemble the blade, shim, and dado-cutter holders.

11 Cut the drawer slides (R) to size, and glue them to the cabinet bottom (D), so they are flush against the cabinet sides (A/B).

Consider a Few Add-Ons

1 For hanging the miter-gauge extension on the side of the cabinet, twice transfer the full-size Holder drawing in **8–34** to ³/₄"-thick stock. Cut the holders (S) to shape, drill a screw hole in each, and screw the holders to the side of the cabinet, where shown on **8–27**.

2 For added security, drill pilot holes and add a pair of 2¹³/₁₆" screw eyes to the cabinet, where shown on **8–27**. Then, crosscut a piece of ¹/₄ x 1" aluminum bar stock to 15" long for the locking bar. (See **8–33** for reference.) Drill a ⁵/₁₆" hole at each end to align with the screw-eye holes. Later, use padlocks to secure the locking bar to the screw eyes.

Sand, Paint, and Finish

1 Remove the rigid casters and screw eyes, and finish-sand the entire cabinet.

2 Apply two coats of clear finish to the trays.

3 Apply a coat of primer to the cabinet. Later, apply two coats of enamel paint to the cabinet. When the paint has dried, you can reattach all of the screw eyes and the casters.

4 Position the tablesaw on the base, drill the mounting holes, and finish by securing the unit with bolts.

8–35.

MOBILE TOOL CABINET

Put your tools within easy reach by rolling this sturdy cabinet (**8–35**) right up to your work area. When you're through, simply close and lock the doors. Then return the cabinet to its storage spot—it occupies only 2 × 2' of floor space. And it's not as tough to make as it looks.

Note: For a natural look and void-free edges, we used Baltic birch plywood for our cabinet, and applied a clear finish. But, as a less expensive alternative, you also can use type AC or BC plywood, particularly if you plan to paint the cabinet.

Start with the Doors

1 From ¾" plywood, cut the door sides (A) and tops and bottoms (B) to the sizes listed in the Materials List. From ½" plywood, cut the shelves (C) to size.

2 On the inside face of the side pieces, rout ½" dadoes ⅜" deep to receive the shelves, where dimensioned on **8–36**. Then, rout ¾" rabbets ⅜" deep across the ends of the sides to accept the tops and bottoms. Now, rout a ½" rabbet ⅜" deep along the back edges of the sides, tops, and bottoms to receive the back (D).

Note: To rout the ½" and ¾" rabbets and dadoes in this project, we used ¼" and ½" straight bits, respectively, making two passes to size the joints exactly to fit the mating plywood parts.

3 Glue and assemble a door's sides, top, bottom, and shelves, as shown in **8–37**, using squaring braces to keep the assembly square. When the glue is dry, sand the door frame's surfaces and edges to 220 grit. Repeat to assemble the other frame.

4 From ½" plywood, cut the backs (D) to size. Sand them to 220 grit, and remove the dust. Glue a back in the rabbeted opening in each door frame, and clamp it securely all around the frame.

5 From ½" plywood, cut the vertical spacers (E) and horizontal spacers (F) to size. Glue and nail the spacers inside each door to the back (D), where shown on **8–36**.

Note: Because actual plywood thicknesses vary from their nominal dimensions, measure all inside dimensions of the tool cabinet and cut the spacers, and later the upper and lower shelves (L, M) and shelf edging (N), to the necessary length for the best fit.

6 Brush or spray the inside and outside of the doors with a finish of your choice. (We brushed on three coats of satin polyurethane, sanding to 220 grit between coats.)

8–36. **DOOR EXPLODED VIEW**

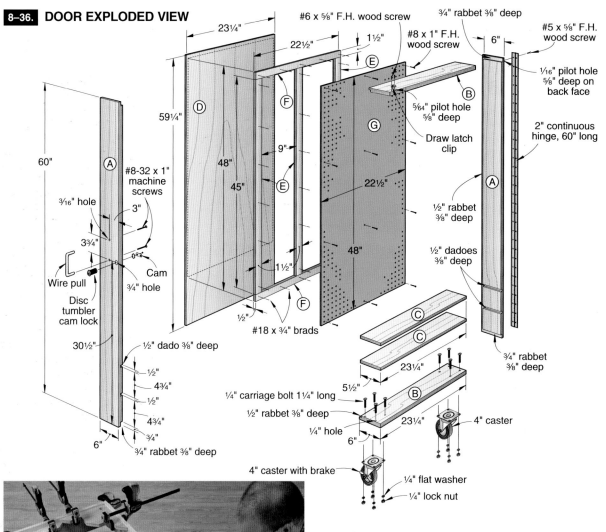

8–37.

With the edges of the door-frame members flush, clamp the assembly together with squaring braces. You can make simple braces from scrap ¾" plywood.

7 Cut the perforated hardboard panels (G) to size. Screw them to each door's spacers, where shown on **8–36**.

Build the Center Cabinet

1 From ¾" plywood, cut the sides (H), bottom and center shelf (I), top (J), and divider (K) to size. On the inside face of the sides, rout ¾" rabbets ⅜" deep to receive the bottom shelf and top and rout a ¾" dado ⅜" deep to accept the center shelf, where dimensioned on **8–38**. Then, rout a centered ¾" groove ⅜" deep in the sides to accept the divider, as shown in **8–39**. Using the same setup, rout the mating grooves for the divider in the bottom and center shelf.

8–38. CENTER CABINET EXPLODED VIEW

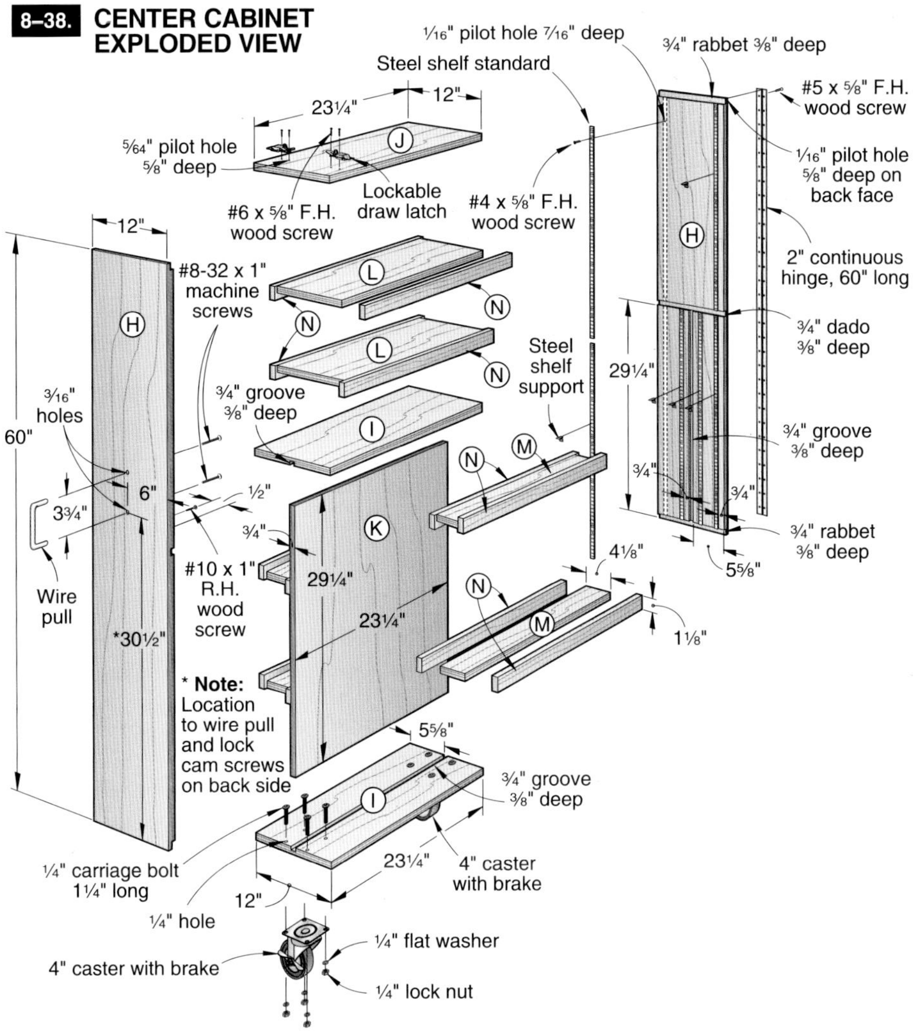
1⁄16" pilot hole 7⁄16" deep
3⁄4" rabbet 3⁄8" deep
Steel shelf standard
#5 x 5⁄8" F.H. wood screw
23¼"
12"
5⁄64" pilot hole 5⁄8" deep
J
1⁄16" pilot hole 5⁄8" deep on back face
Lockable draw latch
#6 x 5⁄8" F.H. wood screw
#4 x 5⁄8" F.H. wood screw
12"
H
2" continuous hinge, 60" long
#8-32 x 1" machine screws
L
N
H
L
N
3⁄4" dado 3⁄8" deep
Steel shelf support
29¼"
3⁄16" holes
3⁄4" groove 3⁄8" deep
I
3⁄4" groove 3⁄8" deep
60"
M
N
6"
½"
3⁄4"
3⁄4"
3¾"
3⁄4"
K
3⁄4" rabbet 3⁄8" deep
#10 x 1" R.H. wood screw
29¼"
4⅛"
5⅝"
Wire pull
N
23¼"
M
*30½"
1⅛"
* Note: Location to wire pull and lock cam screws on back side
5⅝"
3⁄4" groove 3⁄8" deep
I
¼" carriage bolt 1¼" long
23¼"
4" caster with brake
12"
¼" hole
¼" flat washer
4" caster with brake
¼" lock nut

2 Sand the parts to 220 grit, dry-assemble them, and verify they fit correctly. Then, glue and clamp the parts together, again using squaring braces to keep the assembly square. When the glue dries, apply the finish.

3 Measure the openings between the bottom and center shelf (I) and between the center shelf and top (J). (Our openings measured 28½" for the bottom and 29¼" for the top.) From 30"-long steel shelf standards, hacksaw eight pieces to length to fit the bottom openings and four pieces to length to fit the top opening.

4 Position the four standards for the top opening on the sides (H), and mark the mounting-screw hole locations, as shown in **8–40**. Then, drill 1⁄16" pilot holes 7⁄16" deep at the marked locations, and screw the standards to the sides. Following the same process, mount the four outer standards in the bottom openings. Then, mount the four inner standards to the sides, using the 3⁄4" spacers to position them parallel to the divider (K).

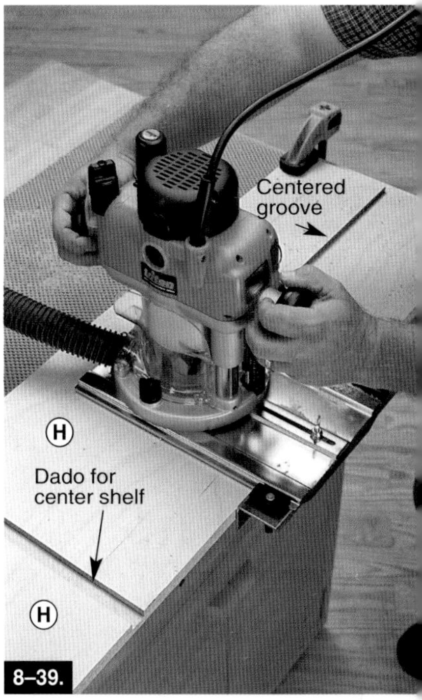
Centered groove
H
Dado for center shelf
H

8–39. *Adjust your router's edge guide to center the bit on the width of a side (H), and rout a groove from the bottom to the center shelf dado for the divider (K).*

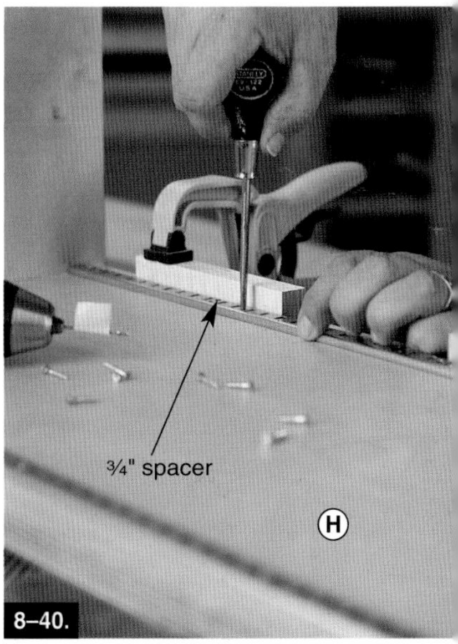
3⁄4" spacer
H

8–40. *Clamp scrap 3⁄4" plywood spacers flush to the edges of each side (H). Position the shelf standard against the spacers, and mark the screw holes with an awl.*

8–41. **CASTER PLACEMENT**

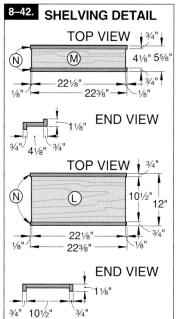

8–42. **SHELVING DETAIL**

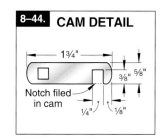

8–43. **DISC TUMBLER CAM LOCK**

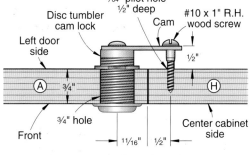

8–44. **CAM DETAIL**

Hinge the Doors to the Cabinet

1 With a hacksaw, cut two 2 × 72" continuous hinges to 60" long. Align the doors and the center cabinet on a flat surface with the back side facing up, and clamp them together. (See the How to Install Continuous Hinges *below*.)

2 Center the hinges along the door/center-cabinet joints. Drill ¹⁄₁₆" pilot holes ⅝" deep in the center of the hinge screw holes, and drive the #5 × ⅝" flat-head woodscrews. Now, set the tool cabinet upright on the casters.

Add the Shelves

1 From ¾" plywood, cut the upper shelves (L), lower shelves (M), and edging (N) to size. (Again, it's a good idea to measure the inside width of the cabinet, both between the sides (H) for the edging and the shelf standards for the shelving, and cut the parts to the exact length for a snug fit.)

5 Position 4" casters on the bottoms (B, I), where dimensioned on **8–41**. Mark the centers of the casters' mounting-bolt holes, and drill ¼" holes

through the bottoms at the marked locations. Bolt the casters to the bottoms, as shown, making sure to locate the casters with a brake where shown.

2 Glue and clamp the edging to the upper and lower shelves, where shown on **8–38** and **8–42**. Center the edging so it overhangs the shelves by ⅛" at each end. When the glue has dried, sand the shelves and edging to 220 grit, and apply the finish. With the finish dry, clip the shelf supports in the standards at the desired locations, and install the shelves on the supports.

SHOP TIP

How to Install Continuous Hinges

When attaching doors with continuous hinges, such as for the tool cabinet, leave a gap between the members to be hinged. The gap will prevent potential hinge binding due to wood movement, which could keep the parts from closing together tightly. To establish the gap, place ¹⁄₁₆"-thick wood spacers between the parts as needed, and then clamp the parts together (8–45).

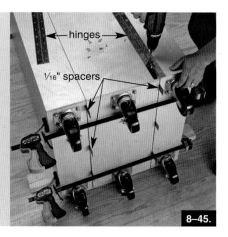

8–45.

Install the Locks and Handles

1 On the front of each door, mark the centerpoint for a ¾" hole to receive a disc tumbler cam lock, where dimensioned on **8–36** and **8–43**. (Note that the two lock installations are mirror images of one another.)

Materials List for Mobile Tool Cabinet

PART	FINISHED SIZE				
	T	W	L	Mtl.	Qty.
DOORS					
A sides	¾"	6"	60"	BB	4
B tops and bottoms	¾"	6"	23¼"	BB	4
C shelves	½"	5½"	23¼"	BB	4
D backs	½"	23¼"	59¼"	BB	2
E vertical spacers	½"	1½"	45"	BB	6
F horizontal spacers	½"	1½"	22½"	BB	4
G perforated panels	¼"	22½"	48"	PH	2
CENTER CABINET					
H sides	¾"	12"	60"	BB	2
I bottom and center shelf	¾"	12"	23¼"	BB	2
J top	¾"	12"	23¼"	BB	1
K divider	¾"	23¼"	29¼"	BB	1
L upper shelves	¾"	10½"	22⅛"	BB	2
M lower shelves	¾"	4⅛"	22⅛"	BB	4
N edging	¾"	1⅛"	22⅜"	BB	12

Materials Key: BB = Baltic birch plywood; PH = perforated hardboard.
Supplies: #18 x ¾" brads; #4 x ⅝", #5 x ⅝", and #8 x 1" flathead wood screws; ¼" carriage bolts 1¼" long (24); ¼" flat washers (24); ¼" lock nuts (24); 4" casters with brake (4); 4" casters without brakes (2); steel shelf standards 30" long (12); steel shelf supports (24); 2 x 72" continuous hinges (2); disc tumbler cam locks (2); #10 x 1" roundhead wood screws (2); 3¾" (96mm) wire pulls with #8-32 x 1" machine screws (3); lockable draw latches (2) with #6 x ⅝" flathead wood screws.
Blades and Bits: ¼" and ½" straight router bits, ¾" Forstner bit.

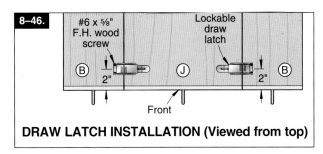

8–46. #6 x ⅝" F.H. wood screw / Lockable draw latch / B / 2" / J / 2" / B / Front

DRAW LATCH INSTALLATION (Viewed from top)

2 Using a ¾" Forstner bit and a backer board to prevent tear-out, drill the lock holes through the doors. Remove the screw attaching the cam to the back of each lock, and remove the cam. Install the locks in the holes as directed in the manufacturer's instructions. File a notch in each cam, where dimensioned on **8–45**. Then screw the cams to the locks.

3 On the back of the center cabinet's front side (H), mark the centerpoints for #10 x 1" round-head wood screws that engage with the locks' cams, where dimensioned on **8–38** and **8–43**. Drill ⁷⁄₆₄" pilot

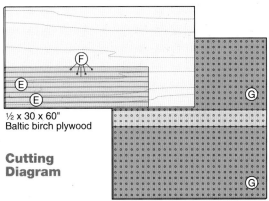

½ x 30 x 60"
Baltic birch plywood

Cutting Diagram

¼ x 48 x 48" Perforated hardboard

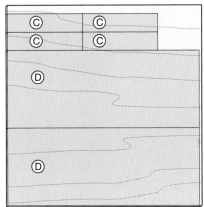

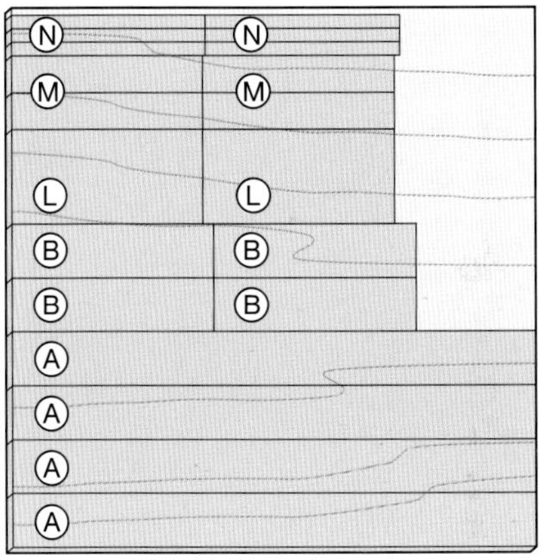

¾ x 60 x 60" Baltic birch plywood

½ x 60 x 60" Baltic birch plywood

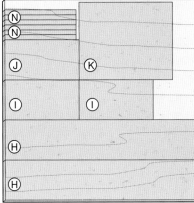

¾ x 60 x 60" Baltic birch plywood

Tool-Hanging and Storage-Bin Options

• **Hooks.** To ensure that your tools' hooks stay securely in place on the perforated hardboard, use locking-type nylon hooks for rock-solid attachment, as shown in 8–47. The hooks come in a variety of shapes, so you can mount virtually any tool or accessory, and are available from woodworking and hardware suppliers.

8–47.

8–48.

• **Storage Bins.** To keep bulk hardware items organized, identified, and easily accessible, it's hard to beat storage bins. We placed bins, available from woodworking and hardware suppliers, on the lower shelves of the doors, as shown in 8–35. The bins hook onto a support strip, as shown in 8–48. A pack of six bins costs about $6.50 and includes the support strip.

holes ½" deep at the centerpoints. Then, drive the #10 × 1" roundhead wood screws to the depth shown.

4 On the front of the doors and center cabinet, mark the screw-hole locations for attaching the wire pulls, where shown on **8–36** and **8–38**. Drill ³⁄₁₆" holes at the marked locations, and mount the pulls using the supplied screws.

5 Clamp the doors snug against the center cabinet at the front. Engage the lockable draw latches in the mating clips, and position the pieces on top of the tool cabinet, where shown on **8–46**. While holding a latch and clip against the cabinet, carefully release the latch lever and mark the mounting-screw hole locations in each piece. Repeat for the other latch. Then, drill ⁵⁄₆₄" pilot holes ⁵⁄₈" deep at the marked spots, and screw the pieces to the top.

6 Now, to mount your tools and organize hardware in the cabinet, see Tool-Hanging and Storage-Bin Options, *above*, for ideas.

8–49.

SWING-OUT CABINET FOR BITS

This wall-hung cabinet (**8–49**) is a real swinger, and it holds a surprising number of assorted drill bits and accessories. To build it, you first assemble the box, and then rip-cut it into three separate sections. It's very easy!

1 Rip and crosscut a piece of ¾" pine to 6¾ × 52". From it, cut two pieces 7¾" long (A) and two pieces 17⅛" long (B). Rabbet the ends of the two longer pieces, where indicated on **8–50**.

2 Glue and clamp the top and bottom pieces (A) to the side pieces (B) to form the frame. Cut the plywood sides (C), and attach them to the frame.

3 After the glue dries, use your tablesaw and rip fence to rip the frame into three sections to the widths shown on **8–51**.

4 Cut shelves (D, E, F) to finished size. Drill holes in the shelves and Sections 1 and 3 for drill bits.

5 Glue and nail the shelves to the frame sections, where shown on the drawing. Attach the cleats (G) to the bottom of Section 2, as shown in **8–51**. Connect the frame with hinges.

6 Cut two pieces of ¾" pine 1½ × 34" and then laminate to form a piece 1½" thick. From this,

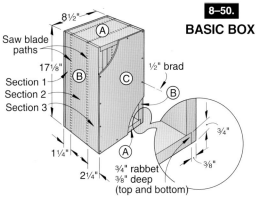

8–50.
BASIC BOX

8½"
Saw blade paths
17⅛"
½" brad
Section 1
Section 2
Section 3
1¼"
2¼"
¾" rabbet ⅜" deep (top and bottom)
¾"
⅜"

cut the mounting bracket pieces (H, I) to finished size. Rabbet both ends of I and one end of H.

7 Assemble the bracket pieces with glue and reinforcing dowels, and then radius the ends and corners. Drill holes in the bracket upright for wood screws or wall anchors.

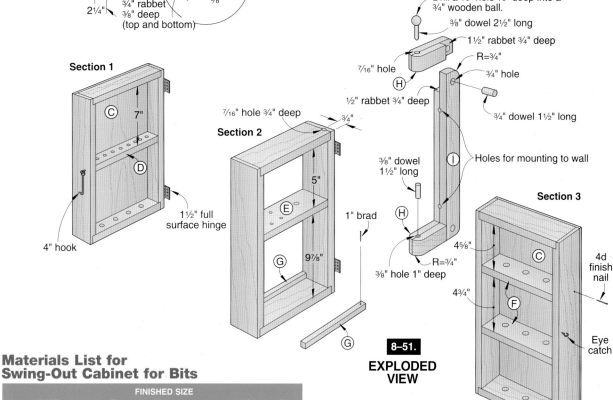

Section 1
C 7"
D
4" hook
1½" full surface hinge

Section 2
7/16" hole ¾" deep
¾"
5"
E
9⅞"
G
1" brad
G

Drill a ⅜" hole ⅜" deep into a ¾" wooden ball.
⅜" dowel 2½" long
1½" rabbet ¾" deep
7/16" hole
R=¾"
¾" hole
½" rabbet ¾" deep
H
⅜" dowel 1½" long
¾" dowel 1½" long
Holes for mounting to wall
I
⅜" hole 1" deep
H
R=¾"

Section 3
C
4⅝"
4¾"
F
4d finish nail
Eye catch

8–51.
EXPLODED VIEW

8 Drill pivot holes in the mounting bracket and in Section 2 for the dowel pivot pins. Cut the bottom pivot pin to length and glue it into the bracket. Cut the removable top pin to length and glue the wooden ball to it. Mount the bracket to the wall and the cabinet to the bracket.

Materials List for Swing-Out Cabinet for Bits

PART	FINISHED SIZE			Mtl.	Qty.
	T	W	L		
A top and bottom	¾"	6¾"	7¾"	P	2
B sides	¾"	6¾"	17⅛"	P	2
C sides	⅛"	8½"	17⅛"	PW	2
D shelf	¾"	1⅛"	7"	P	1
E shelf	¾"	3¼"	7"	P	1
F shelves	¾"	2⅛"	7"	P	2
G cleats	⅜"	⅜"	7"	P	2
H* brackets	1½"	1½"	6¼"	P	2
I* bracket	1½"	1½"	20⅝"	P	1

* Laminated from two pieces of ¾" pine.

Materials Key: PW = plywood; P = pine.
Supplies: ½" brads and 4d finish nails; ⅜" dowel; ¾" wooden ball; two pairs of 1½" full surface hinges; screws for mounting to wall; 4" hook-and-eye latch; polyurethane.

8–52.

SCROLLSAW BLADE DRAWER

Tired of constantly fussing around with your scrollsaw blades? With this clever helper made from common cove moldings (**8–52** and **8–53**), you can retrieve blades easily by reaching in with a finger and sweeping the blades up along the curved edge of each hollow.

To build the drawer, glue and clamp the cove pieces together. Then, screw the guides underneath your bench or scrollsaw stand. A string keeps the drawer from sliding out too far, and a brad nailed to one guide stops the drawer flush with the front of the bench. Clever, huh?

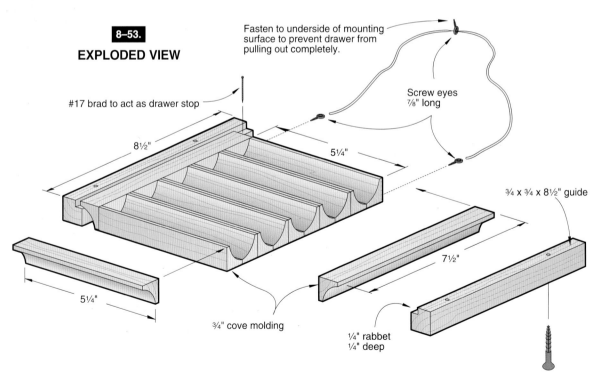

8–53.

EXPLODED VIEW

Fasten to underside of mounting surface to prevent drawer from pulling out completely.

Screw eyes ⅞" long

#17 brad to act as drawer stop

8½"

5¼"

¾ x ¾ x 8½" guide

7½"

5¼"

¾" cove molding

¼" rabbet ¼" deep

DROP-DOWN TOOL TRAY

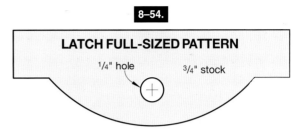

8–54.

LATCH FULL-SIZED PATTERN

¹/₄" hole   ³/₄" stock

H ere's a great way to squeeze a lot of tool-storage space into a small area (**8–54** to **8–56**). The storage tray shown here holds nine measuring and marking tools and quickly folds up into an often-unused space. You could just as easily customize your tray to hold chisels, screwdrivers, wrenches, or other tools.

To build the tray, you must first measure the underside of the cabinet. The length of the tray should equal the inside measurement between the two sides of the cabinet minus ¹/₈" for clearance. The width of the tray must equal the measurement from the wall to the inside edge of the cabinet face frame minus the thickness of the wall cleat and minus ¹/₈" for clearance.

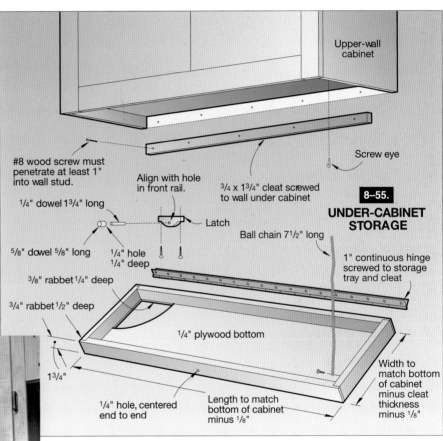

Upper-wall cabinet

Screw eye

#8 wood screw must penetrate at least 1" into wall stud.

Align with hole in front rail.

³/₄ x 1³/₄" cleat screwed to wall under cabinet

¹/₄" dowel 1³/₄" long

8–55.

UNDER-CABINET STORAGE

Ball chain 7¹/₂" long

Latch

⁵/₈" dowel ⁵/₈" long

¹/₄" hole ¹/₄" deep

1" continuous hinge screwed to storage tray and cleat

³/₈" rabbet ¹/₄" deep

³/₄" rabbet ¹/₂" deep

¹/₄" plywood bottom

1³/₄"

¹/₄" hole, centered end to end

Length to match bottom of cabinet minus ¹/₈"

Width to match bottom of cabinet minus cleat thickness minus ¹/₈"

For accurate positioning, center and screw the wooden latch to the bottom edge of the cabinet face frame after you have installed the tray. Then, hold the tray in the closed position, and drill the hole for the dowel through the latch and into the front rail of the tray.

Finally, cut tool-holding blocks and glue them to the tray bottom to prevent your tools from sliding around.

8–56.

WALL HANGING SANDING-SUPPLY ORGANIZER

Wouldn't it be great to store all of your sanding supplies neatly in one place. This *WOOD*® magazine original, featuring easy-to-build projects (**8–57**), will help you do just that. Create these efficient holders for belts, sanding discs and accessory sanders, even rolls of adhesive-backed sandpaper. You'll be glad you did!

8–57.

GUIDE TO SIX GREAT ORGANIZERS

1 **Sanding-stick and sandpaper-strip organizer**

2 **5" sanding disc holder**

3 **Hand-sander rack and tote**

4 **Disc-sander center**

5 **Roll-sandpaper dispenser**

6 **Belt-sander belt divider**

1 Organize Your Sanding Sticks and Sandpaper Strips

When that handcrafted project requires precision sanding, this trio of sanding sticks and adhesive-backed sandpaper strips perform admirably. Build the solid-maple organizer to divide the coarse-, medium-, and fine-grit sandpaper strips, using **8–59** and the Materials List. Refer to **8–58** to make the contoured sticks.

Materials List for Sandpaper-Strip Organizer

PART	FINISHED SIZE			Mtl.	Qty.
	T	W	L		
A	½"	2"	5¾"	M	2
B	½"	5¾"	15½"	BP	1
C	¼"	1½"	15½"	M	4
D	¼"	1½"	20"	M	3

Materials Key: M = maple; BP = birch plywood.
Supplies: 3M adhesive-backed 1½ x 14⅝" sandpaper strips (fine, medium, and coarse); ¼" birch dowel.

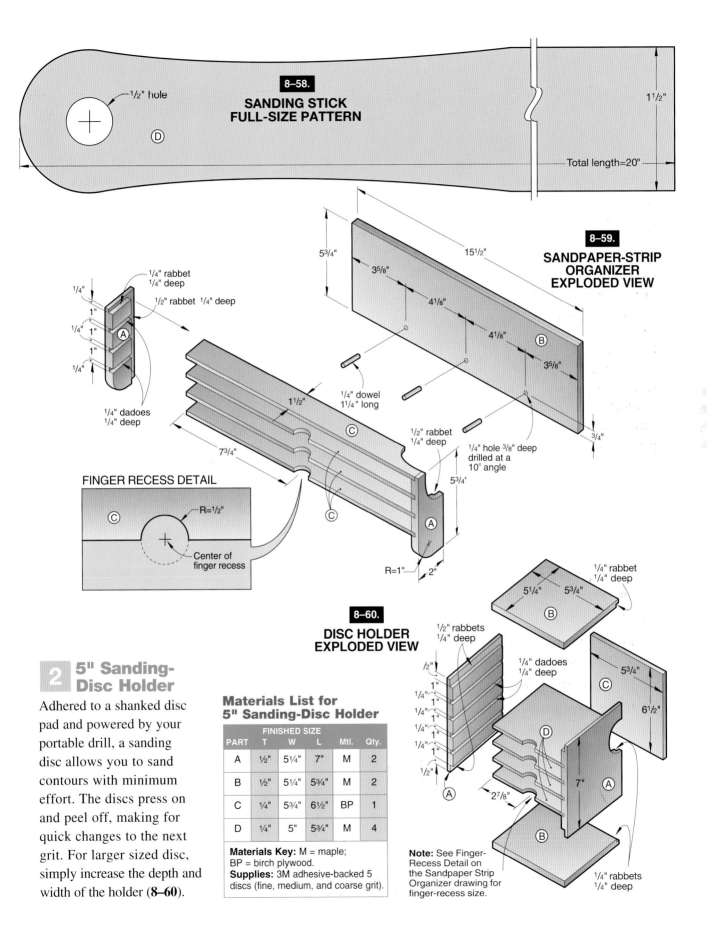

8–58.
SANDING STICK
FULL-SIZE PATTERN

¹/₂" hole

D

1¹/₂"

Total length=20"

8–59.
SANDPAPER-STRIP
ORGANIZER
EXPLODED VIEW

5³/₄"

15¹/₂"

3⁵/₈"

4¹/₈"

4¹/₈"

B

3⁵/₈"

³/₄"

¹/₄" dowel
1¹/₄" long

¹/₄" hole ³/₈" deep
drilled at a
10° angle

¹/₄" rabbet
¹/₄" deep

¹/₂" rabbet ¹/₄" deep

¹/₄"

1"

¹/₄" 1"

A

1"

¹/₄"

¹/₄" dadoes
¹/₄" deep

1¹/₂"

C

7³/₄"

¹/₂" rabbet
¹/₄" deep

5³/₄"

A

FINGER RECESS DETAIL

C

R=¹/₂"

Center of
finger recess

C

R=1"

2"

2 | 5" Sanding-Disc Holder

Adhered to a shanked disc pad and powered by your portable drill, a sanding disc allows you to sand contours with minimum effort. The discs press on and peel off, making for quick changes to the next grit. For larger sized disc, simply increase the depth and width of the holder (**8–60**).

Materials List for 5" Sanding-Disc Holder

PART	FINISHED SIZE			Mtl.	Qty.
	T	W	L		
A	¹/₂"	5¹/₄"	7"	M	2
B	¹/₂"	5¹/₄"	5³/₄"	M	2
C	¹/₄"	5³/₄"	6¹/₂"	BP	1
D	¹/₄"	5"	5³/₄"	M	4

Materials Key: M = maple;
BP = birch plywood.
Supplies: 3M adhesive-backed 5
discs (fine, medium, and coarse grit).

8–60.
DISC HOLDER
EXPLODED VIEW

¹/₂" rabbets
¹/₄" deep

¹/₄" rabbet
¹/₄" deep

5¹/₄"   5³/₄"

B

¹/₄" dadoes
¹/₄" deep

5³/₄"

C

6¹/₂"

/2"

1"

¹/₄"   1"

¹/₄"   1"

¹/₄"   1"

¹/₄"   1"

¹/₂"

D

A

7"

A

2⁷/₈"

B

¹/₄" rabbets
¹/₄" deep

Note: See Finger-
Recess Detail on
the Sandpaper Strip
Organizer drawing for
finger-recess size.

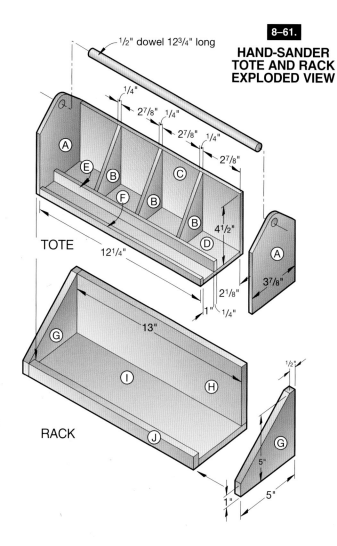

8–61.

**HAND-SANDER
TOTE AND RACK
EXPLODED VIEW**

1/2" dowel 12 3/4" long

1/4"

2 7/8" 1/4"

2 7/8" 1/4"

2 7/8"

4 1/2"

TOTE

12 1/4"

2 1/8"

1" 1/4"

3 7/8"

A

E

B

C

F

B

B

D

A

RACK

13"

G

I

H

J

1/2"

5"

1"

5"

G

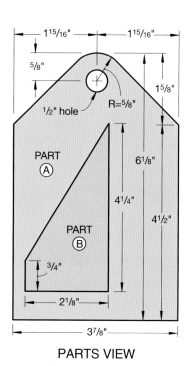

1 15/16" — 1 15/16"

5/8"

1/2" hole

R = 5/8"

1 5/8"

PART
Ⓐ

PART
Ⓑ

6 1/8"

4 1/4"

4 1/2"

3/4"

2 1/8"

3 7/8"

PARTS VIEW

Hand-Sander Tote and Rack

Carry a quartet of hand sanders—each having a different grit—to your workbench with this handy tote (**8–61**). When you're done, just return the tote to the rack. You'll also find room for rolls of adhesive-backed sandpaper behind the hand sanders in the tote.

Materials List for Hand-Sander Tote and Rack

PART	FINISHED SIZE			MTL.	QTY.
	T	W	L		
TOTE					
A	1/4"	3 7/8"	6 1/8"	M	2
B	1/4"	2 1/8"	4 1/4"	M	3
C	1/4"	4 1/2"	12 1/4"	M	1
D	1/4"	3 3/8"	12 1/4"	M	1
E	1/4"	3/4"	12 1/4"	M	1
F	1/4"	1/2"	12 1/4"	M	1
TOTE RACK					
G	1/2"	5"	5"	M	2
H	1/2"	4 1/2"	13"	M	1
I	1/2"	4 1/2"	13"	M	1
J	1/2"	1"	13"	M	1

Material Key: M = maple.
Supplies: 3M Hand-Ease sander;
sandpaper roll refills (very fine birch dowel 2 1/2" x 90",
fine 2 1/2" x 80", medium 2 1/2 x 55").

4 Disc-Sander Center

You can't beat small-diameter disc sanders for sanding the contours of shapely projects and the inside and outside of turned bowls. Here's a slick way to get your sanding-adhesive drill attachments and various discs in order and divided by grit (**8–62**).

Materials List for Disc-Sander Center

PART	FINISHED SIZE T	W	L	Mtl.	Qty.
A	½"	3½"	15½"	M	2
B	½"	3½"	7¼"	M	2
C	½"	3¼"	7¼"	M	4
D	¼"	3¼"	2½"	M	4
E	¼"	3¼"	1¾"	M	1
F	¼"	3¼"	3½"	M	4
G	¼"	7¼"	15	BP	1

Material Keys: M = maple; BP = birch plywood.

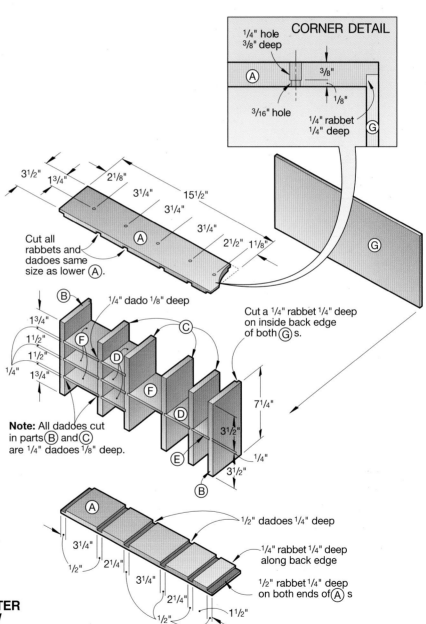

8–62.
DISC-SANDER CENTER EXPLODED VIEW

5 Roll-Sandpaper Dispenser

This handy dispenser (**8–63**) lets you store up to four 4½"-wide rolls of sandpaper. It loads from the bottom, and is dimensioned to let you tear off 4" pieces, perfect for a palm sander.

To remove a needed piece, pull the end of the sandpaper roll flush with the front top edge of part C.

Fold a crease in the sandpaper along the back edge of part C (4" from the front). Pull the paper out until the crease aligns with the front edge of C. Adhere the paper back to part C so the crease is flush with the front edge. Just pull down on the paper to tear off a 4"-long piece

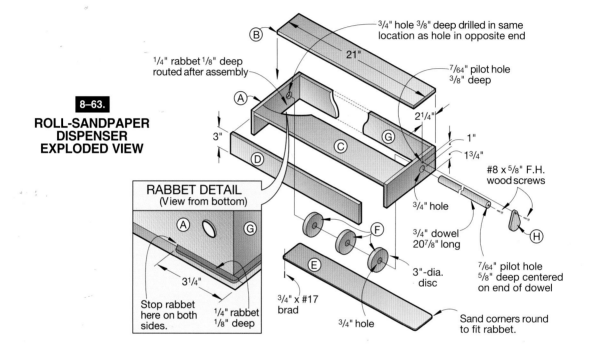

8–63.
ROLL-SANDPAPER DISPENSER EXPLODED VIEW

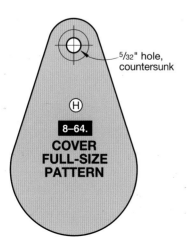

8–64.
COVER FULL-SIZE PATTERN

Materials List for Roll-Sandpaper Dispenser

PART	FINISHED SIZE			Mtl.	Qty.
	T	W	L		
A	½"	3½"	9"	M	2
B	½"	4½"	21"	M	1
C	½"	4"	20"	M	1
D	½"	3"	20"	M	1
E	⅛"	3"	20½"	M	1
F	½"	3" dia.		M	1
G	½"	3½"	20"	M	1
H	¼"	1⅜"	2¼"	M	1

Material Key: M = maple.
Supplies: 3M adhesive-backed sandpaper rolls 4½"-wide by 10 yards long (80-grit, 100-grit, 120-grit, 220-grit); ¾" dowel; #8 x ⅝" flathead wood screws; ¾" x #17 brads; ⅜" dowel.

6 Belt-Sander Divider

The divider was sized for holding 3"-wide sanding belts (**8–65**). For wide belts, just increase the width of pieces B and C accordingly. For longer belts, just increase the pieces A, B, and C to compensate for the length difference.

❖❖

Materials List for Belt-Sander Divider

PART	FINISHED SIZE T	W	L	Mtl.	Qty.
A	½"	10"	20"	M	2
B	½"	3¾"	10"	M	2
C	½"	3¾"	9¾"	M	5
D	¼"	3¾"	19½"	BP	1

Materials Key: M = maple; BP = birch plywood.

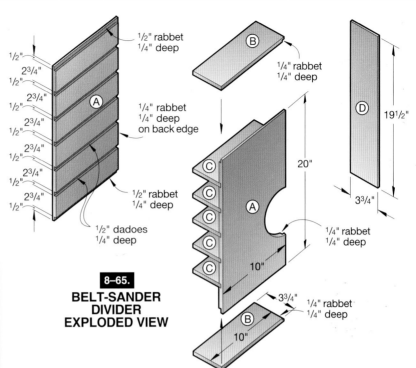

8–65.
BELT-SANDER DIVIDER EXPLODED VIEW

DRUM-SANDER HOLDER

Tired of sanding drums and sleeves rolling around your shop? Solve the problem with our wall-mounted holder (**8–66** and **8–67**). Drumshafts fit in holes in the top, while sleeves fit below.

1 Cut the back (A), storage box tops and bottom (B), and dividers (C) to size. Laminate the two

8–66.

Materials List for Drum-Sander Holder

PART	FINISHED SIZE T	W	T	Mtl.	Qty.
A back	¾"	11"	17"	PB	1
B tops & bottoms	¾"	7½"	17"	PB	3
C dividers	¾"	4"	7½"	PB	4
D drawer front	½"	3¹⁵⁄₁₆"	6⁷⁄₁₆"	P	1
E drawer sides	½"	3¹⁵⁄₁₆"	7¼"	P	2
F drawer back	½"	3¹⁵⁄₁₆"	5⁵⁄₁₆"	P	1
G drawer bottom	¼"	5¹⁵⁄₁₆"	7¼"	HB	1
H trim	¾"	¾"	18½"	P	2
I trim	¾"	¾"	11"	P	2

Materials Key: P = pine; PB = particleboard; HB = hardboard.
Supplies: #8 x 1¼" flathead wood screws; #8 x 2" flathead wood screws; wooden knob; 4d finish nails; masking tape; paint.

top pieces. Lay out and drill 1"-deep holes in the top to house the shanks of your drum sanders.

2 Glue and clamp the storage box together. Drill shank and pilot holes, and drive the screws.

3 Cut the drawer front (D), sides (E), back (F), and bottom (G) to size. Cut or rout a ¼" groove ¼" deep ½" from the bottom in the drawer front and sides.

Cut a ½" dado ¼" deep 1½" from the back edge of each side.

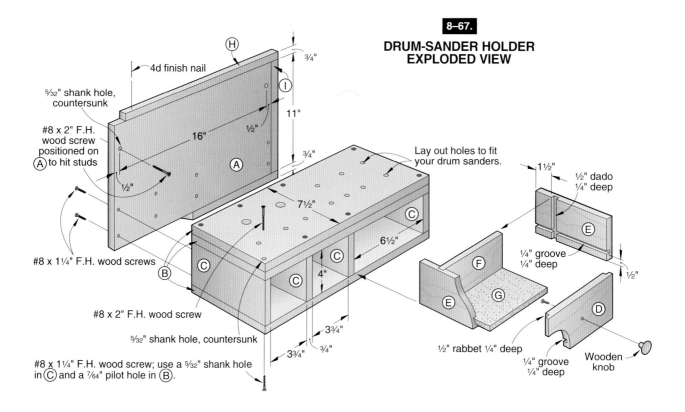

8–67.

DRUM-SANDER HOLDER
EXPLODED VIEW

4d finish nail

⁵⁄₃₂" shank hole, countersunk

#8 x 2" F.H. wood screw positioned on Ⓐ to hit studs

16"

11"

7½"

6½"

4"

1½"

½" dado ¼" deep

Lay out holes to fit your drum sanders.

¼" groove ¼" deep

½"

¾"

½"

#8 x 1¼" F.H. wood screws

#8 x 2" F.H. wood screw

⁵⁄₃₂" shank hole, countersunk

3¾"

3¾"

¾"

½" rabbet ¼" deep

¼" groove ¼" deep

Wooden knob

#8 x 1¼" F.H. wood screw; use a ⁵⁄₃₂" shank hole in Ⓒ and a ⁷⁄₆₄" pilot hole in Ⓑ.

Finally, cut a ½" rabbet ¼" deep along both ends of the front piece.

4 Glue and clamp the drawer together, checking for square. Drill a hole through the drawer front; attach a knob.

5 Cut trim pieces (H, I) to size; glue and nail them to the back (A).

6 Glue and clamp the storage box to the back piece. Drill shank and pilot holes from the back side of the back piece into the back of the box, and screw the back to the box.

7 Mask the trim pieces and paint the storage box and back. Remove the masking tape and apply a clear finish to the trim and drawer.

8 Drill mounting holes through the back piece, and fasten the unit to a wall.

11-IN-ONE SANDER CABINET

8–68.

You'll spend less time gritting your teeth and more time sanding when you store everything you need—sanders, sandpaper, and sanding belts—in one convenient wall-hung cabinet (**8–68** and **8–69**). The cabinet accommodates a portable belt sander and a palm-grip sander.

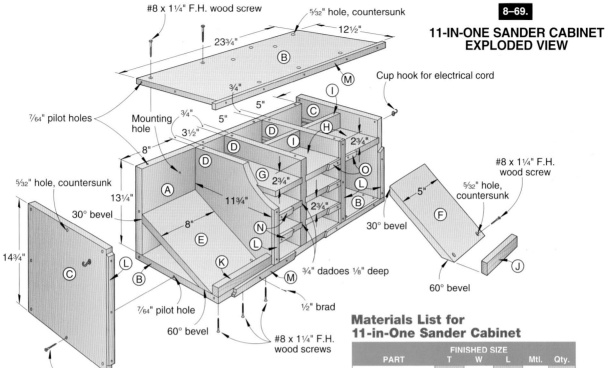

8–69.

11-IN-ONE SANDER CABINET EXPLODED VIEW

From a sheet of ¾" particleboard, lay out and cuts parts A through I to finished size. (We cut E and F to size plus 2" in length to allow for ease in bevel-cutting these pieces later.)

Bevel-cut the front end of the shelves (E, F) for the two sanders to a 60° angle, and the back end to 30° angle. Cut dadoes in the shelf dividers (D) and inside the right-hand side (C), where shown on **8–69**.

Glue and screw the back (A), bottom (B), and sides (C) together. Then, glue and screw the cleats (J, K) and the slanted shelves (E, F) in position. Drill two holes in the back for screws or wall anchors, according to wall-stud spacing.

Loose-fit the shelf dividers (D) and shelves (G, H), starting with the left-hand divider and working toward the right-hand one. As you slide in the center and right-hand shelves (H), glue in the false back (I) for each shelf, where shown on **8–69**.

. Glue and screw the top (B) on the cabinet. Sand all edges flush. Mask off the pine cleats, and then

paint all exposed surfaces to the desired color. Cut the pine trim strips (L, M, N, O) to finished size. Finish all the pine pieces with polyurethane or another clear finish. Attach the trim strips to the cabinet face. Attach cup hooks for electrical cords. Finally, hang the cabinet.

Materials List for 11-in-One Sander Cabinet

PART	FINISHED SIZE			Mtl.	Qty.
	T	W	L		
A back	¾"	13¼"	23¾"	PB	1
B top & bottom	¾"	12½"	23¾"	PB	2
C sides	¾"	12½"	14¾"	PB	2
D dividers	¾"	11¾"	13¼"	PB	3
E* shelf	¾"	8"	12¾"	PB	1
F* shelf	¾"	5"	12¾"	PB	1
G shelves	¾"	3¾"	11¾"	PB	3
H shelves	¾"	5¼"	6¼"	PB	4
I false back	¾"	2¾"	5"	PB	5
J cleat	¾"	1¼"	5"	P	1
K cleat	¾"	1¼"	8"	P	1
L trim	¼"	¾"	13¼"	P	5
M trim	¼"	¾"	25¼"	P	2
N trim	¼"	¾"	3½"	P	3
O trim	¼"	¾"	5"	P	4

*Cut these pieces oversize, and then bevel cut to final lengths.

Materials Key: PB = particleboard; P = pine.
Supplies: #8 x 1¼" flathead wood screws; ½" brads; 1½" cup hooks (2); latex paint; wall anchors or screws for wall mounting.

CHAPTER 9

Storing Lumber and Sheet Goods

BY NOW YOU HAVE A PRETTY *good notion of the many different ways a woodworking shop can go together. You've also seen plenty of workbench options, several ideas for base and wall cabinets, plus dozens of specialized storage units for tools and supplies. But as you probably will realize, you have to provide storage for materials—lumber and sheet goods— in your shop, too. Even if you buy only what you need, you'll still be greatly surprised how leftovers stack up!*

That's why, in this chapter, you'll find some simple, yet highly practical, storage ideas for wood—from the space-saving lumber and sheet goods rack shown on the next page *to a pair of sturdy carts to keep your stock mobilized. And just for good measure, we've thrown in a catchall for cutoffs and a simple, yet effective, sheet goods mover/lifter. You're sure to find them all to be great ideas for your shop.*

QUICK-AND-EASY MOBILE STORAGE

This easily built project (**9–1**) will serve all your wood-working needs. Use it for lumber storage in the shop or as a catchall in the garage or basement. Plus, using ½" plywood and 2 × 4s for its construction, you'll find this project very affordable.

Note: Our unit measures 99" inches long and rests on four casters for mobility. Size the unit to suit your needs and omit the casters if mobility is not one of your requirements.

Building Steps

1 Cut the 2 × 4, plywood, and perforated hardboard parts A, B, C, D, E, and F to the sizes listed in the Materials List.

2 On a flat surface, drill countersunk mounting holes and screw the shelf supports (A, B) together to form four rectangular 2 × 4 frames. Note that the bottom frames uses four Bs and the other frames use just two.

3 Glue and screw the four plywood shelves (C) to the 2 × 4 frames. Check each for square. Sand or rout slight round-overs to break the sharp edges along the top edges of the ½" plywood shelves.

4 Glue and screw the four uprights (D) to what will be the bottom shelf assembly

9–1.

(A, B, C). Use a framing square to ensure squareness and plumb of the uprights to the shelf assembly.

5 Screw the bottom four vertical supports (E) to the inside faces of the uprights (D).

6 Position the next shelf assembly on top of the shelf supports, as shown in **9–3**.

7 Repeat Steps 5 and 6 to secure all the supports and shelf assemblies to the uprights. Before screwing the top four supports (E) in place, make sure the top of the top shelf sits flush with the top ends of the uprights (D). Trim the top four vertical supports if necessary.

Materials List for Quick-And-Easy Mobile Storage

PART	FINISHED SIZE			Mtl.	Qty.
	T	W	L		
A horiz. front & back supports	1½"	3½"	96"	C	8
B end supports	1½"	3½"	20⅞"	C	10
C shelves	½"	23⅞"	96"	P	4
D uprights	1½"	3½"	71⅞"	C	4
E vertical supports	1½"	3½"	18⅝"	C	12
F end panel	¼"	16⅞"	71⅞"	PH	1

Materials Key: C = choice of pine or fir 2 x 4; P = plywood; PH = perforated hardboard.
Supplies: ⅜" lag screws; 1½" long (16); 4" heavy-duty swivel-lock casters (4); #12 x 1" panhead screws; #8 x 3" flathead wood screws; #8 x 2½" flathead wood screws; #8 x 1½" flathead wood screws.

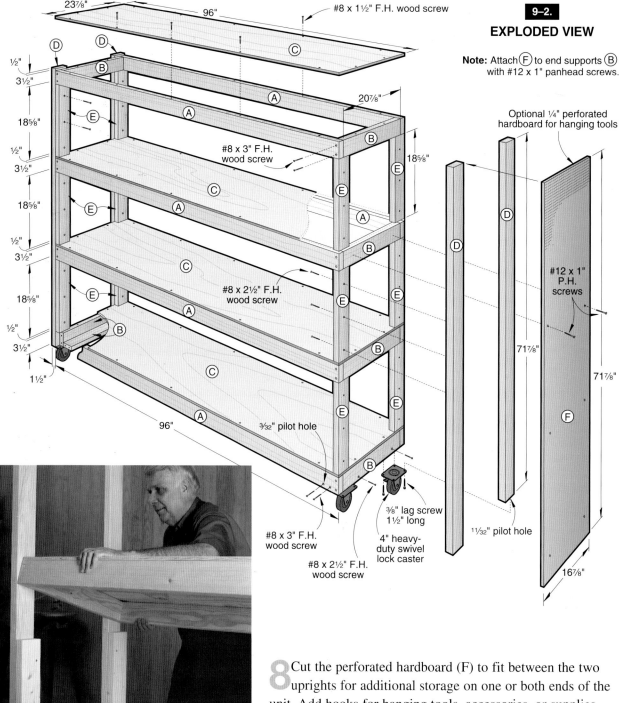

9–2.

EXPLODED VIEW

Note: Attach (F) to end supports (B) with #12 x 1" panhead screws.

23⅞"

96"

#8 x 1½" F.H. wood screw

½"
3½"

18⅝"

½"
3½"

18⅝"

½"
3½"

18⅝"

½"
3½"

1½"

96"

20⅞"

18⅝"

18⅝"

#8 x 3" F.H. wood screw

#8 x 2½" F.H. wood screw

³⁄₃₂" pilot hole

Optional ¼" perforated hardboard for hanging tools

#12 x 1" P.H. screws

71⅞"

71⅞"

¹¹⁄₃₂" pilot hole

16⅞"

⅜" lag screw 1½" long

4" heavy-duty swivel lock caster

#8 x 3" F.H. wood screw

#8 x 2½" F.H. wood screw

9–3.

One level at a time, glue and screw the 2 x 4 vertical supports into position, and add a shelf assembly. Continue the process to the top of the cabinet.

8 Cut the perforated hardboard (F) to fit between the two uprights for additional storage on one or both ends of the unit. Add hooks for hanging tools, accessories, or supplies.

9 Lay the unit on its side, and attach 4" heavy-duty swivel-lock casters to the bottom if desired. Paint the completed project if desired.

❖

ROLL-AROUND PLYWOOD CART

Carrying plywood any distance is difficult, but can be made considerably easier when you use this mobile storage rack (**9–4** to **9–6**). All you need is plywood, dimensional lumber, casters, and wood screws.

Swiveling casters allow you to turn the cart in any direction and back it up to a wall for space-saving storage. Full 4 × 8' sheets fit on the front of the rack. Inside the slats, you can store cut-off pieces.

The construction sequence is as follows: Cut out the two end panels (A) and attach them to the end supports (B). Assemble the base (C, D, E, F) and attach the end panels to that. Complete the rack by attaching the rails (G).

9–4.

Materials List for Roll-Around Plywood Cart

PART	FINISHED SIZE			Mtl.	Qty.
	T	W	L		
A end panels	1½"	11¼"	51"	2 x 12	2
B end supports	1½"	3½"	15½"	2 x 4	2
C base sides	1½"	3½"	93"	2 x 4	2
D base supports	1½"	3½"	12½"	2 x 4	4
E base faces	¾"	5½"	96"	1 x 6	2
F base top	¾"	15½"	96"	P	1
G rails	¾"	3½"	96"	1 x 4	4

Material Key: Plywood
Supplies: #8 x 2" and #8 x 3½" flathead wood screws; 2½" swivel casters; #12 x ¾" panhead sheet-metal screws.

9–5.
EXPLODED VIEW

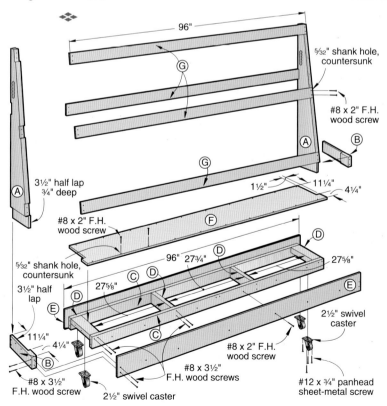

9–6.
SIDE VIEW

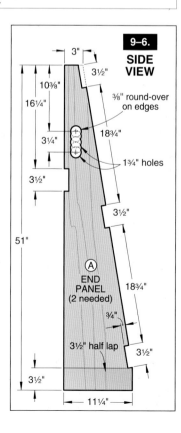

SHEET GOODS MOVER/LIFTER

Few shop chores rival the awkwardness of moving large sheet goods by yourself. And it's almost impossible to get them up on sawhorses without help (or the horses skittering across the floor). You'll never again have to beg reluctant family members to help you if you use this system (**9–7** and **9–8**).

Build the dolly from ¾" plywood or medium-density fiberboard (MDF) and solid stock, as shown in **9–8**. Carefully position the casters so they run parallel to one another—otherwise, the dolly can shimmy like a bad shopping cart.

Each lifter consists of a pair of hinged 2 × 4s. One attaches temporarily to the top of your sawhorse; the other guides the stock from vertical to horizontal.

To work the system, slide the angle iron of the dolly under the edge of the sheet you want to move. (It works best if you store your sheet goods standing on their long edge, raised up on 2 × stickers.) Grab the top of the sheet and tip it toward you. This lifts the material off the stickers and shifts its weight to the dolly. Wheel the sheet to your work area, steering it by tilting and pivoting the dolly on one caster.

To lay the sheet on sawhorses, roll it into position over the angle iron of the lifters, and tip the sheet toward the sawhorses to free the dolly. (The sheet should be resting only on the lifters.) Pull the dolly away and set it aside. Raise the material by tilting the top toward the sawhorses—the lifters will drop to the floor once the weight of the sheet transfers to the horses.

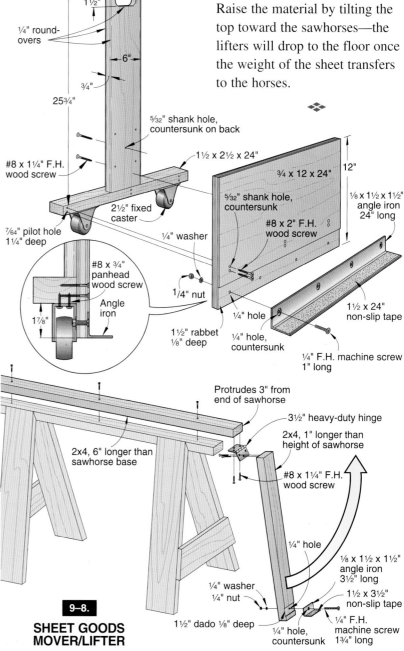

9–7.

9–8.

SHEET GOODS MOVER/LIFTER

LUMBER STORAGE RACK

Looking for versatility in a lumber storage rack? This one's got it! Our rack (**9–9**) features adjustable supports that attach to vertical 2 × 4s for holding loads of boards. The unique sheet-goods bin lets you easily sort through heavy sheets and slide out the one you want. There's even between-the-studs storage for short stock and dowels.

Note: We built our rack to fit against an existing drywalled wall. If you have an exposed stud wall and would like to use it, skip this section and build only the storage rack components covered in the later sections.

Start with the Wall Framework

Measure the distance between your floor and ceiling. The overall height of the rack should be ¼" less than the measured distance.

9–9.

From 2 × 4 stock, crosscut the top and bottom plates (A), uprights (B), and spaces (C) to length. Lay out the holes for the board supports on the uprights (B), where dimensioned on **9–12**. Bore the 9⁄16" holes where marked.

Position the pieces on the floor, and screw the framework together in the configuration shown on **9–12**. Square the uprights with the bottom plate.

Caution: Considering the amount of weight that this unit can hold, you must securely anchor the framework to wall studs and ceiling joists. If the joists run perpendicular to the top plate, screw through the top plate and directly into them. If the joists are parallel to the top plate, install 2 × 4 blocking between the joists, and screw the top plate to the blocking.

With a helper, lift the wall framework into position. Shim the top plate against the ceiling, and firmly secure the top plate to the joists in your shop's ceiling. (We used ¼" lag screws that were 3½" long.) If you can hit wall

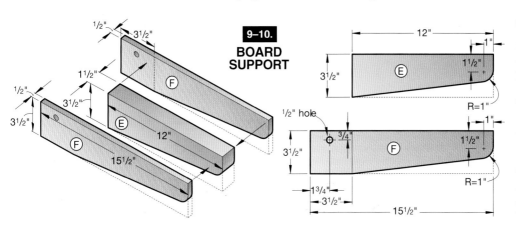

9–10.
BOARD SUPPORT

Cutting Diagram

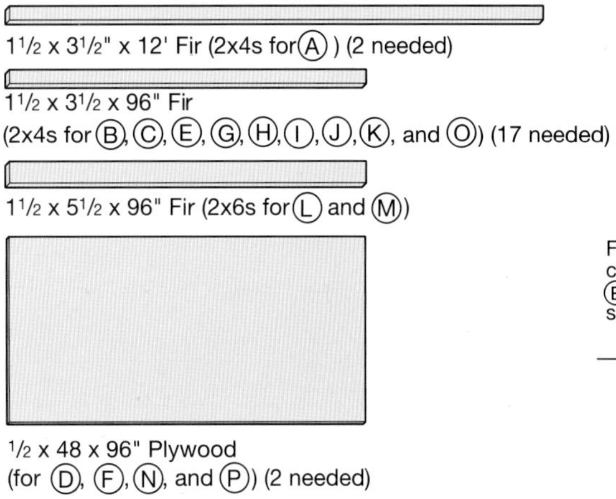

1¹⁄₂ x 3¹⁄₂" x 12' Fir (2x4s for Ⓐ) (2 needed)

1¹⁄₂ x 3¹⁄₂ x 96" Fir
(2x4s for Ⓑ, Ⓒ, Ⓔ, Ⓖ, Ⓗ, Ⓘ, Ⓙ, Ⓚ, and Ⓞ) (17 needed)

1¹⁄₂ x 5¹⁄₂ x 96" Fir (2x6s for Ⓛ and Ⓜ)

¹⁄₂ x 48 x 96" Plywood
(for Ⓓ, Ⓕ, Ⓝ, and Ⓟ) (2 needed)

studs, drill 2" deep counterbored holes
through the uprights (B) or spacers (C),
and use 3"-long screws to further secure
the framework.

5 Secure the bottom plate to your floor.
(We drilled holes in the concrete, and
used plastic concrete anchors and lag
screws; masonry screws would work.)

6 From ¾" plywood or 1 x 4s, cut the
short-stock bin stops (D) to size, and
screw them in place.

Build the Board Supports

1 Referring to **9–10,** on *page 211,* cut
21 center sections (E) and 42 plywood
side pieces (F) to size.

2 Spread an even coat of glue on both
faces of each center section (E), and
clamp it between two of the side pieces (F),
with the top edges and outside end flush.

3 Transfer the profile of one of the supports onto one of the laminations.
Bandsaw the support to shape, and sand the
cut edges smooth to remove the saw marks.
Use this as a template to mark the profile
onto the rest of the supports. Cut and sand
them to shape.

Materials List for Lumber Storage Rack

PART		FINISHED SIZE			Mtl.	Qty.
		T	W	L		
WALL FRAMEWORK						
A	top and bottom plates	1½"	3½"	12"	C	2
B*	uprights	1½"	3½"	93"	C	7
C	spacers	1½"	3½"	22½"	C	4
D	stops	½"	3½"	47¼"	P	3
BOARD SUPPORTS						
E	centers	1½"	3½"	12"	C	21
F	sides	½"	3½"	15½"	P	42
SHEET-GOODS BIN						
G	angled support	1½"	3½"	32"	C	5
H	floor supports	1½"	3½"	18"	C	5
I	cleats	¾"	1½"	4¼"	C	8
J	spacers	1½"	3½"	22½"	C	3
K	spacer	1½"	3½"	19½"	C	1
L	steps	1½"	5½"	22½"	C	3
M	step	1½"	5½"	19½"	C	1
N	floor	½"	11"	94½"	P	1
O	cleat	1½"	1½"	94½"	C	1
P	back	½"	48"	96"	P	1

*Length will depend on distance from floor to
ceiling at chosen rack location.

Materials Key: C = choice (fir, pine, spruce);
P = plywood.
Supplies: 3½" deck screws, #8 x 1¼" flathead
wood screws, #8 x 1½" flathead wood screws,
#8 x 2½" flathead wood screws, #8 x 3" flathead
wood screws, ½" carriage bolts 2" long with flat
washers and nuts, ½" carriage bolts 3" long with
flat washers and nuts, ½" carriage bolts 3 ½"
long with flat washers and nuts, paraffin wax,
masonry screws.

9–11.

SIDE VIEW

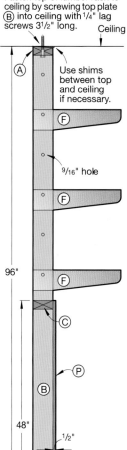

Fasten top of lumber rack to
ceiling by screwing top plate
Ⓑ into ceiling with ¼" lag
screws 3½" long.   Ceiling

Use shims
between top
and ceiling
if necessary.

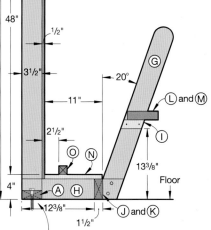

Secure bottom of lumber rack to floor
by screwing bottom plate Ⓐ to floor
with masonry screws or concrete
anchors and lag screws.

Add the Sheet-Goods Bin

1 Using **9–13**, cut the angle supports (G), floor supports (H), and cleats (I) to size. Cut 20° half-lap joints

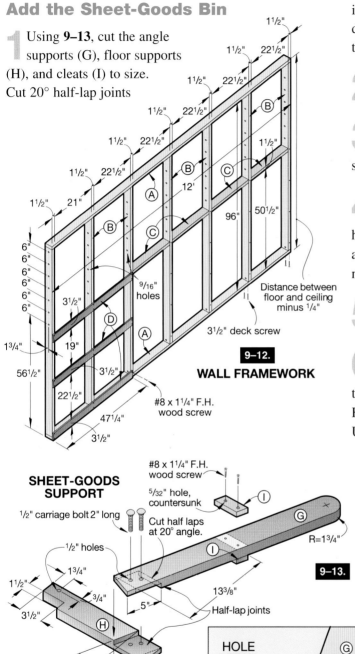

SHEET-GOODS SUPPORT

#8 x 1¹/4" F.H. wood screw

5/32" hole, countersunk

¹/2" carriage bolt 2" long

¹/2" holes

1³/4"

1¹/2"

³/4"

3¹/2"

Cut half laps at 20° angle.

R=1³/4"

13³/8"

5"

Half-lap joints

20°

5"

¹/2" flat washer

¹/2" nut

9–13.

CLEAT DETAIL

20°

5/32" holes, countersunk

1¹/2"

³/4"

4¹/4"

20°

1¹/2"

22¹/2"

1¹/2"

22¹/2"

1¹/2"

22¹/2"

1¹/2"

22¹/2"

1¹/2"

22¹/2"

1¹/2"

1¹/2"

21"

12'

96"

50¹/2"

6"
6"
6"
6"
6"
6"

3¹/2"

9/16" holes

1³/4"

19"

3¹/2"

56¹/2"

22¹/2"

47¹/4"

3¹/2"

Distance between floor and ceiling minus ¹/4"

3¹/2" deck screw

#8 x 1¹/4" F.H. wood screw

9–12.

WALL FRAMEWORK

HOLE DETAIL

20°

1³/4"

1¹/2"

1³/4"

¹/2" holes

Waste

in the H and I pieces, where shown on the drawing. Referring to the Cleat Detail, miter the ends of the step cleats (I) at 20°.

2 Cut a 1½ x 3½" notch in the back of each floor support (H), where shown on the drawing.

3 Glue and clamp the supports together. After the glue dries, cut off the waste areas, where shown on the Hole Detail in **9–13**.

4 Mark the centerpoints, and drill a pair of ½" holes in each glued half lap and a single ½" hole above each notch. To strengthen the joints, add a pair of carriage bolts with flat washers and nut to each of the half-lap joints.

5 Drill the mounting holes, and glue and screw the stop cleats to the angled supports (G).

6 Clamp the support assemblies (G/H/I) to the uprights (B), where shown on **9–14**. Using the previously drilled holes above the notches in H as guides, drill the holes through the uprights. Using carriage bolts, secure the support assemblies to the 2 x 4 uprights.

7 From 2 x 4 and 2 x 6 stock, cut the floor spacers (J, K) and steps (L, M) to length. Drill mounting holes (some need to be angled), and glue and screw the pieces in place. Position each step (L, M), so the back edge won't protrude into the plywood bin and possibly damage any of the stored sheet goods when pressed against the angled supports, as shown on **9–11**.

8 Cut the bin floor (N) and the floor cleat (O) to size. Position the pieces; drill mounting holes, where shown on **9–14**; and attach them to the floor supports (H). To make the sheet goods slide in and out even easier, rub the bin floor with paraffin wax.

9 Lay out and drill pilot holes on the bin back (P), and attach it with wood screws to wall framework.

10 Using ½" carriage bolts, 3" long with washers and nuts, hang the board supports (E/F) at desired weights on the uprights (B).

❖

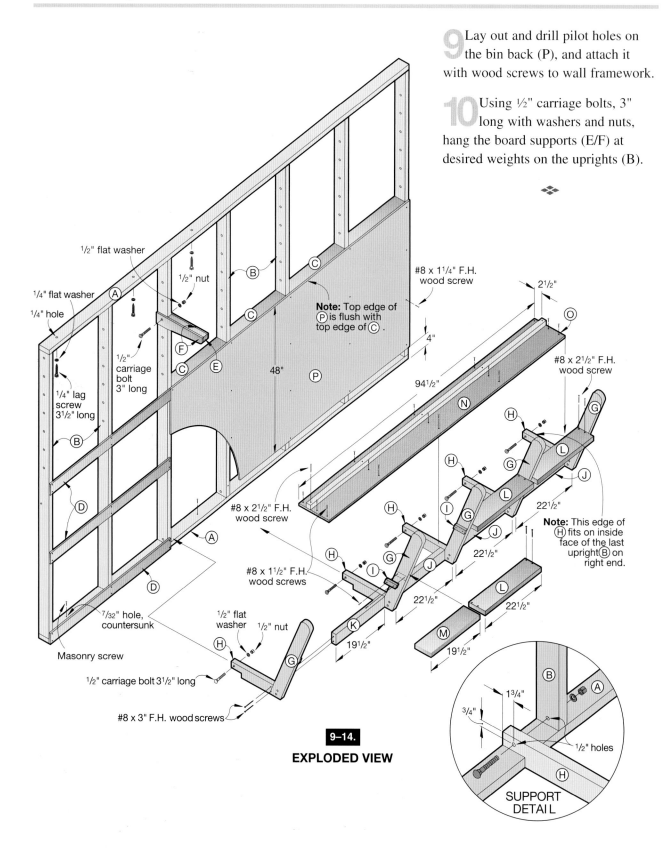

½" flat washer

¼" flat washer

¼" hole

½" nut

½" carriage bolt 3" long

¼" lag screw 3½" long

¼" hole, countersunk

Masonry screw

½" carriage bolt 3½" long

#8 x 3" F.H. wood screws

½" flat washer

½" nut

#8 x 1½" F.H. wood screws

#8 x 2½" F.H. wood screw

Note: Top edge of P is flush with top edge of C.

#8 x 1¼" F.H. wood screw

48"

4"

94½"

2½"

#8 x 2½" F.H. wood screw

22½"

22½"

22½"

22½"

19½"

19½"

Note: This edge of H fits on inside face of the last upright B on right end.

9–14.
EXPLODED VIEW

SUPPORT DETAIL

1¾"

¾"

½" holes

DOUBLE-DUTY CLAMP/LUMBER RACK

Lumber inventories in the typical home woodworking shop always vary. If you stock up on material to make a big project or a lot of holiday gifts, the stacks of lumber on the floor can become a traffic hazard. But if you build a permanent home for those planks, the space sits idle when your wood supply dips.

Here's one solution—foldable racks that can keep pipe clamps tucked against the wall, but drop for instant lumber storage when you need it (**9–15** and **9–16**).

With sides of ¾" plywood and center blocks of 2 × material, these brackets are strong and easy to make. The rear pivot point rotates around a ½" bolt that's placed in the wall stud or, if the wall studs are covered, in a 2 × 4 upright secured to the top and bottom of the wall. The series of holes (on 6" centers) in the stud or upright lets you position the brackets right where you need them, and also provides a locking bolt location when they're folded.

The clamp-holding feature requires just three ¼ × 4¼" screw hooks threaded into each center block. They'll hold ½" and ¾" pipe clamps, as well as clamp styles with square or rectangular bars. Find all the needed hardware at your local hardware retailer or home center.

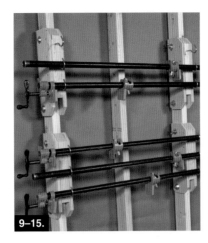

9–15.

9–16.

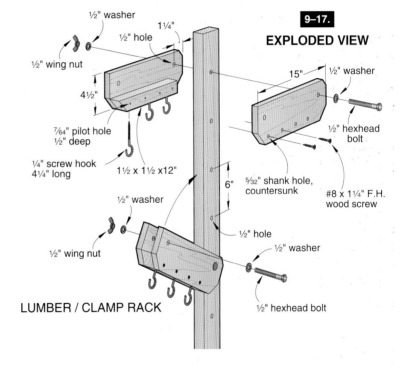

9–17.
EXPLODED VIEW

½" washer
1¼"
½" hole
½" wing nut
4½"
7/64" pilot hole ½" deep
¼" screw hook 4¼" long
1½ x 1½ x12"
15"
½" washer
½" hexhead bolt
5/32" shank hole, countersunk
#8 x 1¼" F.H. wood screw
6"
½" hole
½" washer
½" hexhead bolt
½" washer
½" wing nut

LUMBER / CLAMP RACK

9–18.
PART VIEW

15"
½" hole
12"
1¼"
1¼"
¾"
4½"
12"
½" hole
1½"
1½"
1½"
3"
3"
3"
3/16" hole 1" deep for screw hook

A CATCHALL FOR CUTOFFS

This upright organizer (**9–19** and **9–20**) is designed to hold cutoffs and other short pieces that otherwise might be lost or tossed. And, you can see at a glance what stock you have on hand. Plus, its wall-hugging profile takes good advantage of limited shop space.

9–19.

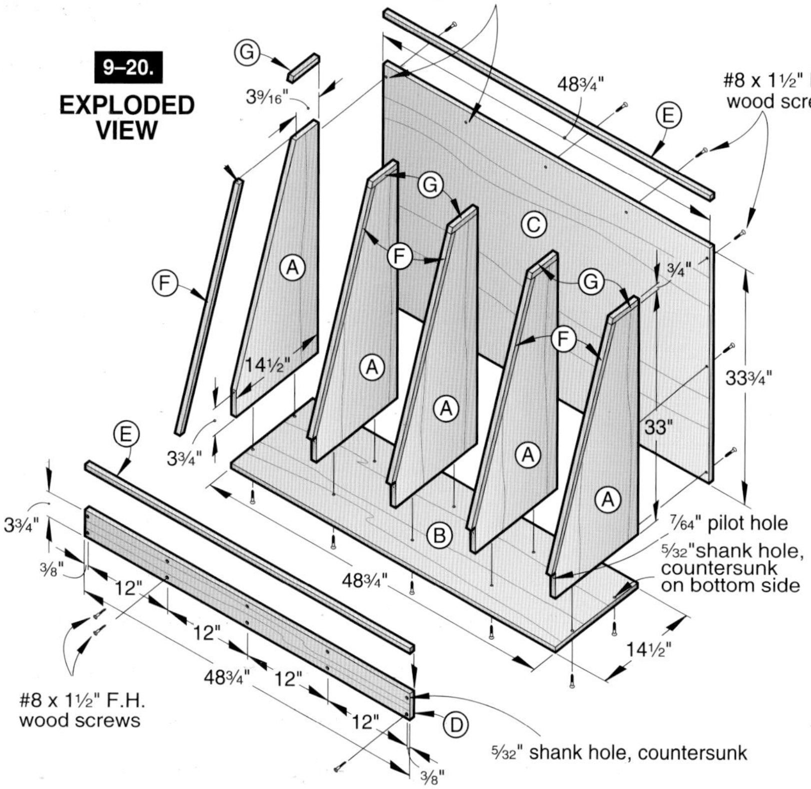
5⁄32" shank holes, countersunk from back side
G
3 9⁄16"
48 3⁄4"
#8 x 1 1⁄2" F.H. wood screws
E
G
C
F
A
F
G
3⁄4"
A
A
F
14 1⁄2"
A
33 3⁄4"
33"
E
A
3 3⁄4"
A
3 3⁄4"
7⁄64" pilot hole
B
5⁄32"shank hole, countersunk on bottom side
3⁄8"
12"
48 3⁄4"
12"
14 1⁄2"
48 3⁄4"
12"
#8 x 1 1⁄2" F.H. wood screws
12"
D
5⁄32" shank hole, countersunk
3⁄8"

9–20. EXPLODED VIEW

Materials List for A Catchall for Cutoffs

	FINISHED SIZE				
PART	T	W	L	Mtl.	Qty.
A dividers	3⁄4"	14 1⁄2"	33"	P	5
B bottom	3⁄4"	14 1⁄2	48 3⁄4"	P	1
C back	3⁄4"	33 3⁄4"	48 3⁄4"	P	1
D front	3⁄4"	3 3⁄4"	48 3⁄4"	P	1
E trim	3⁄4"	3⁄4"	48 3⁄4"	B	2
F trim	3⁄4"	3⁄4"	31 1⁄2"	B	5
G trim	3⁄4"	3⁄4"	4 3⁄8"	B	5

Materials Key: P = plywood; B = birch.
Supplies: #8 x 1 1⁄2" flathead wood screws; paint; clear finish.

CHAPTER 10

Easy Racks, Holders, and Organizers

WRENCHES, SCREWDRIVERS AND *tips, router and drill bits, blades, clamps, and a multitude of other necessities seem to accumulate. Often, they'll find homes in drawers and shelves, turning the immediate need for a small tool into a major search.*

It doesn't have to be that way. As you'll see by the racks, hangers, and organizers in this chapter, it's easy to follow an orderly track. You'll even have fun building those you find fitted to your needs. Check out the simple router bit shelf on page 218. *Then, add it to the router organizer on* page 224. *Or, if you'd rather store router bits in a drawer, you'll like the clever storage system on* page 219.

In this chapter, you'll find solutions for organizing and storing tablesaw and scrollsaw blades, saws and other hand tools, plus racks for clamps of all types. There are even options to keep chisels from chipping, lathe tools from getting dinged, and for keeping glue handy. There is plenty to keep you busy, so go ahead, and kick up some sawdust.

AT-THE-READY BIT SHELF

Here's an easy-to-build organizer (**10–1** and **10–2**) that you can customize to hold as many bits and shanked accessories as you need by simply varying the length and number of holes in this handy project. Angled for better accessibility and visibility of your accessories, our shelves hold router bits, sanding drums, flap sanders, and other drill-press and shanked accessories. Use either thick hardwood or 2 × 6 stock for the angled shelf and ¾" stock for the backboard. We routed a ⅜" round-over on all but the back edges of the shelf, where shown at *right* and finished the project with a clear finish.

❖

10–1.

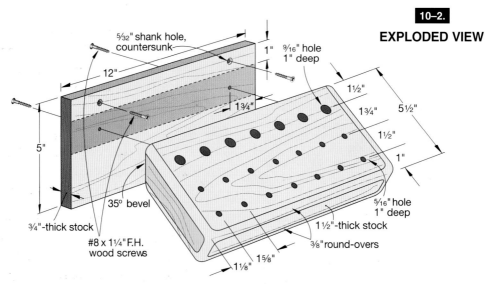

10–2.

EXPLODED VIEW

⁵⁄₃₂" shank hole, countersunk

12"

1"

⁹⁄₁₆" hole 1" deep

1½"

1¾"

5½"

1½"

1"

1¾"

5"

35° bevel

¾"-thick stock

#8 x 1¼" F.H. wood screws

⁵⁄₁₆" hole 1" deep

1½"-thick stock

⅜" round-overs

1⅛"

1⅝"

EASY ROUTER BIT STORAGE

10–3.

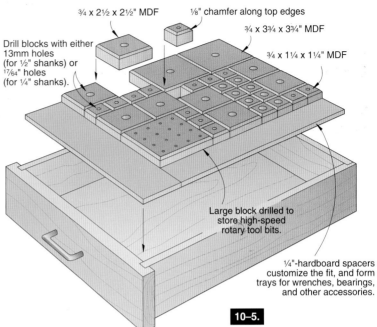

¾ x 2½ x 2½" MDF

⅛" chamfer along top edges

¾ x 3¾ x 3¾" MDF

¾ x 1¼ x 1¼" MDF

Drill blocks with either 13mm holes (for ½" shanks) or ¹⁷⁄₆₄" holes (for ¼" shanks).

Large block drilled to store high-speed rotary tool bits.

¼"-hardboard spacers customize the fit, and form trays for wrenches, bearings, and other accessories.

10–5.

EXPLODED VIEW

(Right). To add versatility to this system, simply bore holes into the module(s) of your choice and glue in craft magnets. You'll find that they hold small steel parts just great.

10–4.

Most router-bit storage systems force you to guess how many bits you'll add to your arsenal in the coming years. *WOOD®* magazine's Dave Campbell neatly sidestepped that dilemma by designing this modular storage system that fits in any drawer and easily grows to meet your expanding bit collection (**10–3** to **10–5**). A 1-2-3 progression of block sizes maximizes the number of possible arrangements. Dave set aside one large square, and drilled it to hold rotary-tool bits. To build your modular storage, do the following:

1 Rip ¾" MDF (medium-density fiberboard) into 1¼"-, 2½"-, and 3¾"-wide strips, and then cross-cut them into squares. Drill centered, slightly over-sized holes for easy bit removal: ¹⁷⁄₆₄" and 13-mm holes for ¼"- and ½"-shank router bits; and ⁷⁄₆₄" and ⁹⁄₆₄" holes for ³⁄₃₂"- and ⅛"-shank high-speed rotary tool bits.

Note: Finding a ³³⁄₆₄" bit to drill the oversized holes for ½" shanks is nearly impossible. Commonly used to install metric hardware, a 13-mm bit is a readily available substitute.

2 Slightly countersink the holes' edges. Then chamfer the blocks' top edges on your table-mounted router. Hold the parts with a padded jointer pushblock to keep your fingers safely away from the router bit.

3 Pour some Danish oil-type finish into a small container, and dip the blocks. After wiping off the excess finish with a rag, dry the blocks on a window screen propped on sawhorses.

4 With the finish dry, arrange the blocks in your drawer. Fill in the extra space in the drawer with snug-fitting pieces of ¼" hardboard. You can use these tray-like spaces for storing accessories.

BANDSAW-BLADE HOLDER

Now you can rid your shop of tangled coils by protecting and storing your blades in this handy organizer made from ¼" birch plywood (**10–6** and **10–7**). After cutting the parts and gluing them together, we mounted the holder to our bandsaw base where they're close at hand when needed. Or, if you prefer, you can fasten the organizer to a wall or cabinet side, keeping your blades nearby when you're ready to work.

10–6.

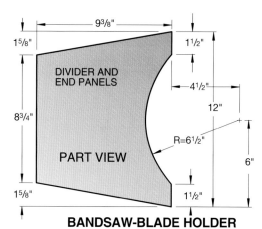

DIVIDER AND END PANELS

9³⁄₈"

1⁵⁄₈"

8³⁄₄"

PART VIEW

1⁵⁄₈"

1¹⁄₂"

4¹⁄₂"

12"

R=6¹⁄₂"

6"

1¹⁄₂"

BANDSAW-BLADE HOLDER EXPLODED VIEW

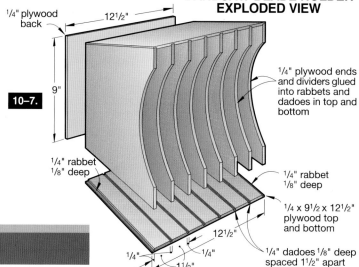

10–7.

¼" plywood back

12¹⁄₂"

9"

¼" plywood ends and dividers glued into rabbets and dadoes in top and bottom

¼" rabbet ⅛" deep

¼" rabbet ⅛" deep

¼ x 9¹⁄₂ x 12¹⁄₂" plywood top and bottom

12¹⁄₂"

¼"

¼"

¼" dadoes ⅛" deep spaced 1¹⁄₂" apart

1¹⁄₂"

SAW-BLADE HOLDER

Wall-mount this handy holder (**10–8** and **10–9**) near the stationary saws in your shop, or, if you like, place it on a convenient benchtop. You'll find it the ideal storage project for organizing an assortment of saw blades and a dado-blade set.

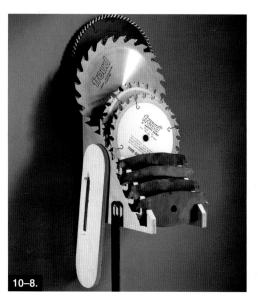

10–8.

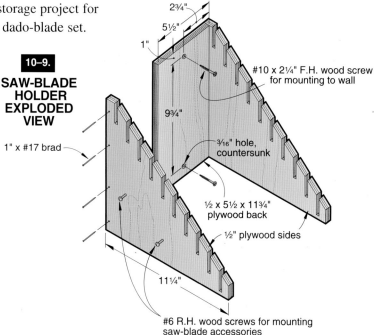

10–9.

SAW-BLADE HOLDER EXPLODED VIEW

2³⁄₄"

5¹⁄₂"

1"

9³⁄₄"

#10 x 2¹⁄₄" F.H. wood screw for mounting to wall

1" x #17 brad

³⁄₁₆" hole, countersunk

¹⁄₂ x 5¹⁄₂ x 11³⁄₄" plywood back

¹⁄₂" plywood sides

11¹⁄₄"

#6 R.H. wood screws for mounting saw-blade accessories

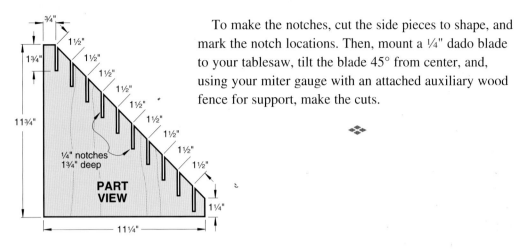

10–9 continued.

SAW-BLADE HOLDER SIDE

¾"
1½"
1¾"
1½"
1½"
1½"
1½"
1½"
1½"
1½"
1½"
11¾"
¼" notches
1¾" deep
PART VIEW
1¼"
11¼"

To make the notches, cut the side pieces to shape, and mark the notch locations. Then, mount a ¼" dado blade to your tablesaw, tilt the blade 45° from center, and, using your miter gauge with an attached auxiliary wood fence for support, make the cuts.

❖

SCROLLSAW-BLADE ORGANIZER

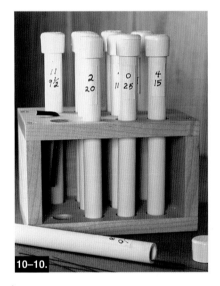

10–10.

Scrollsawers know that laying their hands on the right blade can be tricky and time-consuming, especially if these tiny cutting tools get mixed together. This handy little organizer (**10–10** and **10–11**), designed to separate and store your blades, comes from *WOOD*® magazine reader John M. Turok of Coon Rapids, Minnesota. To build it, all you need is a small bit of scrap stock

and some ½"-diameter (⅝"-O.D.) CPVC pipe and caps.

For still more convenience, drill a few extra holes in a rack top to hold scrollsaw tools, such as Allen wrenches. Consider labeling the tubes for easy reference. You can hang the unit on a wall or set it on a flat surface near your scrollsaw.

❖

CPVC cap

½" (⅝" O.D.) CPVC tube 6" long

Paper label taped to tube

½"-diam. wood plug ½" long epoxied into bottom of CPVC hole

10–11.

BLADE ORGANIZER EXPLODED VIEW

#6 x 1" F.H. wood screw

½"
3¾"
½"
¼"
6"
1⁵⁄₁₆"
1⅛"

Hole sized to fit scrollsaw tool

¾" holes

⁹⁄₆₄" shank hole, countersunk

¾"
1⅛"
1⅛"

⁵⁄₃₂" shank hole, countersunk

½ x 3¾ x 6" back (optional) used to hang organizer

6"

⅜⁄₃₂" pilot hole ½" deep

#8 x 2" F.H. wood screw

⅝" holes ¼ deep
Note: Hole locations are the same as those in the top pieces.

#6 x 1" F.H. wood screw

All stock is ½" thick.

2¾"

3¾"

⁹⁄₆₄" shank hole, countersunk from the bottom

#6 x 1" F.H. wood screws

CORDLESS DRILL RECHARGING STATION

This handy workshop accessory (**10–12** and **10–13**) goes together quickly, and your cordless drill always will be close to your work and fully charged when you're ready to use it.

You'll have to customize the support to accommodate your particular drill and charger. (We built the support for the handle style shown.) For instance, if you have a pistol-grip style drill, you may need to shift the grip opening to the right and modify or reposition one or both cradle brackets to hold the drill.

The charger's feet fit over short lengths of dowel to keep the charger from shifting on the support. To position the dowels correctly, touch an inkpad lightly to the charger's feet, and then set the charger on a sheet of paper to make a pattern. Transfer the pattern to the support, and drill holes for the appropriate dowels. If your charger's feet won't accept a dowel peg, drill holes in the top of the support into which the feet will fit snugly.

10–12.

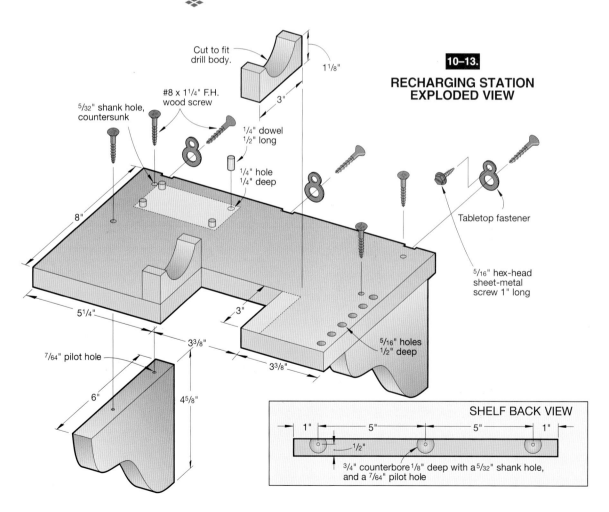

10–13.

**RECHARGING STATION
EXPLODED VIEW**

Cut to fit drill body.

1 1/8"

3"

#8 x 1 1/4" F.H. wood screw

5/32" shank hole, countersunk

1/4" dowel 1/2" long

1/4" hole 1/4" deep

Tabletop fastener

8"

5/16" hex-head sheet-metal screw 1" long

5 1/4"

3"

3 3/8"

3 3/8"

7/64" pilot hole

5/16" holes 1/2" deep

6"

4 5/8"

SHELF BACK VIEW

1"   5"   5"   1"

1/2"

3/4" counterbore 1/8" deep with a 5/32" shank hole, and a 7/64" pilot hole

WALL-HUNG PLANE HOLDER

Like all precision tools in the shop, your planes need and deserve safe, sturdy storage. Consider this adaptable wall-hung organizer (**10–14** and **10–15**) the answer, regardless of which size planes you may own. Using the design shown and the construction notes below, you can build these holders from scrapwood. Then, secure them to your shop wall or to a piece of plywood attached to the wall. We included a dado in the base (A) to protect the plane blade from damage.

10–14.

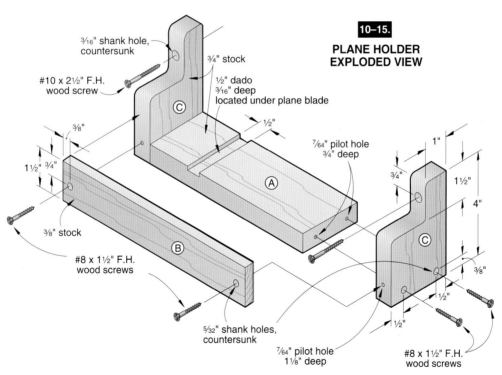

10–15.

PLANE HOLDER EXPLODED VIEW

3⁄16" shank hole, countersunk

#10 x 2½" F.H. wood screw

3⁄4" stock

½" dado 3⁄16" deep located under plane blade

½"

7⁄64" pilot hole 3⁄4" deep

3⁄8"

1½" 3⁄4"

©

3⁄8" stock

B

#8 x 1½" F.H. wood screws

5⁄32" shank holes, countersunk

7⁄64" pilot hole 1⅛" deep

1"

3⁄4"

1½"

4"

©

3⁄8"

½"

½"

#8 x 1½" F.H. wood screws

A

Construction Notes
Length of (A) equals length of plane plus ½".
Length of (B) equals length of (A), plus 1½".
Width of (A) and (C) equals width of plane plus ½".
Locate the dado in (A) under the plane blade.

WALL-HUNG ROUTER ORGANIZER

When you want to use your router, you don't want to waste time routing through drawers and toolboxes for router bits, wrenches, and other accessories. This compact, wall-mounted organizer (**10–16** and **10–17**) keeps everything you need, including your router, in plain sight.

Building Instructions

1 Cut the mounting board (A) to the finished size shown in the Materials List. Lay the board on a flat surface, and position the accessories you want to hang on the board. See **10–17** for the dowel sizes and locations for typical accessories—wrenches,

10–16.

router edge guide, and one or more trammel bases for cutting circles. Once you've established locations for your accessories, drill ⅜"-deep holes in the board for the dowels and cut the dowels to length. To make the 15° hole for the router dowel, use an adjustable drill guide.

Materials List for Wall-Hung Router Organizer

PART	FINISHED SIZE T	W	L	Mtl.	Qty.
A mounting board	¾"	16"	18"	PB	1
B trim	¾"	¾"	19½"	P	2
C trim	¾"	¾"	16"	P	2
*D bit shelf	1½"	5"	10"	P	1

*Laminated from two pieces of ¾" pine.
Material Keys: PB = particleboard; P = pine.
Supplies: 1½" cup hook; 3d finish nails; #10 x 2½" flathead wood screws; 1" dowel; ⅛", ¼" dowel.

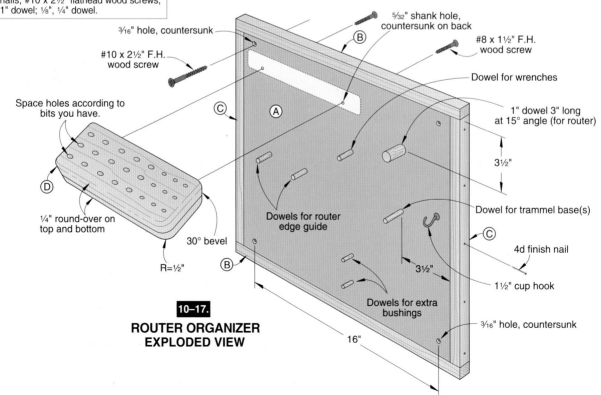

¾₆" hole, countersunk

#10 x 2½" F.H. wood screw

⁵⁄₃₂" shank hole, countersunk on back

Ⓑ

#8 x 1½" F.H. wood screw

Dowel for wrenches

1" dowel 3" long at 15° angle (for router)

3½"

Space holes according to bits you have.

Ⓒ Ⓐ

Ⓓ

¼" round-over on top and bottom

30° bevel

R=½"

Ⓑ

Dowels for router edge guide

Dowel for trammel base(s)

Ⓒ

4d finish nail

1½" cup hook

3½"

Dowels for extra bushings

¾₆" hole, countersunk

16"

3½"

10–17.
**ROUTER ORGANIZER
EXPLODED VIEW**

2 Cut the ¾ × ¾" trim strips (B, C) to finished lengths. To make the bit shelf (D), cut two pieces of ¾" stock slightly over-size and laminate them together, face to face, with woodworker's glue. When the glue has dried, cut the piece to finished length and width, beveling the back edge to a 30° angle, where shown on **10–17**. Radius the front corners and rout a ¼" round-over on all edges except the back.

3 Lay out and drill shank holes in the bit organizer for your router bits. You can leave extra space on the bit organizer for future acquisitions. Our organizer holds 24 bits of various sizes.

4 Paint the mounting board the desired color. After the paint dries, nail the trim strips to the edges of the board. Apply clear polyurethane or varnish to the pine trim pieces, dowels, and bit organizer. Glue the dowels into the board, and attach the cup hooks and bit organizer to the mounting board.

Mount the board to the wall with screws, making sure to align the board's mounting holes with the wall studs.

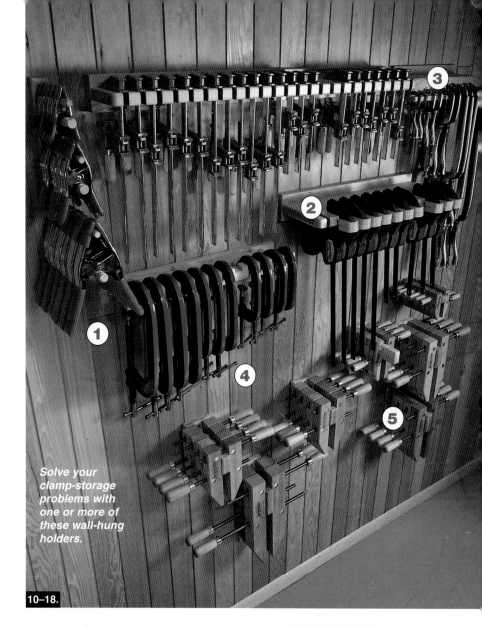

Solve your clamp-storage problems with one or more of these wall-hung holders.

10–18.

FIVE GREAT CLAMP ORGANIZERS

GUIDE TO FIVE GREAT CLAMP ORGANIZERS (10–18)

1 **Spring Clamp Holder**
2 **Bar-Clamp Rack**
3 **Locking C-Clamp Support**
4 **C-Clamp Rack**
5 **Handscrew-Clamp Organizer**

1 Simple, But Sturdy Spring-Clamp Holder

A backboard with protruding dowels does the job for supporting 4", 6", and 9"-long spring clamps (**10–19** and **10–20**). If you've got a similar assortment of these clamps, the rack shown here should suffice. If you've got quite a collection, extend the board or make two or more holders as needed.

We don't recommend extending the dowels longer than dimensioned on **10–20**. Extended too far, the dowels can get bumped and broken.

10–19.

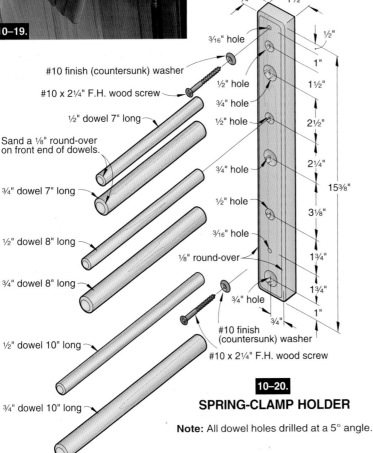

#10 finish (countersunk) washer

#10 x 2¼" F.H. wood screw

½" dowel 7" long

Sand a ⅛" round-over on front end of dowels.

¾" dowel 7" long

½" dowel 8" long

¾" dowel 8" long

½" dowel 10" long

¾" dowel 10" long

¾" — 1½"

³⁄₁₆" hole — ½"

½" hole — 1"

¾" hole — 1½"

½" hole — 2½"

¾" hole — 2¼"

½" hole — 15⅜"

³⁄₁₆" hole — 3⅛"

⅛" round-over — 1¾"

¾" hole — 1¾"

#10 finish (countersunk) washer — 1"

#10 x 2¼" F.H. wood screw — ¾"

10–20.

SPRING-CLAMP HOLDER

Note: All dowel holes drilled at a 5° angle.

10–21.

2 A Home for Bar Clamps

The slots in this quick-and-easy holder (**10–21** and **10–22**) allow you to store your sliding-head clamps in perfect order by simply resting the head on the horizontal support. Without the slots, you'd have to tighten the jaws to hold a clamp in place, and then loosen the jaws when you want to remove the clamp from the holder.

To make the slots, cut the horizontal support to shape, rout ⅜" round-overs along the edges and corners, and mark the notches 1⅝" on center. Mount a ⅜" dado blade to your tablesaw, and raise the blade 2" above the saw table. Mount a wood extension to your miter gauge, and make the cuts where marked. As shown in

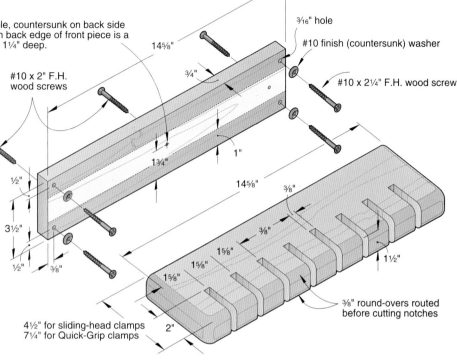

10–22.
BAR-CLAMP RACK

5/32" shank hole, countersunk on back side. Mating hole in back edge of front piece is a 7/64" pilot hole 1¼" deep.

3/16" hole

#10 finish (countersunk) washer

14⅝"

3/4"

1"

#10 x 2¼" F.H. wood screw

#10 x 2" F.H. wood screws

1¾"

14⅝"

3/8"

½"

3½"

½"

3/8"

1⅝"

1⅝"

1⅝"

3/8"

3/8"

1½"

3/8" round-overs routed before cutting notches

4½" for sliding-head clamps
7¼" for Quick-Grip clamps

2"

10–18, we built and placed three racks end to end. You could also lengthen the two parts, and cut the number of slots needed to match your supply of clamps.

For Quick-Grip clamps, we found that an extended version (2¾" more from front to back) of the rack used to support our sliding-head clamps works wonderfully. The front support with its numerous slots holds the clamps upright and keeps them from dinging the wall.

3 **Nifty Support for Locking C-Clamps**

Projects often require clamping pressure applied several inches in from the outside edges for a good bond. When this happens, we turn to our locking C-clamps, the largest of which has a throat depth of more than 15".

To hang our collection of C-clamps, we found this screw-together support is an organizer's dream (**10–23** and **10–24**).

10–23.

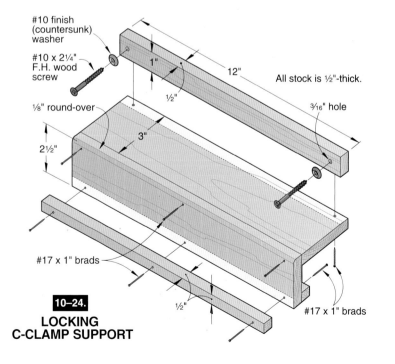

#10 finish (countersunk) washer

#10 x 2¼" F.H. wood screw

⅛" round-over

2½"

1"

½"

3"

12"

All stock is ½"-thick.

3/16" hole

#17 x 1" brads

½"

#17 x 1" brads

10–24.
LOCKING C-CLAMP SUPPORT

4 C-Clamps in a Row

When the job calls for plenty of clamping pressure, no-nonsense C-clamps provide the answer. To hang and organize this type of clamp, build our four-piece wall-mounted hanger with its notched front support (**10–25** and **10–26**). The notches allow you to hang the clamps on the rack without having to tighten them.

10–27.

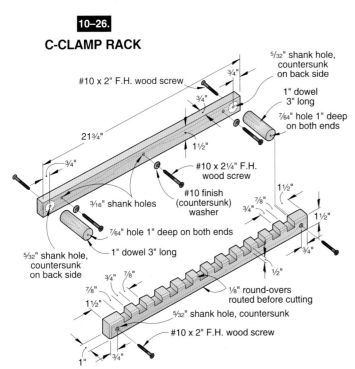

10–25.

5 Handscrew-Clamp Hangout

This is a very simple organizer for handscrew clamps (**10–27** and **10–28**). Just cut the support extension (A) to fit between the threaded rods of the clamp, chamfer the outside end of the support, and securely mount it to the backboard (B). We've found that about four clamps per support is a full load.

10–26.

C-CLAMP RACK

#10 x 2" F.H. wood screw
3/4"
3/4"
5/32" shank hole, countersunk on back side
1" dowel 3" long
7/64" hole 1" deep on both ends
21¾"
1½"
#10 x 2¼" F.H. wood screw
3/4"
#10 finish (countersunk) washer
3/16" shank holes
7/64" hole 1" deep on both ends
5/32" shank hole, countersunk on back side
1" dowel 3" long
3/4"
7/8"
7/8"
7/8"
1½"
1½"
1" round-overs... 1/8" round-overs routed before cutting
5/32" shank hole, countersunk
1½"
3/4"
7/8"
3/4"
1/2"
3/4"
1½"
#10 x 2" F.H. wood screw
1"
3/4"

Width of clamp when hung on (A)
#10 x 2" F.H. wood screws
1/8" round-over
3/16" hole
#10 finish (countersunk) washer
#10 x 2¼" F.H. wood screw
Length equals thickness of 4 clamps.
1"
(A)
Distance between threaded rods minus ¼"
⅔ of overall length of clamp
(B)
3/16" hole
5/32" holes, countersunk on back side
¼" chamfer
3/4"
#10 finish (countersunk) washer
#10 x 2¼" F.H. wood screw

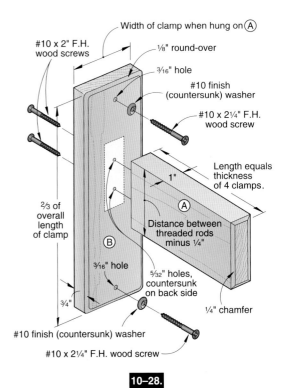

10–28.

HANDSCREW-CLAMP ORGANIZER

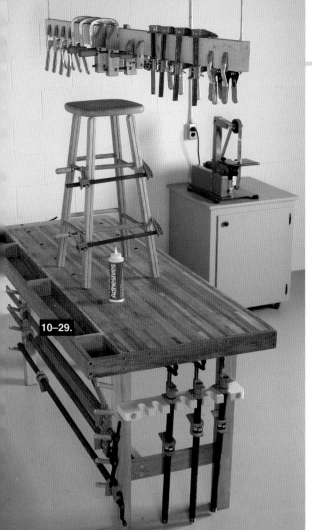

10–29.

spacing. Or, you can nail support blocks between the joists and suspend rods from them.

2 Use a hacksaw to cut the threaded rods to a length that will position the rack about 6" lower than the final anticipated height. Attach the threaded rods to the 2 × 4 rack, and then to ceiling joists, using the fasteners specified on the drawing.

3 Hang a few clamps on the rack, and then adjust the rack to a convenient height. (We positioned our rack about 3½' above the table surface so taller projects, such as a stool, would fit underneath it.) Cut off the rod ends extending beneath the 2 × 4.

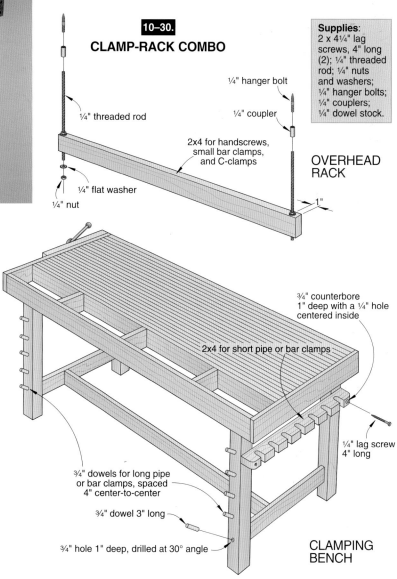

10–30.

CLAMP-RACK COMBO

¼" threaded rod

¼" hanger bolt

¼" coupler

¼" flat washer

¼" nut

2x4 for handscrews, small bar clamps, and C-clamps

OVERHEAD RACK

1"

Supplies: 2 x 4¼" lag screws, 4" long (2); ¼" threaded rod; ¼" nuts and washers; ¼" hanger bolts; ¼" couplers; ¼" dowel stock.

¾" counterbore 1" deep with a ¼" hole centered inside

2x4 for short pipe or bar clamps

¼" lag screw 4" long

¾" dowels for long pipe or bar clamps, spaced 4" center-to-center

¾" dowel 3" long

¾" hole 1" deep, drilled at 30° angle

CLAMPING BENCH

The best place to hang your clamps is at the clamping bench with overhead rack, shown in (**10–29**). The rack easily holds several dozen small clamps of any type. You can hang your longer bar clamps and pipe clamps directly on the bench itself for convenient, easy access.

Building Instructions

1 Cut the 2 × 4 for the overhead rack to the desired length. (We cut ours slightly shorter than the length of the workbench.) Then drill holes for the two ¼" threaded hanger rods, where shown on **10–30**. If you want to hang the rack crosswise to the joists, adjust the length of the 2 × 4 and/or rod-hole locations to correspond with joist

Bench Pegs and Rack

1 Drill the 30°-angled holes for the pegs, using an adjustable drill guide or the homemade drill guide, shown on **10–31**. Space the holes 4" apart.

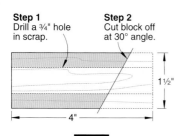

Step 1
Drill a ¾" hole in scrap.

Step 2
Cut block off at 30° angle.

1½"

4"

10–31.
30° ANGLE DRILL GUIDE

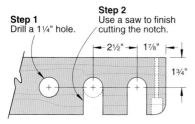

Step 1
Drill a 1¼" hole.

Step 2
Use a saw to finish cutting the notch.

2½" — 1⅞"

1¾"

10–32.
NOTCH

2 Cut the 2 × 4 rack for the end of the bench to the desired length. To make the notches, use a 1¼" hole saw or drill bit, and cut the holes 1½" center to center; then complete the cutout with a handsaw (see **10–32**). Attach the rack to the bench, where shown on **10–30**.

❖

SAFE CHISEL STORAGE

You can create a custom drawer insert (**10–33** and **10–34**) that will protect chisel edges.

Plan the Layout

Add up the handle diameters of the chisels you want to store, and subtract the total from the inside width of the drawer. Divide that number by the spaces that you'll need to get an approximate spacing distance. Our set includes 10 chisels, so we needed to account for 11 spaces. You can vary the spacing by making the end spaces bigger or smaller than the others.

Make the Supports

1 Place your longest chisel with the handle flat on your workbench, and measure the gaps beneath the ferrule and the blade. Also measure the length of the blade.

2 Now, make blade and ferrule supports to match those dimensions. For the blade support, plane a board to ⅝" thickness, rip it to 4¾", and then cut it to a length 1/16" less than the width of the drawer. For the ferrule support, rip a strip 1" wide. Plane it to ½" thickness, and cut it to the width of the drawer. Mark the handle center points on the edge of each support. Extend the points across the ferrule support.

3 Make a customized recess for each chisel by marking on the edge of the blade support the width of each blade plus 1/16", centered on the previously marked points. At your tablesaw, set a dado

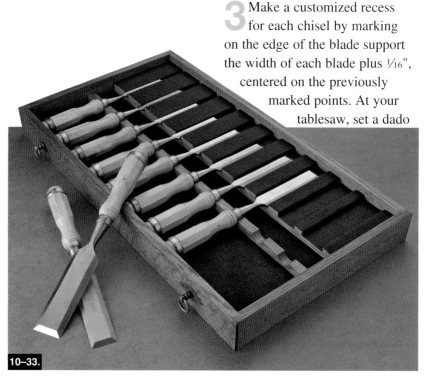

10–33.

Keep your hand chisels safe from damage, well-organized, and looking good with a simple series of felt-lined dadoes.

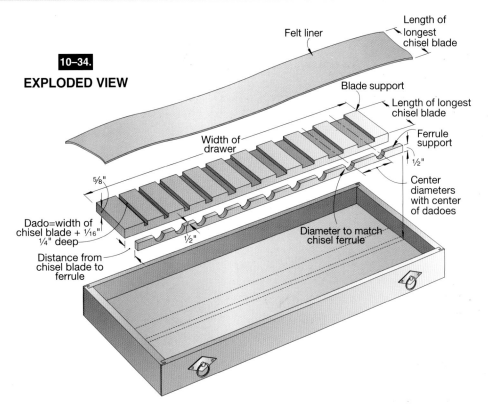

10–34.
EXPLODED VIEW

Felt liner

Length of longest chisel blade

Blade support

Length of longest chisel blade

Width of drawer

Ferrule support

½"

Center diameters with center of dadoes

5/8"

Dado=width of chisel blade + 1/16"
¼" deep

½"

Diameter to match chisel ferrule

Distance from chisel blade to ferrule

---

blade to cut ¼" deep, and cut a dado between each set of marks. Use the appropriate chisel and a scrap of felt to test for a snug fit.

4 Using a drill bit of the same diameter as the ferrule (we used a ¾" Forstner), drill a hole at each line on the ferrule support stock. To do this, clamp a fence on your drill-press table to keep the holes lined up, and locate the fence so that the holes will hold the ferrules at the desired height. (In our case, we centered the holes ½" from the edge to leave 1/8" of wood under each ferrule.) After drilling the holes, sand both faces. Rip the stock through the center of the holes to leave a series of notches.

Add Finish and Felt

1 Apply a stain or clear finish to match the tool chest or storage unit in which you'll place the drawer. Coat the top and sides of the ferrule support, but only the front edge of the blade support.

2 Now cut an oversized piece of felt from the fabric store. Mask the edges of the blade support with tape, and then spray adhesive on the top and ends. Remove the tape, and apply the felt, fitting it tightly into the dadoes and around the ends. After the adhesive sets, turn the support upside-down, and trim the felt along the edges and the ends with a knife.

3 If your drawer bottom is unlined, cut a piece of poster board 1/8" smaller in length and width than the inside dimensions of the drawer. Apply double-faced tape to the bottom of the poster board around its perimeter. Cut felt 2" larger in length and width than the poster board. Center the poster board, top down, on the felt, trim off the corners of the felt, and fold it onto the tape. Flip the poster board over, and press it into the drawer.

4 Place strips of cloth-backed, double-faced tape on the blade support bottom, and press it into the drawer. Finally, apply double-faced tape to the bottom of the ferrule support, and press it into place.

CHISEL STAND

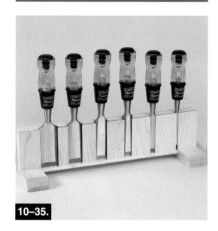

10–35.

You can use this practical stand (**10–35** and **10–36**) as a freestanding unit, or remove the feet and fasten it to your workshop wall. Either way, this project displays your chisels proudly and protects their finely honed edges.

Note: We built our stand for a set of Stanley no. 60 chisels. The dimensions might vary slightly for your chisel set.

❖

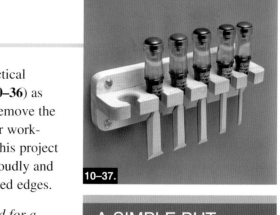

10–37.

A SIMPLE BUT TOP-NOTCH CHISEL RACK

Sometimes, the best ideas are also the simplest. For this handy little shop project (**10–37** and **10–38**, *opposite page*), we went to our scrap pile for the material and invested about a half hour of shop time. Now, we have a top-notch rack for our chisel set.

Note: All stock is ¾" thick. Hole sizes may vary with different brands of chisels.

❖

WRENCH RACK

We built our racks (**10–39** and **10–40**) to handsomely hold a 16-piece (¼" to 1¼") Stanley combination wrench set. As described here and on the Notch Detail in **10–40**, you may need to change a few dimensions for your particular set. Also, depending upon how much space you have

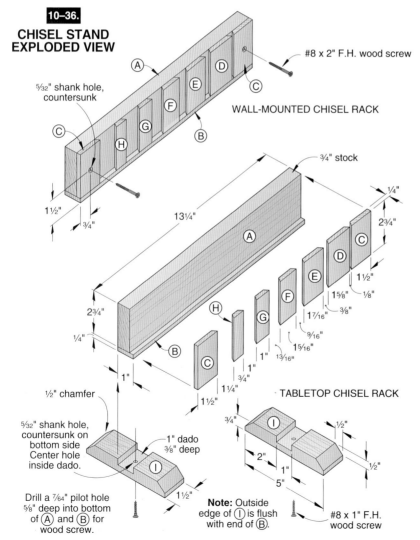

10–36.

CHISEL STAND EXPLODED VIEW

⁵/₃₂" shank hole, countersunk

#8 x 2" F.H. wood screw

WALL-MOUNTED CHISEL RACK

¾" stock

1½"
¾"
13¼"
2¾"
¼"
¼"
2¾"
1½"
1⅝"
⅛"
1⁷/₁₆"
⅜"
9/₁₆"
15/₁₆"
1"
13/₁₆"
1"
¾"
1¼"
1½"
1"
½" chamfer

⁵/₃₂" shank hole, countersunk on bottom side Center hole inside dado.

1" dado ⅜" deep

TABLETOP CHISEL RACK

¾"
½"
2"
1"
½"
5"
1½"

Drill a ⁷/₆₄" pilot hole ⅝" deep into bottom of Ⓐ and Ⓑ for wood screw.

Note: Outside edge of Ⓘ is flush with end of Ⓑ.

#8 x 1" F.H. wood screw

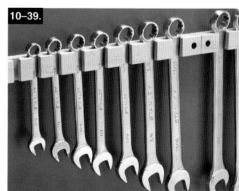

10–39.

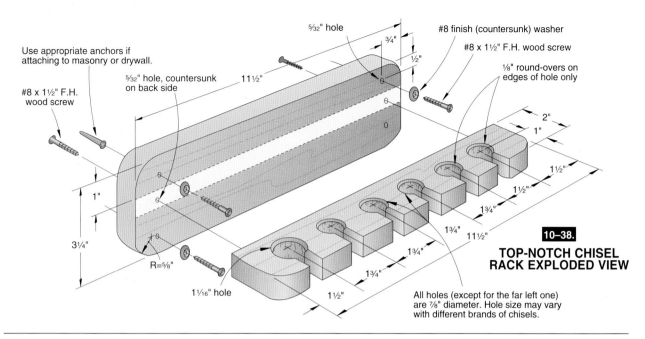

Use appropriate anchors if attaching to masonry or drywall.

#8 x 1½" F.H. wood screw

⁵⁄₃₂" hole, countersunk on back side

⁵⁄₃₂" hole

11½"

¾"

½"

#8 finish (countersunk) washer

#8 x 1½" F.H. wood screw

⅛" round-overs on edges of hole only

2"

1"

1½"

1½"

1¾"

1¾"

1¾"

1¾"

11½"

1"

3¼"

R=⅝"

1¹⁄₁₆" hole

1½"

All holes (except for the far left one) are ⅞" diameter. Hole size may vary with different brands of chisels.

10–38.
TOP-NOTCH CHISEL RACK EXPLODED VIEW

for hanging your wrenches, you may want to place the racks end to end, as shown in **10–39**, or hang one rack under the other.

Note: In our research, we discovered that several manufacturers offer slightly different wrench designs and sell sets containing varying numbers of wrenches. For this reason, the size and number of openings in the racks you make may need to differ from the ones shown here.

To build the racks, cut the front and back pieces to size. Measure the width (A) of each wrench, and to this width add ¹⁄₁₆" on the front face of the front pieces. The width of the dadoes in the back piece will be the width of the wrench (A) plus ⁵⁄₁₆". Now, measure the thickness (B) of each wrench, and cut the dadoes to this depth plus ¹⁄₁₆" on the front face of the back pieces.

Glue the pieces together, and then cut away the waste (shaded areas) from the front pieces, where shown on **10–40**.

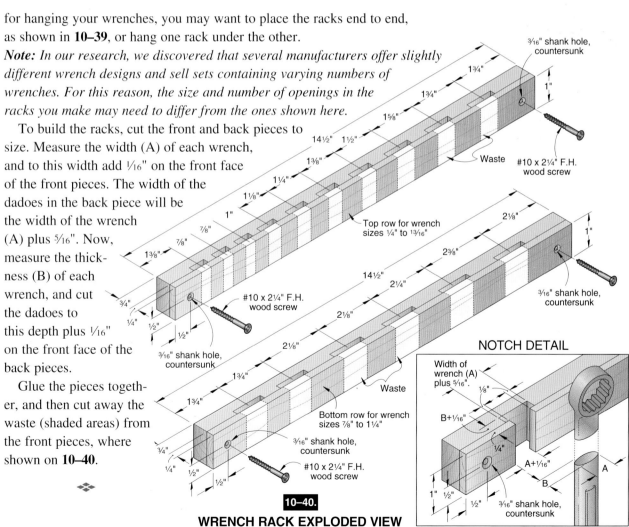

³⁄₁₆" shank hole, countersunk

1¾"

1¾"

1⅝"

1"

14½"   1½"

1⅜"

1¼"

1⅛"

1"

⅞"

⅞"

1⅜"

¾"

¼"   ½"

½"

³⁄₁₆" shank hole, countersunk

Waste

#10 x 2¼" F.H. wood screw

Top row for wrench sizes ¼" to 1³⁄₁₆"

2⅛"

1"

2⅜"

2¼"

#10 x 2¼" F.H. wood screw

14½"

2¼"

2⅛"

2⅛"

³⁄₁₆" shank hole, countersunk

1¾"

1¾"

¾"

¼" ½"

½"

³⁄₁₆" shank hole, countersunk

Waste

Bottom row for wrench sizes ⅞" to 1¼"

³⁄₁₆" shank hole, countersunk

#10 x 2¼" F.H. wood screw

10–40.
WRENCH RACK EXPLODED VIEW

NOTCH DETAIL

Width of wrench (A) plus ⁵⁄₁₆".

⅛"

B+¹⁄₁₆"

¼"

A+¹⁄₁₆"

A

B

1"   ½"

½"

³⁄₁₆" shank hole, countersunk

SIMPLE SAW RACK

Tired of hanging your handsaws on perforated hardboard hooks or worse, nails? If so, dress up your shop with this sturdy maple organizer (**10–41** and **10–42**). As dimensioned on **10–42**, we glued ⅛"-thick spacers between the supports for regular handsaws and ⅜"-thick spacers for backsaws. Then, we glued the top between the ends and on top of the support strip. For safety, hang your saws with the teeth facing the wall.

10–43.

READY-FOR-WORK TOOL HOLDERS

Wall-mounted perforated hardboard provides ideal tool storage—as long as you can get the hanger "pegs" to stay put. Faced with that challenge, we designed tool holders that attach firmly, but take only a minute or two to move to a new, more convenient location on the perforated hardboard (**10–43** and **10–44**, *opposite page*).

To make the pliers holder, mark slots, where shown on **10–44**, and drill ½" holes at each end of the marked slots. Drill slightly over-lapping holes to remove the stock between the end holes, and smooth up the sides of the slots with a chisel. Then, attach the wrought steel desktop fasteners, available from woodworking suppliers, as shown.

The 5/0 handscrew-clamp holder, shown in **10–44**, mounts to the perforated hardboard with a short length of ¼ × 1" aluminum bar stock and two hexhead sheet-metal screws. To store larger size handscrew clamps, make the wooden block longer, and wider.

By using varying lengths of all-thread rod, you can create multi-purpose hangers for bar clamps or individual tools, such as the adjustable wrenches in **10–43**. We dipped the all-thread rod in plastic coating for protection from sharp edges and to provide a no-slip surface to keep tools in place.

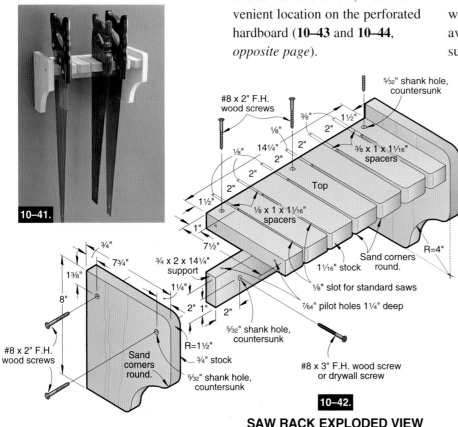

10–41.

10–42.

SAW RACK EXPLODED VIEW

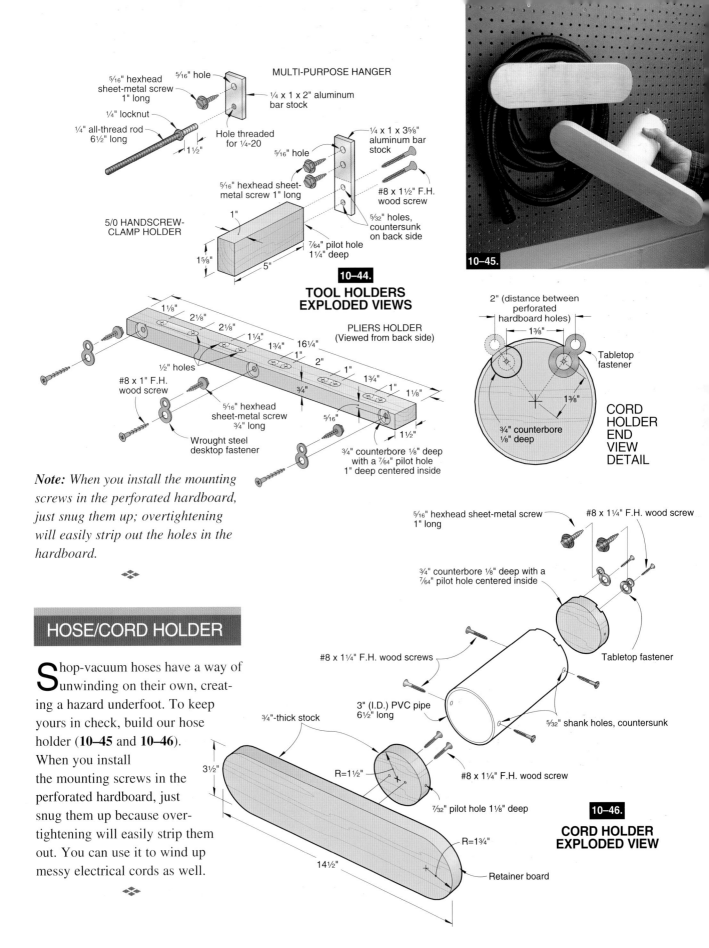

MULTI-PURPOSE HANGER

5/16" hexhead sheet-metal screw 1" long

5/16" hole

1/4 x 1 x 2" aluminum bar stock

1/4" locknut

1/4" all-thread rod 6½" long

1½"

Hole threaded for 1/4-20

1/4 x 1 x 3⅝" aluminum bar stock

5/16" hole

5/16" hexhead sheet-metal screw 1" long

#8 x 1½" F.H. wood screw

5/32" holes, countersunk on back side

5/0 HANDSCREW-CLAMP HOLDER

1"

1⅝"

5"

7/64" pilot hole 1¼" deep

10–44.

TOOL HOLDERS EXPLODED VIEWS

10–45.

PLIERS HOLDER
(Viewed from back side)

1⅛"   2⅛"   2⅛"   1¼"   1¾"   16¼"   1"   2"   1"   1¾"   1"   1⅛"

¾"

½" holes

#8 x 1" F.H. wood screw

5/16" hexhead sheet-metal screw ¾" long

Wrought steel desktop fastener

5/16"

1½"

¾" counterbore ⅛" deep with a 7/64" pilot hole 1" deep centered inside

2" (distance between perforated hardboard holes)

1⅜"

Tabletop fastener

1⅜"

¾" counterbore ⅛" deep

CORD HOLDER END VIEW DETAIL

Note: When you install the mounting screws in the perforated hardboard, just snug them up; overtightening will easily strip out the holes in the hardboard.

❖❖

HOSE/CORD HOLDER

Shop-vacuum hoses have a way of unwinding on their own, creating a hazard underfoot. To keep yours in check, build our hose holder (**10–45** and **10–46**). When you install the mounting screws in the perforated hardboard, just snug them up because over-tightening will easily strip them out. You can use it to wind up messy electrical cords as well.

❖❖

5/16" hexhead sheet-metal screw 1" long

#8 x 1¼" F.H. wood screw

¾" counterbore ⅛" deep with a 7/64" pilot hole centered inside

#8 x 1¼" F.H. wood screws

Tabletop fastener

3" (I.D.) PVC pipe 6½" long

5/32" shank holes, countersunk

¾"-thick stock

3½"

R=1½"

#8 x 1¼" F.H. wood screw

7/32" pilot hole 1⅛" deep

10–46.

CORD HOLDER EXPLODED VIEW

14½"

R=1¾"

Retainer board

GLUE CADDY

If you want to get your shop better organized, make this handy holder (**10–47** to **10–49**) for your supplies.

Forming the Disks

1 Cut two pieces of ¾" stock (we used a pine 1 × 12) to 11" square. If you don't have stock this wide, laminate narrower pieces together.

2 Find the center of one of the squares by marking diagonals. Mark a 10"-diameter circle at the center.

3 Stick the two squares together, marked circle up, with double-faced tape. Cut the discs to shape.

4 Arrange your glue bottles and accessories on top of the discs. Trace around each container. (We added 35-mm film canisters for holding cotton swabs, brushes, and other items used to apply glue into those hard-to-get-at places. We also included a foam cup for soaking the glue brushes after each use.)

5 To prevent drilling into your drill-press table, mount a piece of scrap material to it. Drill a ¾" hole through the center of both discs. Then, if you plan on using the film canisters, as shown in **10–47**,

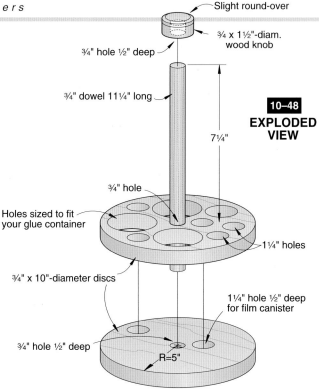

Slight round-over

¾ x 1½"-diam. wood knob

¾" hole ½" deep

¾" dowel 11¼" long

10–48
EXPLODED VIEW

7¼"

¾" hole

Holes sized to fit your glue container

1¼" holes

¾" x 10"-diameter discs

1¼" hole ½" deep for film canister

¾" hole ½" deep

R=5"

We used a circle cutter for cutting the large-diameter holes.

10–49.

switch bits, and drill 1¼" holes through the first disc and ½" deep into the second (We used a Forstner bit.) Separate the discs, and remove the tape.

6 With the 1¼" bit still chucked into your drill press, drill holes through the top disc for additional film canisters. Now, using a circle cutter or holesaw, bore the larger holes, as shown in **10–48**. Bore through the top disc and just a fraction into the scrap top mounted to your drill-press table. Sand both discs and the openings smooth.

Finishing Up

1 Cut a piece of ¾" dowel stock 11¼" long for the handle.

10–47.

2 Slide the top disk onto the ¾" dowel, and position its top face 7¼" down from the top end of the dowel. Mark the position of the disc (top and bottom) on the dowel. Slide the disc away from the marked area, and apply glue between the marked lines on the dowel. Now, slide the disc back onto the glued area, turn it upside down, and run a small bead of glue around the dowel below the top disc. Later, after the top disc is firmly glued in place, glue the dowel into the ¾" hole in the base disc, aligning the holes in the discs.

3 Mark a 1½" circle on a piece of ¾" stock for the knob. Bore a ¾" hole ½" deep at the centerpoint. Cut the knob to shape on a bandsaw, sand it smooth, and glue it to the top end of the dowel. Finally, apply a clear finish to all the parts.

❖❖❖

LATHE TOOL RACK

Long-handled tools can be cumbersome, especially when your try to store them in vertical racks. As an alternative to vertical storage, this wall-hung organizer (**10–50** and **10–51**) makes infinite sense. It even features a tray at the bottom to hold your turning accessories. Get organized, while protecting your tool investment, with this easy-to-make unit.

❖❖

10–50.

Although it looks like a fine gun rack, this storage unit is designed to keep long-handled turning tools at the ready.

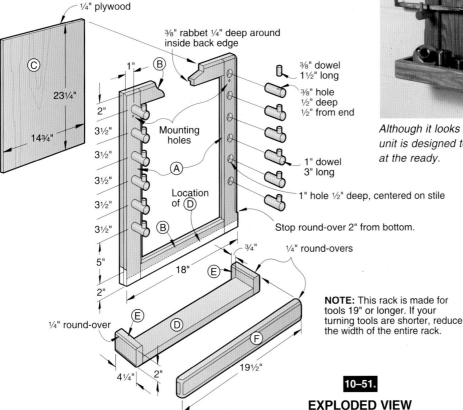

¼" plywood

Ⓒ
23¼"
14¾"

3⁄8" rabbet ¼" deep around inside back edge

1"
Ⓑ
2"
3½"
Mounting holes
3½"
3½"
Ⓐ
3½"
Location of Ⓓ
3½"
3½"
Ⓑ
5"
18"
Ⓔ
2"

¼" round-over

Ⓔ
Ⓓ
Ⓕ
4¼"
2"
19½"

3⁄8" dowel 1½" long

3⁄8" hole ½" deep ½" from end

1" dowel 3" long

1" hole ½" deep, centered on stile

Stop round-over 2" from bottom.

¾"
¼" round-overs

NOTE: This rack is made for tools 19" or longer. If your turning tools are shorter, reduce the width of the entire rack.

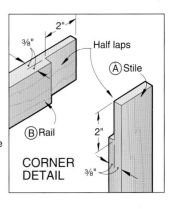

2"
3⁄8"
Half laps
Ⓐ Stile
Ⓑ Rail
2"
CORNER DETAIL
3⁄8"

10–51.
EXPLODED VIEW

LATHE TOOL BOARD

You will be able to quickly grab needed tools and accessories from this convenient lathe-tool board (**10–52** and **10–53**).

1 Cut the mounting board (A), tool holder (B), magnet support (C), and slanted accessory holders (D) to size, bevel-ripping the front edge of C at 10° and the back edge of each D at 30°. Cut the trim pieces (E, F) to size.

2 Drill ¼"-deep holes in the tool holder (B), ¼" larger in diameter than the butt of your lathe-tool handles. Drill ¼"-deep holes in the holders (D) for your circular sanding attachments and live centers. Drill holes the same diameter as the shanks of your accessories. Radius the front corners of all the holders, and rout a ¼" round-over along all edges except the back.

3 Paint the board and let dry. With the board flat, position the tool holder, magnet support, and accessory holders; then position your turning accessories on the open area. Mark and drill dowel holes ⅜" deep to hang these items, and cut dowels to length.

4 Nail the trim pieces (E, F) to the board. Apply polyurethane to all trim except the front edge of the magnet support. Affix the adhesive-backed magnetic strip, available at crafts supply stores, to the front of the magnetic support.

❖

10–52.

Your lathe work will take a turn for the better when you can quickly grab needed tools and accessories from this convenient lathe-tool board.

Materials List for Lathe-Tool Board

PART	FINISHED SIZE			Mtl.	Qty.
	T	W	L		
A mounting board	¾"	36"	48"	PL	1
B tool holder	1½"	3"	46"	F	1
C support	¾"	1"	34¾"	F	1
D holders	1½"	3"	6"	F	2
E trim	¾"	¾"	36"	P	2
F trim	¾"	¾"	49½"	P	2

Material Keys: PL = plywood; P = pine; F = fir.
Supplies: # 10 x 1¼" flathead wood screws; ⅜" dowel stock; ¾" dowel stock; 4d finish nails; polyurethane; flexible magnetic strip.

10–53.
EXPLODED VIEW

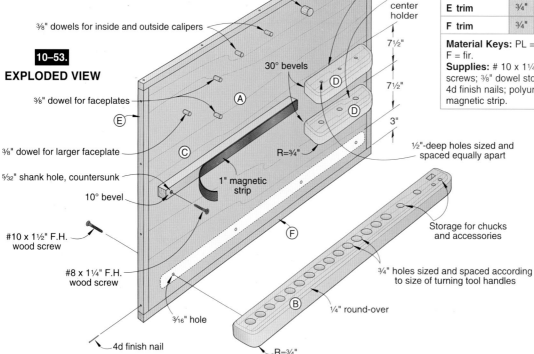

¾" dowel for dust mask

¾" dowel for shield

⅜" dowels for inside and outside calipers

⅜" dowel for faceplates

⅜" dowel for larger faceplate

5/32" shank hole, countersunk

10° bevel

#10 x 1½" F.H. wood screw

#8 x 1¼" F.H. wood screw

3/16" hole

4d finish nail

30° bevels

R=¾"

1" magnetic strip

Circular-sander holder

Live-center holder

7½"

7½"

3"

½"-deep holes sized and spaced equally apart

Storage for chucks and accessories

¾" holes sized and spaced according to size of turning tool handles

¼" round-over

R=¾"

CHAPTER 11

Stands, Stools, and Supports

A WOODWORKING WORKSHOP IS *never quite complete, it seems. Even if well-planned and laid out from the very beginning, the addition of new and different tools frequently means building something to put them on or stash them in. That's why in this chapter we'll show you how to craft three clever options, each of which you can adapt to other machines. Then, because machine mobility is often critical in a shop, we have a plan for a mobile base that'll save you big bucks over the ready-made variety.*

Of course, sitting yourself down in the woodworking shop is also important. So to that end, we've included two clever shop-seating options. Then, for the times when you need extra height, or when a youngster comes by for a woodworking lesson, you'll find helpful step-by-step instructions for a great-looking step stool.

Finally, to help you out when cutting sheet goods or crosscutting long lengths, we added a couple of work supports and sawhorses that hide away when not needed. To round out these final pages, you'll find plans for a pair of rugged, but basic, sawhorses that'll last for many years. Have fun building any or all of the great projects you'll see on the following pages.

LEG STAND FOR STATIONARY TOOLS

11–1.

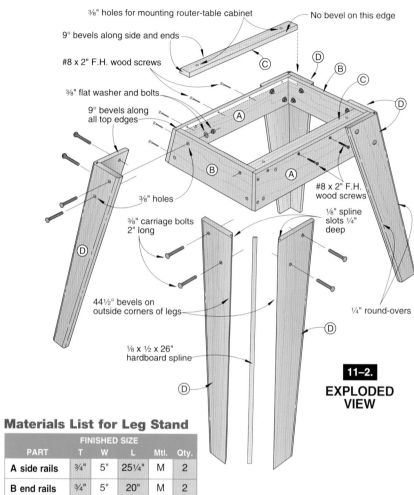

³⁄₈" holes for mounting router-table cabinet

No bevel on this edge

9° bevels along side and ends

#8 x 2" F.H. wood screws

³⁄₈" flat washer and bolts

9° bevels along all top edges

#8 x 2" F.H. wood screws

³⁄₈" holes

¹⁄₈" spline slots ¹⁄₄" deep

³⁄₈" carriage bolts 2" long

44¹⁄₂° bevels on outside corners of legs

¹⁄₄" round-overs

¹⁄₈ x ¹⁄₂ x 26" hardboard spline

11–2.
EXPLODED VIEW

A leg stand, with its tapered and slightly splayed legs, makes a sturdy, yet good-looking base for any machine. Use the dimensions in the Materials List to make one, as shown in **11–1** and **11–2**, or alter the lengths of the parts and custom-make one to fit a specific machine in your shop. To size the parts for any machine, follow these guidelines:

1 Subtract 1⅝" from the length of the machine for the length of the side rails (A).

2 Subtract 3¼" from the width of the machine for the length of the end rails (B).

3 Subtract 3" from the length of the side rails (A) for the length of the cleats (C).

Materials List for Leg Stand

PART	FINISHED SIZE				
	T	W	L	Mtl.	Qty.
A side rails	¾"	5"	25¼"	M	2
B end rails	¾"	5"	20"	M	2
C cleats	¾"	1½"	22¼"	M	2
D leg halves	¾"	6½"	27⅛"	M	8

Material Key: M = Maple.
Supplies: ⅛" tempered hardboard; ³⁄₈" carriage bolts 2" long (16); flat washers and nuts; #8 x 2" flathead wood screws (16); finish.

4 Multiply your desired height by 1.074 for the length of the blanks for the leg halves (D). Screw a piece of plywood to the cleats and bolt your machine to it. Here's how to put your leg stand together.

After cutting the rails (A/B) to the dimensions shown on **11–3**, screw them together to form a rectangular frame. Fit the cleats (C) into the rail frame and screw them in place. Set the frame aside.

Cut blanks for the leg halves (D) to size and, before forming the legs, drill the holes where shown. Remember to make mirrored pairs. Bevel-rip the mating edges. Then, with the saw blade tilted at the same angle, cut the spline slots in the bevels. Next, make the angled and beveled cuts at the top and bottom of each leg half. Do not cut the leg tapers until the halves have been glued together.

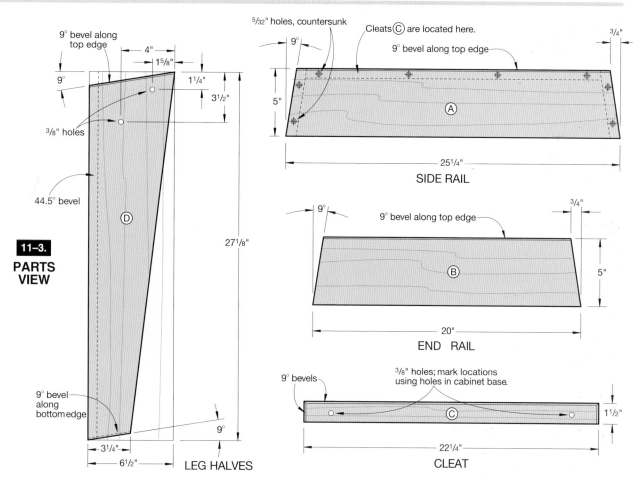

11–3.

PARTS VIEW

9° bevel along top edge

4"

1⁵⁄₈"

9°

1¹⁄₄"

3¹⁄₂"

³⁄₈" holes

44.5° bevel

Ⓓ

27¹⁄₈"

9° bevel along bottom edge

9°

3¹⁄₄"

6¹⁄₂"

LEG HALVES

⁵⁄₃₂" holes, countersunk

Cleats Ⓒ are located here.

9°

9° bevel along top edge

³⁄₄"

5"

Ⓐ

25¹⁄₄"

SIDE RAIL

9°

9° bevel along top edge

³⁄₄"

Ⓑ

5"

20"

END RAIL

9° bevels

³⁄₈" holes; mark locations using holes in cabinet base.

Ⓒ

1¹⁄₂"

22¹⁄₄"

CLEAT

11–4.

Clamping the legs (D) to the frame (A/B/C) holds them at the proper angle while you finish clamping their length.

SUPER-STURDY SCROLLSAW STAND

E levate your benchtop scroll-saw to new heights with this sturdy shop-made stand (**11–5** to **11–7**). It provides solid support directly beneath the saw to help minimize vibration.

We recommend that you elevate the scrollsaw table to a comfortable height, typically at elbow level. So, you may have to adjust the height of the uprights (C) slightly. We dry-clamped our pieces together before drilling the mounting holes to verify a comfortable working height.

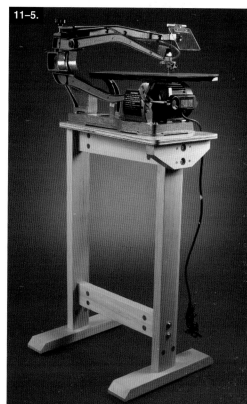

11–5.

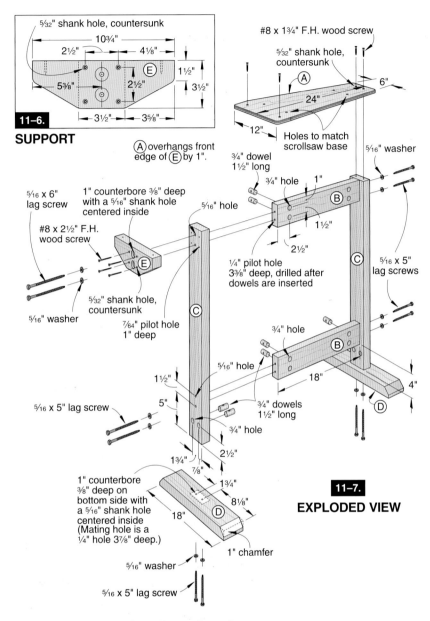

11–6.

SUPPORT

Ⓐ overhangs front edge of Ⓔ by 1".

11–7.

EXPLODED VIEW

Materials List for Scrollsaw Stand

PART	FINISHED SIZE			Mtl.	Qty.
	T	W	L		
A top	¾"	12"	24"	P	1
B cross members	1½"	3½"	18"	C	2
C uprights	1½"	3½"	31½"	C	2
D feet	1½"	3½"	18"	C	2
E support	1½"	3½"	10¾"	C	1

Materials Key: P = plywood; C = choice of fir or pine 2 x 4 stock (handpicked from defect-free pieces).
Supplies: ⁵⁄₁₆ x 5" lag screws (10); ⁵⁄₁₆ x 6" lag screws (2); ⁵⁄₁₆ x 5" flat washers (12); 38 x 1¾" flathead wood screws; 38 x 2½" flathead wood screws; ¾" dowel stock; clear finish.

For additional strength, drill the mounting holes, and drive the lag screws through the uprights (C) and feet (D) and into the ¾" dowels in the mating pieces.

❖

MOBILE MITERSAW STAND

Finally, the ultimate mitersaw center has arrived. This sturdy plywood unit (**11–8** and **11–9**) rolls around and locks in place where needed. Fold-down table extensions let you manage long pieces for crosscutting. And there's more, including built-in waste disposal for cutoffs and sawdust, and a unique leveling system for parking the unit securely on an uneven floor.

First, Build a Sturdy Carcase

1 Cut the cabinet sides (A), bottom (B), and back (C) to the dimensions found in the Materials List.

Note: We used birch plywood for this and other shop projects for several good reasons: Compared to fir plywood, it's flatter, contains fewer voids and patches, and paints and finishes better.

2 From solid stock (we used birch), cut enough ¼ x ¾" material for the sides' top and front edge banding pieces (D, E), the bottom's front edge banding (F), and the back edge banding (F). (See **11–10** for reference. Then cut the parts to length.)

3 Glue and clamp the side top edge bandings (D) on the sides first; then follow with the remaining bandings. Sand these applied pieces flush.

Materials List for Mobile Mitersaw Stand

PART		FINISHED SIZE			Mtl.	Qty.
		T	W	L		
A	cabinet sides	¾"	24¾"	29⅝"	BP	2
B	cabinet bottom	¾"	24⅜"	29¼"	BP	1
C	cabinet back	¾"	29¼"	29⅝"	BP	1
D	side top edge bandings	¼"	¾"	24¾"	B	2
E	side front edge bandings	¼"	¾"	29⅞"	B	2
F	edge bandings for parts B, C	¼"	¾"	29¼"	B	2
G	top shelf supports	¾"	3½"	23½"	B	2
H*	top shelf	1½"	23½"	28½"	BP	1
I	top shelf edge strip	¾"	1½"	28½"	B	1
J	vertical cleats	¾"	1¼"	23½"	B	2
K	horizontal cleats	¾"	1¼"	23½"	B	2
L	rod support	1½"	3½"	22¼"	P	1
M	knobs	¾"	2½" dia.		B	2
N	door	¾"	27¾"	23½"	BP	1
O	door vertical edge bandings	¼"	¾"	23½"	B	2
P	door top edge banding	¼"	¾"	28¼"	B	1
Q**	saw tables	3/16"	10¹¹⁄₁₆"	3¹³⁄₁₆"	B/L	2
R**	saw table fences	¾"	6⅞"	3⅞"	BP	2
S	saw table fence edge banding	¼"	¾"	3⅞"	B	2
T	folding tables	13/16"	10⁹⁄₁₆"	33"	B/L	2
U	folding table supports	¾"	15½"	32¾"	BP	2
V	folding table support side edge bands	¼"	¾"	4⅜"	B	2
W	folding table support top edge bands	¼"	¾"	33"	B	2

*Cut parts marked with an * oversized. Trim to finished size according to the instructions.
**Parts are cut slightly undersized and reach finished size when plastic laminate is applied
**Part dimensions subject to change due to mitersaw model used. (See 11–12.)
Materials Key: BP = Baltic birch plywood; B = birch; P = pine or fir; B/L = Baltic birch plywood with plastic laminate.
Supplies: 2⅜" T-nuts; ⅜" threaded rod, 34" long (2); #8 x 1¼" flathead wood screws; #8 x 2" flathead wood screws; #8 x 2½" flathead wood screws; #8 x 3" flathead wood screws; 1½" continuous hinge, 33" long (2); 1½" continuous hinge, 10½" long (2); ⅜ x 1½" hexhead bolts; ⅜" washers and nuts; 4" barrel bolts (2); 3½" draw catches (2); 1 pair of 1 x 3 nonmortise hinges; 4" wire pull; magnetic catch and catch plates; 18-gallon storage box; clear finish.
Buying Guide: HD fixed casters (3) and HD double-locking (braking) swivel casters (2), 4"; 1 x 3 door hinges (one pair); 4" wire pull; magnetic catch and strike.

11–8.

With the folding tables down, the mitersaw cabinet stores easily against a shop wall.

11–9.

4 Cut the rabbets on parts A and C, where shown.

5 Referring to **11–10** and **11–11**, drill the ⁷⁄₁₆" counterbored holes for the threaded rod and the ⅜" holes for the casters in the bottom (B). (We used the casters as hole guides.) Now hammer ⅜" T-nuts in the holes for the threaded rods in the bottom side of the carcase bottom.

6 Glue and assemble parts A, B, and C. Then, drill the countersunk holes in the bottom and back, and secure the joining parts with screws.

11–10.
CARCASE

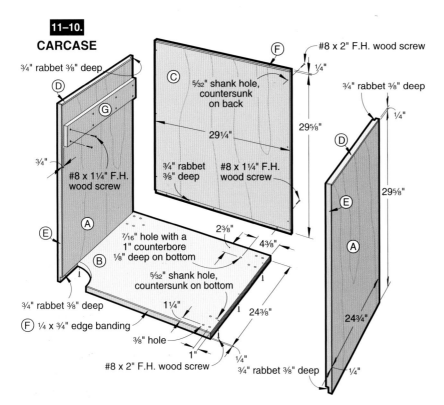

¾" rabbet ⅜" deep

#8 x 2" F.H. wood screw

¼"

⁵⁄₃₂" shank hole, countersunk on back

¾" rabbet ⅜" deep

¼"

29¼"

29⅝"

¾"

#8 x 1¼" F.H. wood screw

¾" rabbet ⅜" deep

#8 x 1¼" F.H. wood screw

29⅝"

⁷⁄₁₆" hole with a 1" counterbore ⅛" deep on bottom

2⅜"

4⅜"

⁵⁄₃₂" shank hole, countersunk on bottom

¾" rabbet ⅜" deep

1¼"

24⅜"

24¾"

① ¼ x ¾" edge banding

⅜" hole

#8 x 2" F.H. wood screw

1"

¼"

¾" rabbet ⅜" deep

¼"

Build and Install a Heavy-Duty Top Shelf

Note: *The top shelf height shown is for a particular 12" mitersaw. Depending on the mitersaw model you use, some dimensions and part sizes will change. These we've indicated in* **11–12.**

1 Cut the top shelf supports (G) to size and drill the counter-sunk holes. We'll install these later.

2 Cut two pieces of ¾" birch plywood to 24 x 29" for the top shelf (H). Glue and clamp these pieces together, keeping the edges and ends flush. Then cut the lamination to size. See **11–13.**

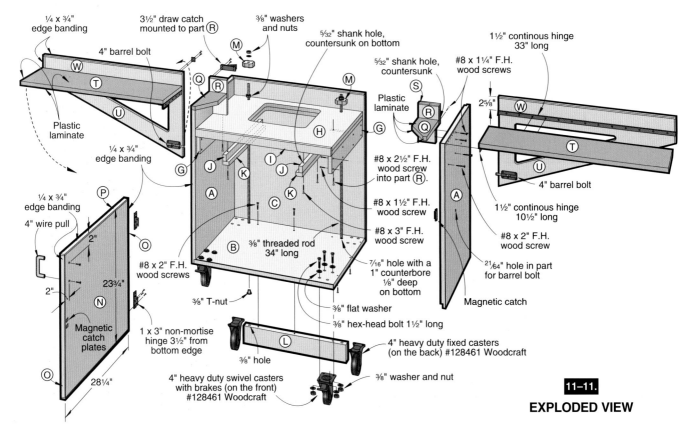

¼ x ¾" edge banding

3½" draw catch mounted to part ®

⅜" washers and nuts

⁵⁄₃₂" shank hole, countersunk on bottom

1½" continous hinge 33" long

4" barrel bolt

⁵⁄₃₂" shank hole, countersunk

#8 x 1¼" F.H. wood screws

Plastic laminate

2⅝"

¼ x ¾" edge banding

Plastic laminate

#8 x 2½" F.H. wood screw into part ®.

4" barrel bolt

¼ x ¾" edge banding

4" wire pull

#8 x 1½" F.H. wood screw

1½" continous hinge 10½" long

2"

#8 x 3" F.H. wood screw

#8 x 2" F.H. wood screw

23¾"

#8 x 2" F.H. wood screws

⅜" threaded rod 34" long

⁷⁄₁₆" hole with a 1" counterbore ⅛" deep on bottom

²¹⁄₆₄" hole in part for barrel bolt

2"

Magnetic catch plates

⅜" T-nut

⅜" flat washer

⅜" hex-head bolt 1½" long

Magnetic catch

28¼"

1 x 3" non-mortise hinge 3½" from bottom edge

⅜" hole

4" heavy duty fixed casters (on the back) #128461 Woodcraft

4" heavy duty swivel casters with brakes (on the front) #128461 Woodcraft

⅜" washer and nut

11–11.
EXPLODED VIEW

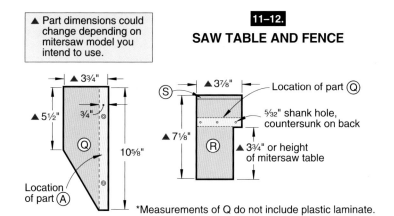

▲ Part dimensions could change depending on mitersaw model you intend to use.

11–12.
SAW TABLE AND FENCE

*Measurements of Q do not include plastic laminate.

3 From ¾" stock, cut the top shelf edge strip (I) to size. Determine the width by measuring the thickness of the top shelf. (Ours measured 1½".)

4 Glue and clamp the top shelf edge strip (I) to the front edge of the shelf top (H). Sand smooth.

5 Place the top shelf face down and lay out the opening on the bottom surface. Also, mark the locations of the ⅜" holes for the all-thread rod and the ⁵⁄₃₂" holes that are countersunk on the bottom (**11–13**).

Next, drill blade start holes in the four corners of the layout. Jigsaw out the opening, cutting just inside the line. Sand to the line using a drum sander or oscillating spindle sander.

6 To locate the shelf top at the exact height for your mitersaw, you'll need to cut two alignment blocks first. To make these, place your mitersaw on a flat surface. Then, take a 6" piece of ¾" birch plywood and glue a piece of scrap plastic laminate to it. This should be a scrap piece of the same plastic laminate you will later glue to the saw table surfaces.

Next, cut two pieces of scrap (ours measured ¾ x 4 x 6") and place one on top of the laminated scrap and alongside your mitersaw. Mark the precise location of the saw's table surface on the scrap and cut a ¾" notch at this location. Now, hold this "alignment block" against the saw, as shown in **11–14**. Once satisfied, cut the second alignment block the same way.

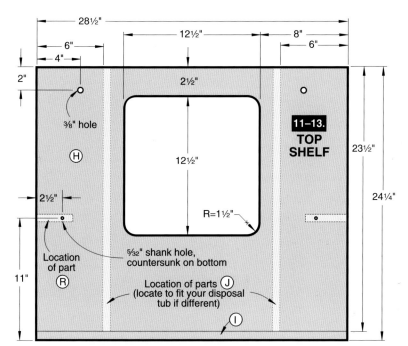

11–13.
TOP SHELF

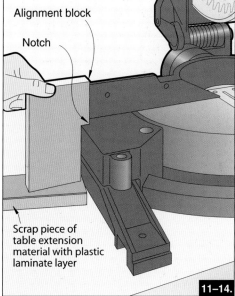

11–14.

Working off laminated scrap and the height of your mitersaw table, mark the notch location in the alignment blocks needed to fix the top shelf height even with the saw.

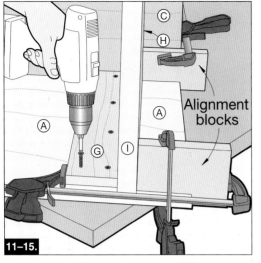

11–15.

Lay the cabinet carcase down on its side; then gather the top shelf supports (G), top shelf (H), and the alignment blocks, as shown in **11–15**, on the inside of the cabinet. Clamp these in place. With the shelf support flush to the cabinet back, screw (don't glue) the top shelf support in place. (By just screwing the supports in place, you later can adjust the top shelf height should you change mitersaws.) Lay the cabinet on the opposite side and repeat the process, securing the other support.

Stand the cabinet upright with the shelf top resting on the top shelf supports. Mark centered screw-hole locations on the sides and back for securing the top shelf. Drill the countersunk holes and screw the top shelf firmly in place.

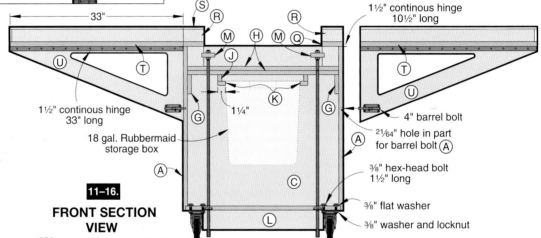

11–16.

FRONT SECTION VIEW

11–17.

TURN KNOB FULL-SIZE PATTERN

Now Add the Cabinet's Special Touches

Cut the vertical cleats (J) and horizontal cleats (K) to size. Glue and clamp parts (J) to (K), referring to **11–11**. (These glued-up runners support a waste tub. We used an 18-gallon storage box.)

Once dry, drill three countersunk holes through each runner. Turn the cabinet upside down and fasten the runners to the bottom of the top shelf (see **11–16**), factoring in the waste tub dimensions.

With the cabinet still upside down, install the casters, using the ⅜" holes drilled earlier. Note that the braking swivel casters mount at the front and the fixed casters at the back of the cabinet.

11–18.
SIDE SECTION VIEW

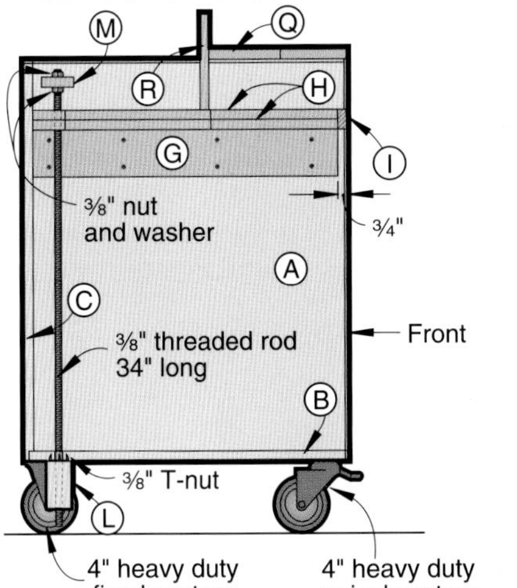

M
Q
R
H
G
I
3⁄8" nut and washer
3⁄4"
A
C
3⁄8" threaded rod 34" long
Front
B
3⁄8" T-nut
L
4" heavy duty fixed casters (on the back) #128461 Woodcraft
4" heavy duty swivel casters with brakes (on the front) #128461 Woodcraft

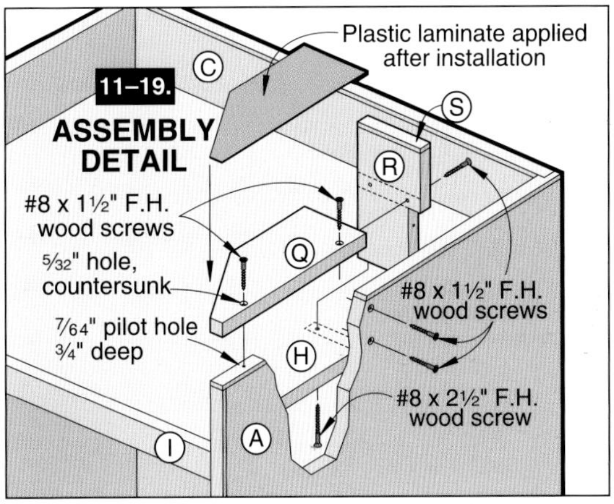

Plastic laminate applied after installation
C
11–19.
ASSEMBLY DETAIL
S
R
#8 x 1½" F.H. wood screws
5⁄32" hole, countersunk
Q
#8 x 1½" F.H. wood screws
7⁄64" pilot hole 3⁄4" deep
H
#8 x 2½" F.H. wood screw
I
A

3 Cut the threaded rod support (L) to size. Mark the 3⁄8" hole locations on (L), spaced the same as the threaded-rod T-nut holes in the cabinet bottom (B). Drill 3⁄8" holes across the width of part L at these locations.

4 To secure (L) to the cabinet bottom, first establish its location by using 5"-long 3⁄8" bolts (or threaded rod), and washers and nuts to temporarily hold (L) firmly to the cabinet bottom. Then drill three evenly spaced countersunk 5⁄32" holes for #8 x 2" flathead wood screws. Glue and screw part L in place. Remove the extra hardware.

5 To make the leveling system turn knobs (M), start with a 3⁄4 x 4 x 8" piece of stock. Attach the knob pattern from **11–17** to one half of the workpiece. Using a 1¼" Forstner bit, bore the 5⁄8"-radiused holes. Next, drill the 3⁄8" center hole, and bandsaw the knob to shape. Sand, and use this knob as your template for the other knob. Now, install the threaded rod and knob in the cabinet using washers and nuts. See **11–11** and **11–18** for reference.

Give Your Cabinet a Door

1 From 3⁄4" birch plywood, cut the door (N) to size. Cut the vertical edge-bandings (O) for the door and glue and clamp these in place. Now, cut and apply edge banding (P) for the door's top edge. Sand the parts flush and smooth.

2 Lay out the locations of the non-mortise hinges, where shown on **11–11**. Drill the screw holes and attach the hinges to the door. Allowing for a 1⁄8" reveal, attach the door to the cabinet.

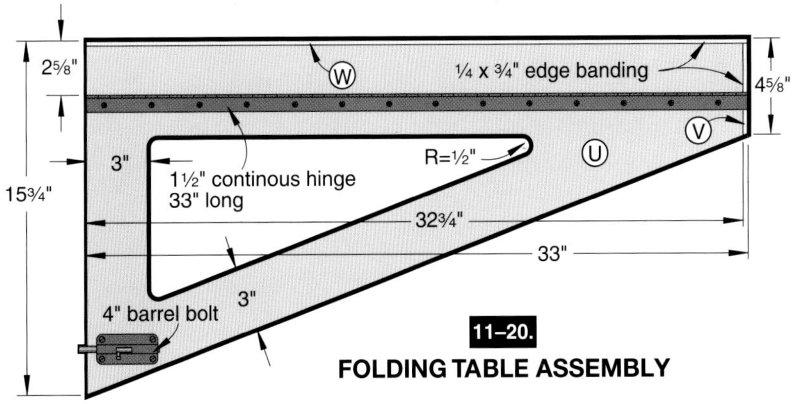

2⅝"
W
¼ x ¾" edge banding
4⅝"
V
3"
R=½"
U
1½" continous hinge 33" long
15¾"
32¾"
33"
3"
4" barrel bolt

11–20.
FOLDING TABLE ASSEMBLY

3 Locate and install the wire pull on the door. Next, screw on the magnetic catch and magnetic catch plates on the cabinet side and door back.

Table Talk, Beginning with the Cabinet

1 Referring to **11–12**, cut the left and right saw tables (Q) to size from ¾" birch plywood. Note that applying the ¹⁄₁₆"-thick plastic laminate will result in the finished sizes in the Materials List.

2 Cut enough plastic laminate to cover the surface and the exposed edges of saw tables (Q). Make the laminate pieces slightly oversized. (Do not laminate the back and long outside edges or bottom of these pieces.) Apply the edge pieces only, using contact cement. File or trim with a router and flush-trimming bit.

Next, glue and screw these pieces to the cabinet sides, as shown in **11–19**, keeping them flush to the sides and side front edges. Secure with #8 x 1½" flat-head wood screws and countersink the holes. Apply the top plastic laminate piece over the screws and trim flush. Smooth any sharp edges with a file.

3 Cut the saw fences (R) to the needed dimensions, making left and right side parts. Cut the saw-table fence edge bandings (S), and glue and clamp the pieces in place. Sand flush. Finally, drill the countersunk holes on the back side, where shown on **11–11**.

4 Clamp and screw the saw fences (R) to the saw tables (Q). Determine the countersunk screw-hole locations on the sides and drill the holes. Now, screw the fences to the sides (A) and top shelf (H).

Add the Sturdy Folding Tables

1 Cut the folding tables (T) to ¾ x 10½ x 32¹⁵⁄₁₆". Label one left and one right. Then glue plastic laminate to the outside end of each piece and trim. Laminate the front edge and trim. Now, laminate the top surfaces and trim.

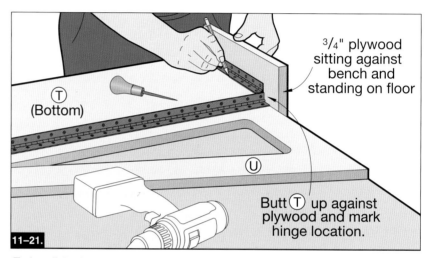

11–21.

To install the hinge flush with the end of the folding table, we butted plywood against the table, marked the hinge location, and then screwed it in place.

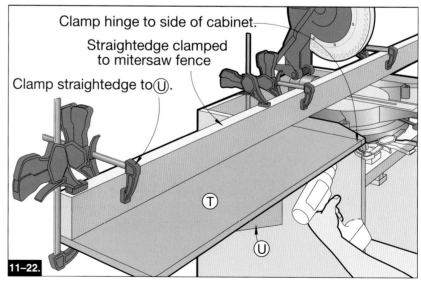

11–22.

To ensure table and fence alignment, clamp a straightedge to the cabinet fences. Clamp the folding table to the straightedge, and secure it to the cabinet.

2 Cut two lengths of 1½" continuous hinge to 33" long. Screw one to each back edge of the folding tables (T). Make sure the hinge barrels are flush with the laminate surface.

3 Refer to **11–20** and cut out the left and right folding table supports (U) to shape.

4 Cut out the end and top edge banding pieces (V, W). Glue and clamp the end pieces to folding table supports (U); then apply the top edge banding.

5 Lay the support flat on your workbench and clamp a folding table (T) on top of support (U) with the hinge edge at the location on **11–20**. Flush the ends, and screw the support to the folding table. Keep in mind that the support also serves as a saw fence; its height above the folding table should match the height of the mitersaw cabinet fence (R) above the saw table (Q). Make minor adjustments as needed. Then, repeat the process for the right folding table and support.

6 Install the barrel bolts on the front faces of folding table supports (U). See **11–16** and **11–23**.

7 Cut two more continuous hinges to fit. Now, screw

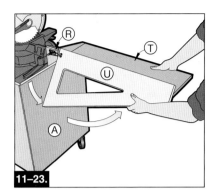

11–23.

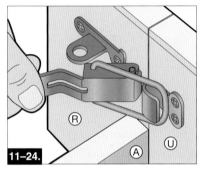

11–24.

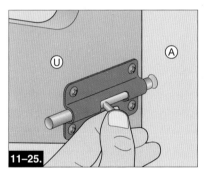

11–25.

these to the bottom of folding tables (T) and flush with their unlaminated inside ends, as shown in **11–24**.

8 To attach the folding tables, place your mitersaw on the cabinet and align its fence with the saw fences (R). Bolt the

mitersaw to the cabinet. Next, cut and place a straight, edge-jointed 2 x 4 on edge and against the aligned fences. Clamp it in position. Align one folding table assembly snugly beneath the 2 x 4 and clamp it in place. Screw it to the cabinet, as shown in **11–25**.

9 Using a square, adjust the folding table support so it angles 90° to the folding table. Mark the hole location for the barrel bolt on the cabinet side (A) and drill the hole. Repeat Steps 8 and 9 for the folding table on the opposite side.

10 Attach draw catches to the fences to keep the tables from sagging. (See **11–22**.)

11 Remove the hardware; then finish. (We used polyurethane.) Once dry, reinstall the hardware.

Raise Your Folding Tables for Action

To use the folding tables, simply lift one table up, swing down the support until it's at a right angle to the table, and lock it in place using the barrel bolt and draw catch, as shown in **11–23** to **11–25**.

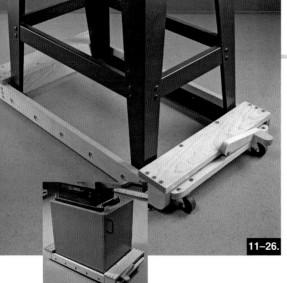

11–26.

Cutting Diagram

(A)   (A)

³/₄ x 3¹/₂ x 96" Maple   (2.7 bd.ft.)

(D)   (F)

(B)   (C)   (C)   (G)   (E)

³/₄ x 7¹/₄ x 96" Maple (5.3 bd.ft.)

Materials List for Roll-Around Tool Base

PART	FINISHED SIZE			Mtl.	Qty.
	T	W	L		
A sides	³/₄"	2³/₄"	TBD	M	2
B ends	³/₄"	2³/₄"	TBD	M	2
C end caps	³/₄"	4"	TBD	M	2
D blocks	1"	2"	3"	LM	2
E caster mount	³/₄"	4"	TBD	M	2
F cam lever	³/₄"	1⁵/₈"	7¹/₄"	M	1
G toekick	³/₄"	1¹/₂"	2"	M	1

TBD: To be determined. Read the instructions for details on sizing a base to fit your machine base.

Materials Key: M = maple; LM = laminated maple.
Supplies: ¹/₈" x 1¹/₂" angle iron crosscut to fit your base (2); ¹/₄ x ³/₄" flathead machine screws with ¹/₄" T-nuts (10); ¹/₄ x 1¹/₄" flathead machine screws with hex nuts and lock washers (8); ¹/₄ x 2¹/₂" flathead machine screws with hex nuts and lock washers (8); 2" fixed casters and two 2" swivel casters (90-lb. plate casters with rubber wheels) (2); 1¹/₂ x 2¹/₂" butt hinges (2) with twelve #10 x ¹/₂" panhead sheet-metal screws (12); ¹/₄ x 2" lag screw with a flat washer (1); #14 x 2" sheet-metal screws (28); ³/₈" threaded inserts with matching leveler glides (2); #8 x 1¹/₂" flathead wood screws (2); clear finish.

ROLL-AROUND TOOL BASE

Struggling to move heavy woodworking tools around your shop can cause headaches, let alone backaches. But you can do the job quickly and easily with a mobile, economical, four-wheeled tool base (**11–26**). It's a triumph of brains over brawn. Once you build one (or more), you'll wonder how you ever got by without it.

Build the Basic Wood Frame

1 From ³/₄" hardwood stock (we used maple), cut the base sides (A) and ends (B) to the lengths determined using the information in Tips on Sizing Your Tool Base and to the width listed in the Materials List.

2 Clamp the ends (B) between the sides (A) in the configuration shown on the side drawing on **11–28** and where dimensioned on **11–31**. Check for square. The opening should measure ¹/₄" longer than your tool base and ⁵/₈" wider. Verify this, and then drill countersunk holes through the sides (A) and centered into the ends of the ends (B). Drive in the screws, but do not glue the joint yet. For extra holding power, we used sheet-metal screws. For ease in driving the screws, add beeswax to the screw threads.

3 Measure from the outside face of one A to the outside face of the opposite A to determine the length of the end caps (C). Cut the end caps to length from stock ripped to 4" wide.

4 Laminate stock to form the 1"-thick caster blocks (D), and then cut them to size. Temporarily clamp the end cap (C) for the fixed casters in place, and glue the caster blocks to the bottom side of the end cap, flush against the end (B) and sides (A). Do not glue the blocks to A and B.

5 Cut the swivel-caster mount (E) to the length of the opening minus ¹/₈". Bevel-rip one edge at 15°, where shown on **11–28**. Mark and cut a pair of 1" radiuses on opposite corners. Drill the mounting holes, and bolt a pair of 2" swivel casters to the bottom side of the mount (E).

6 Mount a pair of 1¹/₂" butt hinges 2¹/₂" long to the bottom side of the caster mount (E). (After twisting off several of the screws supplied with the hinges, we used #10 x ¹/₂" panhead sheet-metal screws to secure the hinges to the caster mount.) Then, drill the pilot holes and screw the hinges to the end (B), where dimensioned on the Swivel Caster drawing in **11–29**.

Add the Angle Iron and Leveling Glides Next

1 Hacksaw a pair of $1/8 \times 1\frac{1}{2}$" angle iron pieces cut to the length of the opening minus $1/4$". With a cloth and some paint thinner, thoroughly clean the angle iron before bringing it in contact with your wood.

2 Drill equally spaced mounting holes through the angle iron. See the Side drawing (A) on the Parts View (**11–30**) for spacing in. Next, transfer the hole centerpoints to the inside face of the sides (A). Using a high-speed steel countersink bit, countersink the holes on the inside face of each angle iron.

3 Disassemble the tool base, and drill the mounting holes through the sides (A) for attaching the angle iron. Counterbore the holes on the outside face of the sides (A) for housing the T-nuts.

4 Cut a $1\frac{1}{2}$" radius on the bottom corners of the sides (A), where shown on the drawings.

5 Mark the centerpoints, and drill the mounting holes for the $3/8$" threaded inserts on the bottom edge of the sides (A), $2\frac{1}{2}$" in from the ends, where shown on the Swivel Caster drawing in **11–29** on *page 252.*

6 For ease in driving the inserts into the sides (A), cut the head off a $3/8 \times 3$" bolt. Double-nut the bolt and thread the insert onto the bottom of the bolt below the nuts. Chuck the assembly into your drill press, and turn the chuck by hand to drive the insert squarely into the mounting hole. Repeat for the second insert.

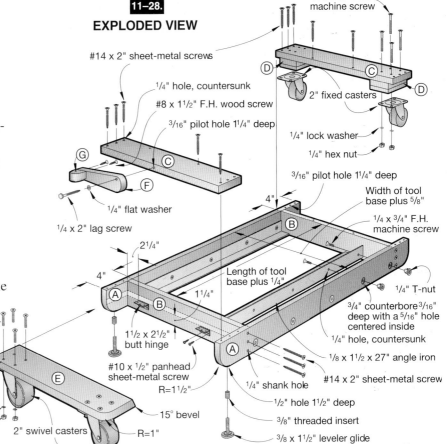

11–28.
EXPLODED VIEW

#14 x 2" sheet-metal screws
$1/4$" hole, countersunk
#8 x $1\frac{1}{2}$" F.H. wood screw
$3/16$" pilot hole $1\frac{1}{4}$" deep
$1/4$" flat washer
$1/4$ x 2" lag screw
$2\frac{1}{4}$"
4"
$1/4$ x $1\frac{1}{4}$" F.H. machine screw
$1/4$" hole, countersunk
$1/4$" lock washer
$1/4$" hex nut
2" swivel casters
$1\frac{1}{2}$ x $2\frac{1}{2}$" butt hinge
#10 x $1/2$" panhead sheet-metal screw
R=$1\frac{1}{2}$"
15° bevel
R=1"

$1/4$ x $2\frac{1}{2}$" F.H. machine screw
2" fixed casters
$1/4$" lock washer
$1/4$" hex nut
$3/16$" pilot hole $1\frac{1}{4}$" deep
Width of tool base plus $5/8$"
$1/4$ x $3/4$" F.H. machine screw
Length of tool base plus $1/4$"
$1\frac{1}{4}$"
4"
$1/4$" T-nut
$3/4$" counterbore $3/16$" deep with a $5/16$" hole centered inside
$1/4$" hole, countersunk
$1/8$ x $1\frac{1}{2}$ x 27" angle iron
#14 x 2" sheet-metal screw
$1/4$" shank hole
$1/2$" hole $1\frac{1}{2}$" deep
$3/8$" threaded insert
$3/8$ x $1\frac{1}{2}$" leveler glide

Tips on Sizing Your Tool Base

To determine the size of the mobile tool base, start by measuring the outside length and width of your tool base. Then, add $8\frac{1}{4}$" to the length to determine the length of the sides (A), and add $5/8$" to the width to determine the length of the ends (B) (**11–27**). Next, adjust the other pieces according to the instructions. The mobile base we built fits the base of our jointer, as shown.

See Supplies at the end of the Materials List for the hardware needed.

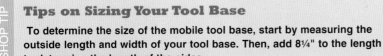

Length of tool base plus $8\frac{1}{4}$"
Width of tool base plus $5/8$"

11–27.

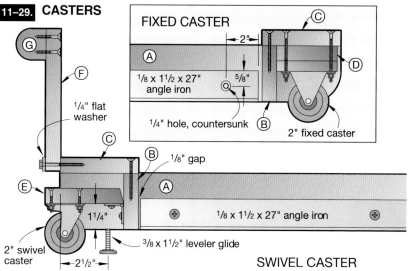

11–29. CASTERS

FIXED CASTER

Ⓐ ⅛ x 1½ x 27"
angle iron

Ⓒ

Ⓓ

2"

⅝"

¼" hole, countersunk Ⓑ

2" fixed caster

Ⓖ

Ⓕ

¼" flat washer

Ⓒ

Ⓔ

Ⓑ ⅛" gap

Ⓐ

1¼"

2" swivel caster

2½"

⅛ x 1½ x 27" angle iron

⅜ x 1½" leveler glide

SWIVEL CASTER

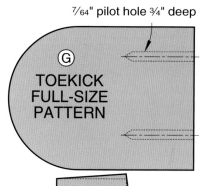

7/64" pilot hole ¾" deep

Ⓖ

TOEKICK
FULL-SIZE
PATTERN

11–30.
FULL-SIZE
PATTERNS

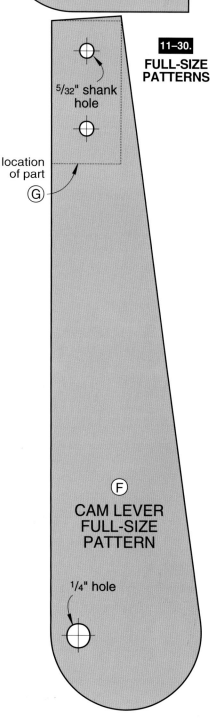

5/32" shank
hole

location
of part
Ⓖ

Ⓕ

CAM LEVER
FULL-SIZE
PATTERN

¼" hole

Add the Cam-Action Foot Lever

1 Transfer the patterns for the full-size cam lever (F) and the toekick (G) in **11–30** to ¾" stock, and bandsaw them to shape. Sand the edges to remove the saw marks.

2 Drill a ¼" hole through the cam lever for attaching it to the end cap (C) later. Mark the centerpoint, and drill a mating pilot hole in the end cap used at the swivel caster end of the base.

3 Drill a pair of mounting holes through the cam lever and into the toekick. Glue and screw the toekick to the cam lever.

Final Assembly and Finishing

1 Remove the casters from the base and the hinges from the end (B) and caster mount (E).

2 Glue and screw the base ends (B) between the sides (A), checking for square. Screw the end caps (C) in place.

3 Add a clear finish or paint the pieces the same color as the machine it will support. Be careful not to get any finish in the threaded inserts used to house the levelers. Paint the angle iron.

4 Drive the panhead sheet-metal screws to hinge the caster mount (E) to the end (B).

5 Use a wrench to drive the ¼" lag screw connecting the cam lever (F) to the end cap (C).

6 Tap the T-nuts in place, and secure the angle iron to the base sides. Now bolt the casters in place.

7 Add the levelers. Raise the levelers so the base rests on the levelers when the cam lever is flipped to the right side, raising the caster mount and swivel casters. When you want to move the tool, flip the cam lever to the left to lower the casters.

❖

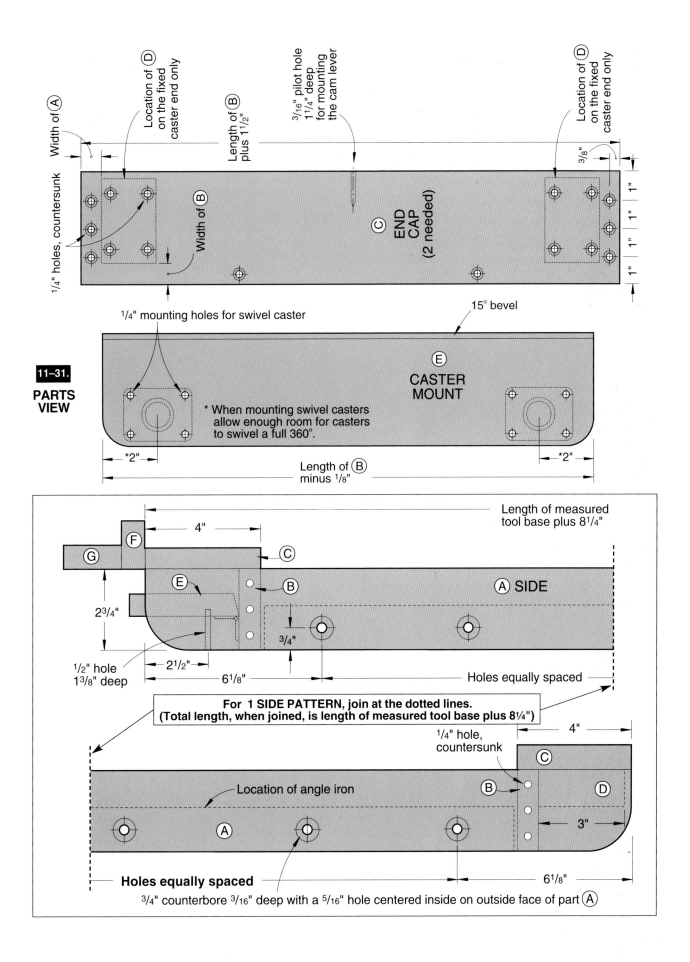

11–31.
**PARTS
VIEW**

Width of (A)

Location of (D) on the fixed caster end only

Length of (B) plus 1½"

³/₁₆" pilot hole 1¼" deep for mounting the cam lever

Location of (D) on the fixed caster end only

3/8"

Width of (B)

(C)
END CAP
(2 needed)

¼" holes, countersunk

1" 1"
1"
1"
1"
1"

¼" mounting holes for swivel caster

15° bevel

(E)
CASTER MOUNT

* When mounting swivel casters allow enough room for casters to swivel a full 360°.

*2"

*2"

Length of (B) minus ⅛"

Length of measured tool base plus 8¼"

4"

(F)
(G)
(C)

(E)
(B)
(A) SIDE

2¾"

3/4"

½" hole 1³/₈" deep

2½"

6⅛"

Holes equally spaced

**For 1 SIDE PATTERN, join at the dotted lines.
(Total length, when joined, is length of measured tool base plus 8¼")**

¼" hole, countersunk

4"

(C)

Location of angle iron

(B)
(A)
(D)

3"

Holes equally spaced

6⅛"

³/₄" counterbore ³/₁₆" deep with a ⁵/₁₆" hole centered inside on outside face of part (A)

WORKSHOP REST STOP

We call our craft woodworking, but for most hobbyists, being in the shop offers a chance to relax while they get immersed in a project. This custom shop stool (**11–32**) can help you avoid the strain of being on your feet or bending over a workbench all day. And to help hone your skills, you'll learn step-by-step how to make the intersecting sliding dovetail joints that help lock the legs in place.

Prepare Stock, and Start with the Platform

1 Start by edge-gluing two wide panels for the platform (A) and seat (C). (We chose 5/4 ash as the material for our shop stool.) After the glue cures, sand the joints, and trim the panels to rough blanks—13½" square and 16" square, respectively. (See the Materials List for the finished sizes.)

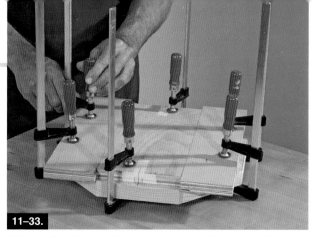

11–33.

After routing the rough grooves, reinstall the setup block and the guides, now spaced ³⁄₁₆" apart.

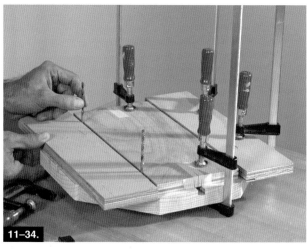

11–34.

Then remove the setup block and rout the dovetail slot.

2 Mark a pair of intersecting diagonal lines (corner to corner) on the underside of the platform blank. Then use a compass to draw a 12½"-diameter circle, centered where the diagonals intersect. Drill a ⅛" hole at the center.

3 Trim the corners off the platform blank to create a rough octagon (which makes for easier clamping and routing access); then set it upside down on the workbench.

Rout Sliding Dovetails for Sturdy Leg Connections

1 The sliding dovetail joint, sometimes called a French dovetail, isn't complicated, but it does require that you fit the mating parts precisely. Cut a 12"-long setup block from ¾" plywood. The block's width should be equal to the router base diameter, and you'll need a centerline at each end.

Cutting Diagram

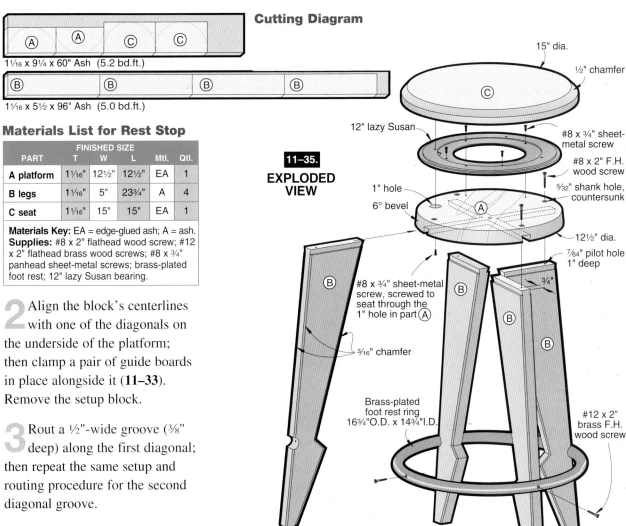

1¹⁄₁₆ x 9¼ x 60" Ash (5.2 bd.ft.)

1¹⁄₁₆ x 5½ x 96" Ash (5.0 bd.ft.)

Materials List for Rest Stop

PART	FINISHED SIZE			Mtl.	Qtl.
	T	W	L		
A platform	1¹⁄₁₆"	12½"	12½"	EA	1
B legs	1¹⁄₁₆"	5"	23¾"	A	4
C seat	1¹⁄₁₆"	15"	15"	EA	1

Materials Key: EA = edge-glued ash; A = ash.
Supplies: #8 x 2" flathead wood screw; #12 x 2" flathead brass wood screws; #8 x ¾" panhead sheet-metal screws; brass-plated foot rest; 12" lazy Susan bearing.

2 Align the block's centerlines with one of the diagonals on the underside of the platform; then clamp a pair of guide boards in place alongside it (**11–33**). Remove the setup block.

3 Rout a ½"-wide groove (³⁄₈" deep) along the first diagonal; then repeat the same setup and routing procedure for the second diagonal groove.

4 Next, add two centered layout marks ½" apart on the

11–35. EXPLODED VIEW

15" dia.
½" chamfer
12" lazy Susan
#8 x ¾" sheet-metal screw
#8 x 2" F.H. wood screw
⁵⁄₃₂" shank hole, countersunk
1" hole
6° bevel
12½" dia.
⁷⁄₆₄" pilot hole 1" deep
¾"
#8 x ¾" sheet-metal screw, screwed to seat through the 1" hole in part Ⓐ
³⁄₁₆" chamfer
Brass-plated foot rest ring 16¾"O.D. x 14¾"I.D.
#12 x 2" brass F.H. wood screw

DOVETAIL DETAIL

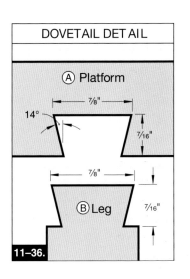

Ⓐ Platform
⅞"
14°
⁷⁄₁₆"
⅞"
Ⓑ Leg
⁷⁄₁₆"

11–36.

ends of the centering block, and reposition it over one of the grooves you just routed. Add a couple of ³⁄₁₆" spacers (drill bits work great) along each edge; then reclamp the outer guide boards in place again (**11–34**). Then, once again, remove the setup block.

5 Install a ½"-diameter, 14° dovetail bit in your router, and adjust the cutting depth to ⁷⁄₁₆" (**11–36**). Start the cut at one

end of the groove, keeping the router base up against one of the guide blocks. Make a return pass along the other guide block, making sure you don't lift the router during the cut. Then change the setup and follow the same procedure on the second diagonal. You now have two perpendicular dovetailed grooves intersecting at the center of the platform.

Sanding and Drilling Complete the Platform

1 Use a bandsaw or jigsaw to trim the platform blank to within ⅛" or less of the circle outline drawn earlier.

2 A disc sander works best for truing the circle. Adjust the sander table to a 6° angle, and use a circle sanding jig to position the platform against the sanding disc. Rotate the platform on the jig's pivot point to true up the entire circumference.

Note: If you don't have appropriate sanding equipment, a router with a trammel guide can be used to trim the circle to its finished diameter.

3 Drill the 1" hole shown in **11–45**. This hole will provide access for mounting the lazy Susan bearing that connects the seat to the platform.

Forming the Legs: From Rough-Cut Blanks to Locking Joinery

1 Start by laying out the shape of one leg (B) on a 24"-long blank (**11–46**). Rough–cut the inside edge to within ⅛" of the lines, using a bandsaw or jigsaw; then trace the outline of the leg on the remaining three 24" blanks.

2 Adjust your miter gauge for a 6° cut, fit it with an auxiliary fence and a stopblock, and trim the top of each leg on the tablesaw (**11–37**).

11–37.

With your miter gauge set at a 6° angle, trim the top of each leg. Then reset the miter gauge to 6° the other direction, reset the stopblock, and cut the bottom ends.

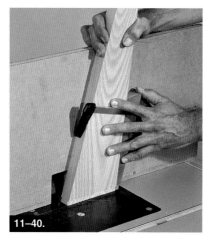

11–38.

Use tape-adhered guide blocks and a flush-trim bit to rout the first leg to final shape. Then use it as your template.

11–39.

To make the half-round notches for the foot ring, clamp legs back-to-back and drill a 1" hole right on the centerline.

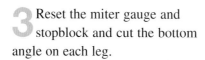

11–40.

Make test cuts in scrap; then rout the dovetail tongue on the top end of each leg. A tall fence and follower block help.

3 Reset the miter gauge and stopblock and cut the bottom angle on each leg.

4 To clean up the cuts on the inside edges, use double-faced tape to adhere guides along the layout lines on the first leg.

11–41.

Two of the legs will have to be tapped in toward the center to make clearance for the foot ring, which installs from the top.

Install a flush-trim bit in your router table and trim the edge to shape (**11–38**). Then use this leg as a template to trim the other three, again by adhering them to the pattern leg with double-faced tape.

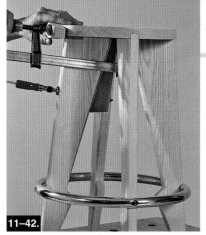

11–42.

After you get the foot ring into position, use clamps to pull the inset legs back to their final position; then fasten them.

11–43.

Center the lazy Susan bearing on the platform, with the wide flange down and the 1" access hole lined up. Then fasten with screws.

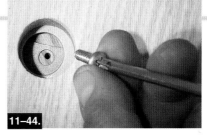

11–44.

Set the base assembly upside down on the seat, and rotate it to expose the screw holes for fastening the seat.

11–45. **PARTS VIEW**

5⁄32" shank hole, countersunk on top face

¾"   ⅞"   6°

Ⓐ

7⁄16"   12½"

PLATFORM
(side view)

5⁄32" shank hole, countersunk on top face

12½" dia.

Grain direction

⅛" hole for seat alignment   ¾"

⅞" dovetail grooves

Ⓐ

1½"

1" hole for access to screw lazy Susan to seat

PLATFORM
(viewed from bottom)

11–46.
LEG PROFILE

5"   6°

7⁄16"

3⁄16" chamfer

23¾"   Ⓑ   23 3⁄16"

3⅛"

1" hole

2⅛"

7¼"

6½"

6°

7"

1"

Next, clamp the legs together in pairs (outside edges back-to-back) and drill the 1" holes that create the notches for the foot ring (**11–39**).

Take the same dovetail bit you used earlier for the platform dovetail grooves and install it in your router table. After tuning the settings and test-cutting a scrap piece—one pass on each face—rout the dovetail tongue on the top end of each leg (**11–40**). Aim for a snug fit—these joints won't have to be glued if the fit is tight, making it easier to assemble everything later.

Rout the 3⁄16" chamfer along the edges, but avoid the ring notches and the corners of the dovetail tongues.

Fashion a Simple Seat

The seat (C) is the simplest component of the whole project. Start with the 16" square blank you glued up and cut to size earlier. Draw diagonal lines on the underside to locate the center, and use a compass to mark the 15" diameter of the finished circle.

If necessary, drill a shallow ⅛"-diameter hole in the bottom of the seat for the pivot pin on your circle-sanding jig.

Use a bandsaw or jigsaw to rough-cut the seat blank close to the outline; then set up the circle jig again to sand to the finished diameter.

Rout a ½" chamfer around the top edge of the seat.

Assemble and Finish the Stool

1 If you achieved a tight fit with your sliding dovetails, prep for assembly by sanding all the parts and by brushing some paste wax into the dovetail grooves in the platform. If the fit is looser, you may instead want to glue the joints to coax a little more strength from them.

2 With the platform clamped vertically in a bench vise, tap the legs into their grooves (**11–41**). Leave two opposing legs in their correct "flush" position, and drive the other two inside the edge of the platform about ½". You'll need this extra (temporary) clearance to install the foot ring.

3 Drop the foot ring into position; then use clamps to draw the two inset legs back to their flush position (**11–42**), and secure them with screws.

4 Turn the foot ring until the screw holes line up with the center of each leg. Drill pilot holes and drive the #12 × 2" brass screws.

5 Drill a countersunk pilot hole through the platform at each leg location, ¾" in from the edge. Drive #8 × 2" flathead screws to secure the legs.

6 Mask off the foot ring and apply two coats of semigloss polyurethane to the base assembly and the seat.

Install the Swivel Bearing and Give Your Seat a Spin

1 Align the lower flange of a 12" lazy Susan swivel with the access hole drilled earlier. Drill pilot holes for the mounting screws and secure the bearing to the platform (**11–43**).

2 Next, set the seat upside down on your workbench and place the base assembly on it. Insert a finish nail through the hole at the center of the platform and into the hole in the center of the seat. Rotate the base assembly so the access hole exposes—one at a time—the screw holes in the top flange of the bearing (**11–44**). Drive in the mounting screws; then pull the alignment nail out of the center hole, turn the stool upright, and take it for a test sit. As you spin around on your newest shop toy, don't worry if you hear strange voices. It's just your back and feet saying "Thank you."

❖

TWO-IN-ONE SEAT/STEP STOOL

This handy seat/step stool (**11–47**) provides additional height to get to those hard-to-reach places. After using its ladder-like capability, fold the top half over for additional seating.

Note: We used southern yellow pine (also called longleaf pine)

11–47.

for this project because it often exhibits wonderful straight grain, and the 12'-long 1 x 10s we bought had no knots. In fact, none of the boards inspected at the lumberyard had any but the slightest defects in grain. This pine also exhibits great strength. A note of caution, though. Yellow pine is notorious for cup and twist, especially in wide boards. Our recommendation is to consider edge-joining 4- to 5"-wide boards to make the wide pieces called for in this project.

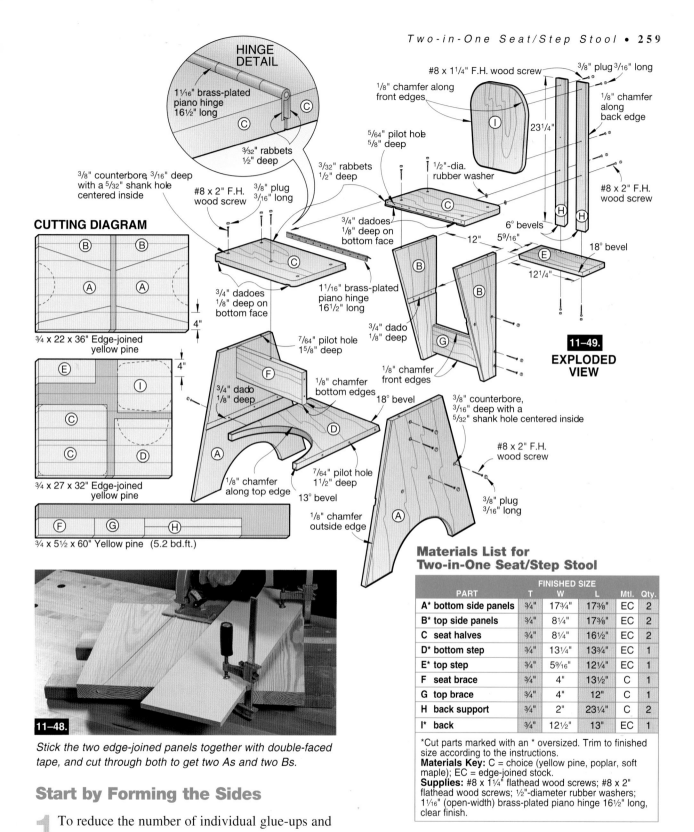

HINGE DETAIL

1¹/₁₆" brass-plated piano hinge 16½" long

³/₃₂" rabbets ½" deep

³/₈" counterbore, ³/₁₆" deep with a ⁵/₃₂" shank hole centered inside

#8 x 2" F.H. wood screw

³/₈" plug ³/₁₆" long

³/₃₂" rabbets ½" deep

³/₄" dadoes ¹/₈" deep on bottom face

¾" dadoes ⅛" deep on bottom face

1¹/₁₆" brass-plated piano hinge 16½" long

⁷/₆₄" pilot hole 1⁵/₈" deep

³/₄" dado ⅛" deep

⅛" chamfer bottom edges

¾" dado ⅛" deep

18° bevel

⅛" chamfer along top edge

13° bevel

⅛" chamfer outside edge

⁷/₆₄" pilot hole 1½" deep

#8 x 1¼" F.H. wood screw

⅛" chamfer along front edges

⁵/₆₄" pilot hole ⅝" deep

½"-dia. rubber washer

³/₈" plug ³/₁₆" long

⅛" chamfer along back edge

23¼"

#8 x 2" F.H. wood screw

6° bevels

12"

5⁹/₁₆"

18° bevel

12¼"

¾" dado ⅛" deep

⅛" chamfer front edges

³/₈" counterbore, ³/₁₆" deep with a ⁵/₃₂" shank hole centered inside

#8 x 2" F.H. wood screw

³/₈" plug ³/₁₆" long

11–49.
EXPLODED VIEW

CUTTING DIAGRAM

¾ x 22 x 36" Edge-joined yellow pine

4"

4"

¾ x 27 x 32" Edge-joined yellow pine

¾ x 5½ x 60" Yellow pine (5.2 bd.ft.)

11–48.

Stick the two edge-joined panels together with double-faced tape, and cut through both to get two As and two Bs.

Start by Forming the Sides

1 To reduce the number of individual glue-ups and for consistent grain patterns on the wide pieces, edge-glue enough ¾" stock for one panel measuring 22 x 36" and the second 27 x 32", after trimming.

Materials List for Two-in-One Seat/Step Stool

PART	FINISHED SIZE			Mtl.	Qty.
	T	W	L		
A* bottom side panels	¾"	17¾"	17⅜"	EC	2
B* top side panels	¾"	8¼"	17⅜"	EC	2
C seat halves	¾"	8¼"	16½"	EC	2
D* bottom step	¾"	13¼"	13¾"	EC	1
E* top step	¾"	5⁹/₁₆"	12¼"	EC	1
F seat brace	¾"	4"	13½"	C	1
G top brace	¾"	4"	12"	C	1
H back support	¾"	2"	23¼"	C	2
I* back	¾"	12½"	13"	EC	1

*Cut parts marked with an * oversized. Trim to finished size according to the instructions.
Materials Key: C = choice (yellow pine, poplar, soft maple); EC = edge-joined stock.
Supplies: #8 x 1¼" flathead wood screws; #8 x 2" flathead wood screws; ½"-diameter rubber washers; 1¹/₁₆" (open-width) brass-plated piano hinge 16½" long, clear finish.

Cut the 22 x 36" panel into two pieces measuring 22 x 17⅜". See Step 1 on **11–50** for reference. Mark one end on each 22 x 17⅜" panel as "bottom."

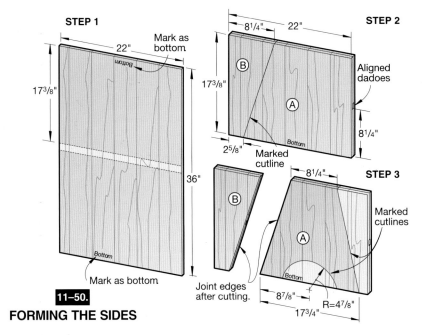

STEP 1

22"

Mark as bottom

17³/₈"

Bottom

36"

Bottom

Mark as bottom

11–50.

FORMING THE SIDES

STEP 2

8¹/₄" — 22"

B

A

Aligned dadoes

17³/₈"

8¹/₄"

Bottom

2⁵/₈"

Marked cutline

STEP 3

8¹/₄"

B

A

Marked cutlines

Joint edges after cutting.

Bottom

8⁷/₈"

R=4⁷/₈"

17³/₄"

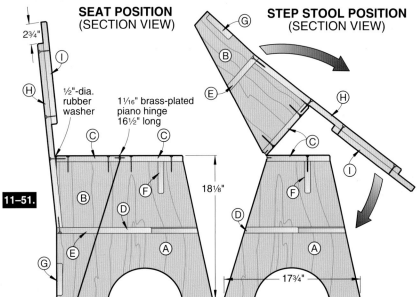

SEAT POSITION (SECTION VIEW)

2³/₄"

I

H

½"-dia. rubber washer

1¹/₁₆" brass-plated piano hinge 16½" long

C

C

B

F

D

E

A

G

11–51.

STEP STOOL POSITION (SECTION VIEW)

G

B

E

H

C

F

I

D

A

18¹/₈"

17³/₄"

11–52.

Use a tablet back to space the seat halves for the proper clearance.

you to get one A and one B from each panel.

3 As shown in **11–48**, use a straightedge and portable circular saw to cut through both pieces. Joint the edges to remove the saw marks. Leave the pieces taped together. Cut a 4⁷/₈" radius on the bottom end of the side panel (A), where dimensioned on Step 3 of **11–50** and **11–53**. Mark the cutline, and make the angled cut along the front edge of the taped-together pieces.

Cut the Rest of the Pieces to Shape

1 Cut the seat halves (C) to size. Cut ¾" and 1½" radii on the outside corners, where shown on **11–53** on *page 261.*

2 Cut a ³/₃₂" rabbet ½" deep along the inside edge of each seat half. The combined width of the rabbets should accommodate the folded thickness of the piano hinge when the edges of the seat halves are pushed together in the seat position shown on **11–51**. The combined depth of the rabbets needs to accommodate the open width of the hinge plus the thickness of the tablet-back cardboard spacers placed between the seat halves when they are face-to-face in the step-stool position, as shown in **11–52**. In case the two seat halves are not perfectly flat, the space provided by the cardboard will prevent stress on the hinge in the step-stool position.

2 Cut a ¾" dado ⅛" deep 8¹/₄" up from the bottom end on the inside face of each panel, where shown on **11–53**, on *page 262.* Using double-faced tape, adhere the two panels face-to-face, with the edges and ends flush and the dadoes aligned, as shown on Step 2 on **11–50**. Mark an angled cutline on the top piece, enabling

3 Cut the bottom step (D) and top step (E) to size plus ½" in width. Then, bevel-rip the mating edges at 18°, where shown on **11–49**. Fit the bottom step into the dadoes in the sides (A), mark its final width, and bevel-rip it to width. Mark and cut an arc on the bottom step, where dimensioned on **11–53**.

4 Fit the top step (E) into the dado in the rear side panel (B) with their front 18° edges flush. Mark a trim line on the back edge of E, and cut it to final width.

5 Cut the seat brace (F), top brace (G), the back supports (H), and back (I) to size and shape. Bevel-cut the bottom ends of the back supports (H) at 6°.

6 Lay out counterbores on all pieces except for the counterbores in v the top step (E) where the back supports (H) are fastened and in the back supports (H) where they are fastened to the rear seat half (C). Drill the holes. Rout ⅛" chamfers on all pieces as indicated on **11–50** and **11–53**. Finish-sand all pieces.

Assemble the Pieces

1 Clamp the bottom step (D) in the dadoes between the bottom side panels (A). Clamp the front seat half (C) onto the top of the sides (A).

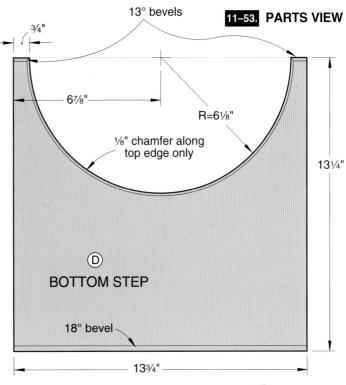

11–53. PARTS VIEW

13° bevels

¾"

6⅞"

R=6⅛"

⅛" chamfer along top edge only

13¼"

Ⓓ BOTTOM STEP

18° bevel

13¾"

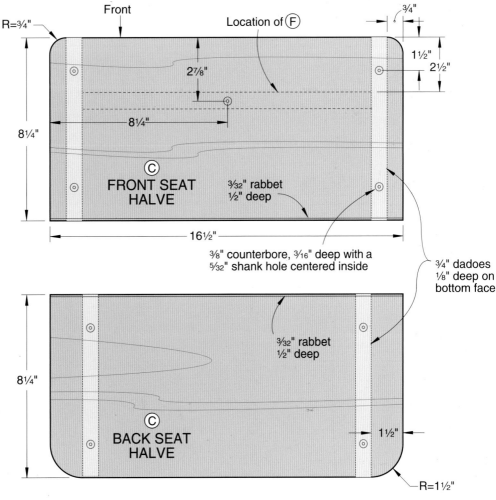

Front

R=¾"

Location of Ⓕ

¾"

1½"

2½"

2⅞"

8¼"

Ⓒ FRONT SEAT HALVE

8¼"

3⁄32" rabbet ½" deep

16½"

3⁄8" counterbore, 3⁄16" deep with a 5⁄32" shank hole centered inside

¾" dadoes ⅛" deep on bottom face

8¼"

3⁄32" rabbet ½" deep

Ⓒ BACK SEAT HALVE

1½"

R=1½"

Fit the brace (F) in place. Check for square. Using the previously drilled counterbored holes in the sides as guides, drill pilot holes and screw the assembly (A, C, D, F) together.

2 Clamp the top step (E) in the dadoes between the top side panels (B). Clamp the rear seat half (C) onto the top of the sides (B). Check for square. Using the counterbored holes in the sides as guides, drill pilot holes, and screw the assembly (B, C, E) together.

3 Glue and clamp the back supports (H) to the back (I), making certain the supports are parallel and that their bottom beveled ends are flush.

4 Position the beveled ends of the back supports (H) so they are centered on the top step (E). To hold the back assembly (H/I) in place, clamp the assembly to the rear seat

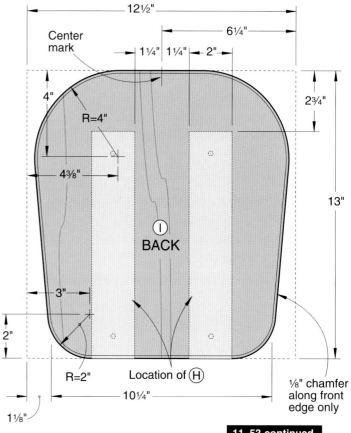

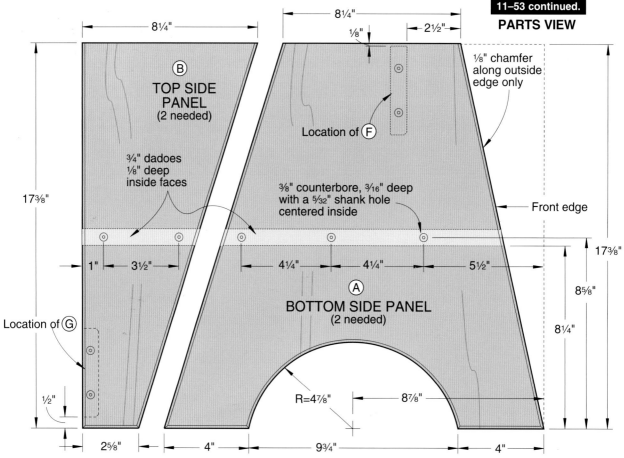

11–53 continued.
PARTS VIEW

half (C). Drill the mounting holes through the top step (E), and fasten the bottom ends of the back supports with screws to the step.

5 Drill holes through the back supports (H) and into the back edge of the rear seat half (C). Insert a ½" rubber washer (a common flat or domed faucet washer works fine) between the back supports and the seat. With the washers between the pieces, drive the screws. The rubber washers fill the angled gap between the two pieces.

6 Check that the top brace (G) fits between the sides (B). Trim if necessary. Then, drill the mounting holes, and screw the brace in place.

7 Place the rear top assembly (B/C/E/G/H/I) on the bottom assembly (A/C/D/F). Drill pilot holes, and screw the piano hinge in place to connect the two assemblies. Check the operation of the chair/step stool. Remove the piano hinge and plug all the counterbores. Trim and sand the plugs flush.

8 Apply two coats of semi-gloss polyurethane, sanding between coats. Screw the hinge back in place, then take a seat.

❖

STURDY STOOL

You'll always have a step stool handy in the shop (or elsewhere) if you build either (or both) of these (**11–54**). The tall one features a storage shelf underneath. Box joints and angled sides make them look so good that you won't have to hide either away between uses. And, our jig for cutting the angled box joint makes both simple to build.

Build the Angled Box-Joint Jig

1 Referring to **11–55**, saw the parts for the jig to size. Assemble them as shown, except for the runners and guide block.

2 Install a ½" dado blade on your tablesaw, and adjust the cutting depth to ½".

3 To install the guide block, first mark a point on your saw table midway between the miter-gauge slots. Then, mark the midpoint of the jig's width on the face near the bottom edge. Place the jig on the saw table, the face toward the back of the saw. Offset the center mark on the jig 1" to the left of the center mark on the table.

Slide the saw's fence up against the right side of the jig. Holding the jig firmly against the fence, saw a dado 3" into the base. Install the guide block in the dado, extending ¾" beyond the face.

4 Next, install the runners. To do this, slide the tablesaw fence exactly 1" to the right. Then, put a strip of double-faced cloth tape on top of each runner. Place the runners in the miter-

11–54.

Cutting Diagram

$3/4$ x $5^{1}/_2$ x 96" Oak (4.0 bd.ft.)

$3/4$ x $11^{1}/_4$ x 60" Oak (5.0 bd.ft.)

$1/4$ x 24 x 24" Oak plywood

gauge slots, taped side up, with a $1/16$"-thick shim strip under each one. (Shimming brings the runner tops flush with the table's surface.)

Holding the jig against the fence to keep it square, press the base down to stick the runners to it. Lift the jig without disturbing the runners' positions on the base, and attach the runners with screws, where shown.

5 Raise the blade to $3/4$" cutting depth. Cut test joints in $3/4$"-thick scrapwood to verify finger spacing. Cut at least eight fingers for a good test. When cutting with the jig, clamp the workpiece to the face. To keep the jig from tipping, which would cut an inaccurate joint, press down on the back of the jig as you push it forward.

Materials List for Sturdy Stool

PART	FINISHED SIZE			Mtl.	Qty.
	T	W	L		
A side	$3/4$"	$14^{5}/_8$"	$16^{1}/_4$"	O	2
A side (optional for short bench)	$3/4$"	$11^{3}/_4$"	8"	O	2
B* top	$3/4$"	$9^{3}/_{16}$"	15"	O	1
C crossbar	$3/4$"	$2^{1}/_4$"	$14^{5}/_{16}$"	O	1
D spreader	$3/4$"	2"	$17^{11}/_{16}$"	O	2
E shelf	$1/4$"	$8^{1}/_4$"	$17^{11}/_{16}$"	OP	1

*Make wider initially, and trim to finished size in accordance with instructions.

Materials Key: O = oak; OP = oak plywood.
Supplies: Woodworker's glue; #8 x 1" and 2" flathead wood screws; clear finish.

11–55.
BOX-JOINT CUTTING JIG

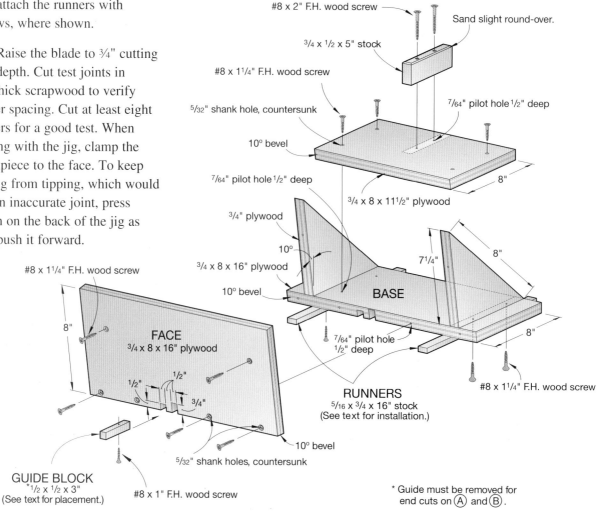

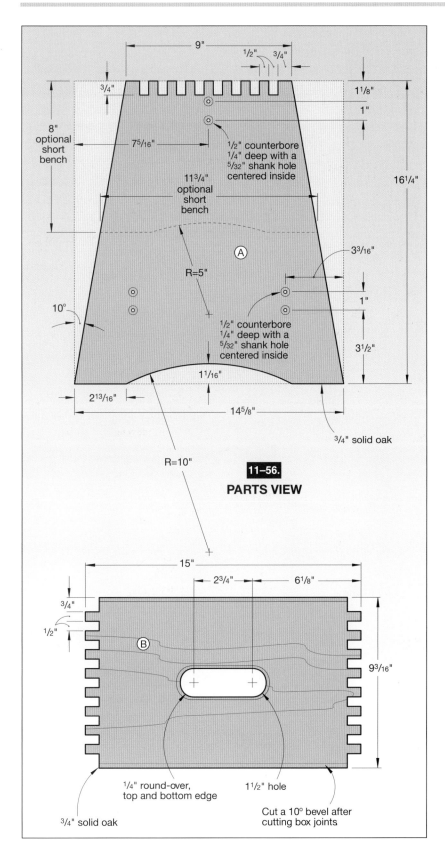

9"

1/2" 3/4"

3/4"

1 1/8"

1"

8"
optional
short
bench

7 5/16"

1/2" counterbore
1/4" deep with a
5/32" shank hole
centered inside

11 3/4"
optional
short
bench

16 1/4"

Ⓐ

3 3/16"

R=5"

1"

10°

1/2" counterbore
1/4" deep with a
5/32" shank hole
centered inside

3 1/2"

1 1/16"

2 13/16"

14 5/8"

3/4" solid oak

R=10"

11–56.
PARTS VIEW

15"

2 3/4"   6 1/8"

3/4"

1/2"

Ⓑ

9 3/16"

1/4" round-over,
top and bottom edge

1 1/2" hole

Cut a 10° bevel after
cutting box joints

3/4" solid oak

If the joint doesn't fit together properly, adjust the jig position on the runners by tapping the jig with a hammer. To increase the distance between the fingers, tap the jig to the right; to reduce it, tap it to the left.

Make the Box-Jointed Stool Sides and Top

1 Glue up stock, and cut the sides (A) to the dimensions shown in the Materials List. (To make the short stool, cut the sides to the optional size shown.) Bevel the ends to 10° as you cut the pieces to length. (Make sure you saw the bevels parallel.)

2 Cut the top (B) to the length shown in **11–56,** but make it 10" wide to start. (You'll cut it to finished width after sawing the box joints.) Bevel both ends to 10°. (On this part, you must saw converging bevels. The long side will be the part's bottom.)

3 Lay out the box joints on sides (A) and top (B), shown on **11–56**. Start from the center on each piece. Locate a finger at the center of each side (A) and a space at the center of each end on the top (B). (We applied masking tape to the face of each part in the joint area, and drew our layout marks on the tape.) Lay out the cutlines for the tapered sides now, to ensure that the joints will be centered.

4 Cut the box joints in the sides (A), using the jig and a 1/2" dado blade set to a 3/4" cutting

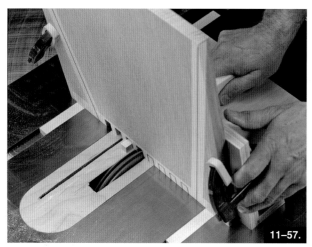

11–57.

Clamp the workpiece to the jig, and press down at the back for stability while sawing.

depth, as shown on **11–56**. To start, remove the guide block from the jig. Align the layout marks for the first space on the right with the dado blade. Clamp the workpiece to the jig, and then saw the space (**11–57**).

Replace the guide block in the jig. Place the dado you just sawed over the guide pin, clamp the part to the jig face, and make the cut. Saw all the fingers on both sides (A) this way.

5 Cut the mating fingers on both ends of the top (B). Again, remove the guide block to make the first cut on each end, and replace it for subsequent ones.

11–58.

Screw the crossbar (C) into place to clamp the joints at the correct angle.

Because of the wider fingers at the outside of the joint, the first and last cuts on both ends of the top (B) will be wider than the dado blade. Make these cuts in two passes.

6 Saw the tapered edges on the sides (A). (Save the waste pieces to cut screw-hole plugs from later.) Drill and counterbore the screw holes, where shown, and bandsaw the arch that forms the feet. Sand both pieces smooth.

7 Dry-assemble the sides (B) and top (B). Mark the width and bevel on part B, and bevel-rip it to width.

8 Lay out the oval handle opening in the top (B). Bore two 1½" holes to form the handle ends, and cut between them with a scrollsaw or jigsaw. Rout a ¼" round-over around the top and bottom of the handle opening.

Put Together a Handy Shelf to Go Underneath

1 Cut the crossbar (C), spreaders (D), and shelf (E) to size, and bevel the ends. Bandsaw the centered handle notch in crossbar (C).

2 Dry-assemble parts A, B, and C to check the fit of C. If the part seems too short, plane or saw a little off the top (notched) edge. If it is too long, trim equal amounts from each end.

3 Form a ⅜" rabbet ¼" deep along the bottom inside edge of each spreader (D). Clamp the spreaders and shelf (E) together, and then drill and countersink screw holes from the bottom. Glue and screw the assembly together.

4 Disassemble parts A, B, and C. Apply yellow glue to the box joints, assemble them, and clamp. (Clamps with rubber or soft plastic pads grip better on the angled sides.) After pulling the joints up snugly, remove the clamps. Install the spreader (D), drill pilot holes into the ends, and drive the screws, as shown in **11–58**.

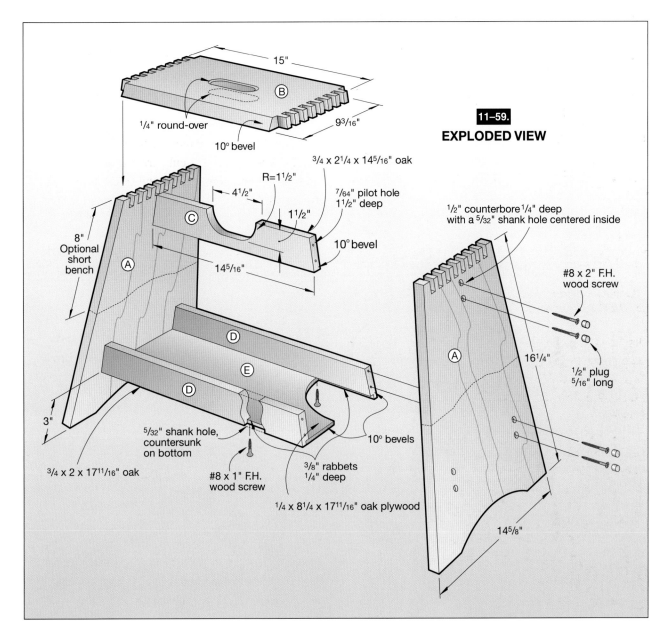

11–59.
EXPLODED VIEW

15"

B

9³/₁₆"

¹/₄" round-over

10° bevel

³/₄ x 2¹/₄ x 14⁵/₁₆" oak

R=1¹/₂"

4¹/₂"

C

1¹/₂"

⁷/₆₄" pilot hole
1¹/₂" deep

8"
Optional
short
bench

A

14⁵/₁₆"

10° bevel

¹/₂" counterbore ¹/₄" deep
with a ⁵/₃₂" shank hole centered inside

#8 x 2" F.H.
wood screw

A

16¹/₄"

¹/₂" plug
⁵/₁₆" long

D

E

D

3"

⁵/₃₂" shank hole,
countersunk
on bottom

10° bevels

³/₄ x 2 x 17¹¹/₁₆" oak

#8 x 1" F.H.
wood screw

³/₈" rabbets
¹/₄" deep

¹/₄ x 8¹/₄ x 17¹¹/₁₆" oak plywood

14⁵/₈"

5 Install the shelf assembly
(D/E). Drill pilot holes into
the shelf, and drive in the screws.

6 Using a plug cutter, cut 12
screw-hole plugs from the
waste. Glue the plugs into the
counterbores, aligning the grain
to make them inconspicuous.
After the glue dries, trim the
plugs flush. Sand the joints
flush, and finish-sand the stool.

Put On a
Long-Lasting Finish

1 Wipe off the sanding dust,
and spray or brush on a coat
of satin polyurethane varnish.
Apply a thin coat to prevent runs.

2 Allow the varnish to dry,
and then sand the stool with
320-grit sandpaper. Dust the
surface, and apply another light
coat of finish.

3 The end grain on the box-joint
fingers absorbs the finish,
making the finger ends look duller
than the rest of the stool. To avoid
that, brush extra finish onto the
end grain to build up the coating.

4 Sand the stool once more
with 320-grit sandpaper. Then,
put on the final coat of finish.

❖

FOLD-OUT WORK SUPPORT

Short on work space? Try this compact project (**11–60** and **11–61**). Spread the legs for a stable support when cutting bulky pieces of sheet goods. Or, position a piece of plywood on its top, and use it as a temporary worktable as needed. Then, when you're done, just fold it up and hang the support on a wall on two ladder hooks.

❖❖❖

11–60.

Bench folded into stored position

Screw into wall stud.

Ladder hooks

HANGING DETAIL

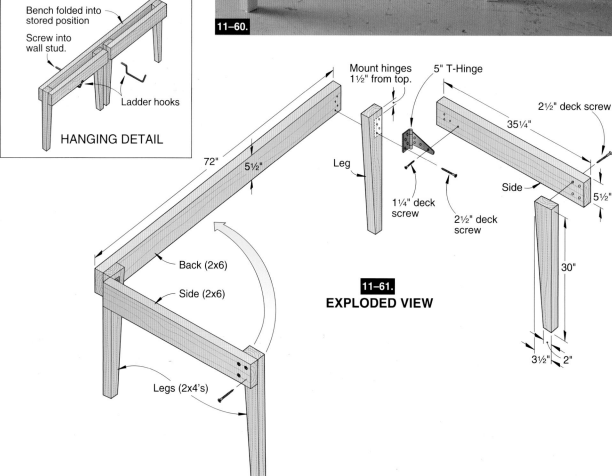

Mount hinges 1½" from top.

5" T-Hinge

2½" deck screw

35¼"

Leg

1¼" deck screw

2½" deck screw

Side

5½"

72"

5½"

Back (2x6)

Side (2x6)

30"

Legs (2x4's)

3½"

2"

11–61.
EXPLODED VIEW

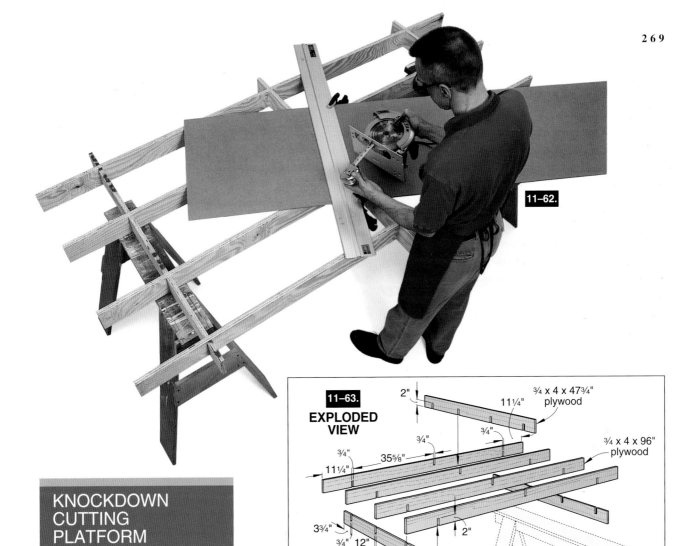

11–62.

11–63.
EXPLODED VIEW

2"
11¼"
¾ x 4 x 47¾" plywood
¾"
¾"
¾ x 4 x 96" plywood
¾"
35⅝"
11¼"
3¾"
¾" 12"
¾"
2"
¾" 12"
¾" 3¾"
4"

KNOCKDOWN CUTTING PLATFORM

Sheet goods often present a real challenge when it comes to cutting them down into project-sized pieces. Even if you have the room in your shop to maneuver a full sheet of plywood, single-handedly balancing it on your tablesaw while cutting it can prove to be impossible.

After struggling with 2 × 4s and sawhorses for too long, we determined that it wasn't the sawhorses, but the shifting, sagging 2 × 4s that were the problem. So we came up with the idea for a knock-down platform that sets up quickly and easily (**11–62** and **11–63**).

Simply rip six 4 × 96" strips from a sheet of ¾" plywood, and then cut three 47¾"-long pieces from two of the strips. Cut the notches, where shown in **11–63**.

To use the platform, position your sawhorses so they support the two short end rails. Then drop the center rail in from the top.

Adjust your saw to cut about ⅛" deeper than the thickness of the sheet being cut. When taken apart for storage, the pieces of the platform make a stack less than 5¼" thick that can be stored easily in the rafters of your shop or garage.

KNOCKDOWN SAWHORSES

A pair of sawhorses come in mighty handy when you cut sheet goods or need to set up a temporary work area. But where do you corral them when you're not using them? Build these sturdy horses (**11–64** to **11–66**), and when you are done using them, you simply break them down and hang them flat against the wall.

With no hardware to fuss with, you can assemble and disassemble the pieces in just seconds. Because they're made completely of wood, you'll never have to worry about catching a saw blade on a metal bracket or fastener.

One full sheet of ¾" plywood yields a pair of sawhorses. The beam ends and stretcher ends are shown in **11–64**. To make them, enlarge the patterns in **11–65**.

❖❖❖

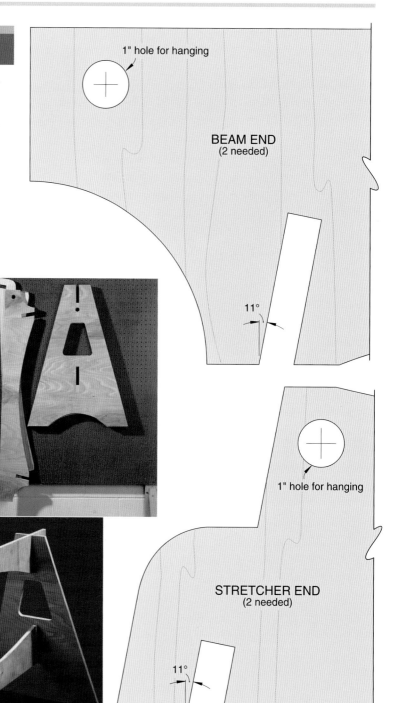

1" hole for hanging

BEAM END
(2 needed)

11°

1" hole for hanging

STRETCHER END
(2 needed)

11°

11–65.

**STRETCHER AND BEAM
HALF-SIZE PATTERNS**

Note: Enlarge to 200%
to make full-size patterns.

11–64.

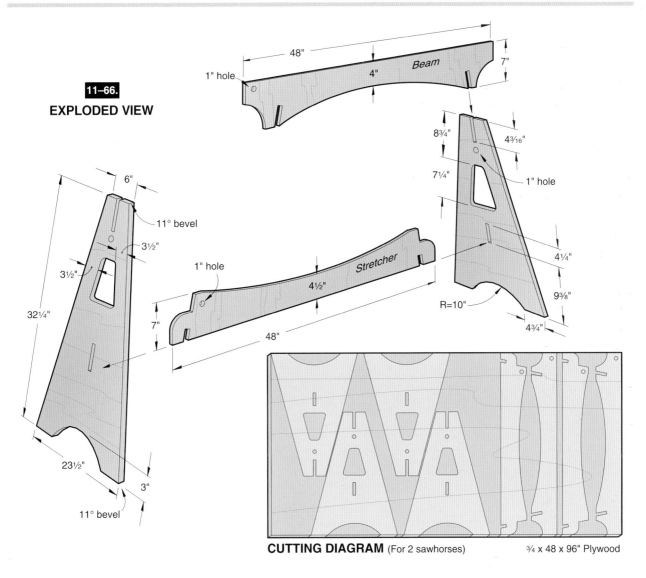

11–66.
EXPLODED VIEW

48"
1" hole
4"
Beam
7"

8¾"
4³⁄₁₆"
7¼"
1" hole
4¼"
9⅜"
R=10"
4¾"

6"
11° bevel
3½"
3½"
32¼"
1" hole
7"
Stretcher
4½"
48"
23½"
3"
11° bevel

CUTTING DIAGRAM (For 2 sawhorses)   ¾ x 48 x 96" Plywood

11–67.

BACK-TO-BASICS SAWHORSES

Our simple, sturdy sawhorses (**11–67** to **11–70**) are only 2' tall, but you can build a pair any size you want and get the help you need. Here's how:

1 To build a 30"-long × 24"-high sawhorse, begin by crosscutting a 30" top beam and four 30" legs. (We used a scrap

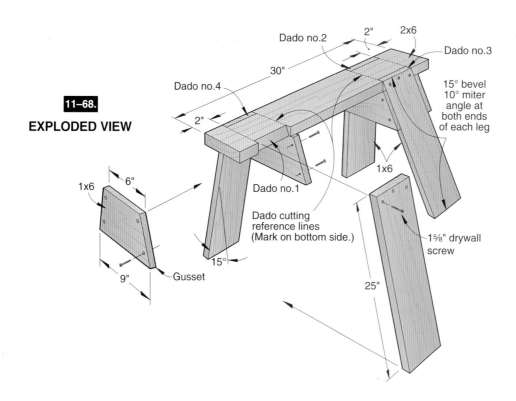

11–68.

EXPLODED VIEW

Dado no.2   2"   2x6

Dado no.3

30"

Dado no.4

15° bevel
10° miter
angle at
both ends
of each leg

2"

1x6

1x6   6"

Dado no.1

Dado cutting
reference lines
(Mark on bottom side.)

1⅝" drywall
screw

9"

15°

Gusset

25"

construction-grade 2 × 6 for the
top and 1 × 6s for the legs.)

2 Tilt the tablesaw blade to 15°,
and bevel one edge of a 12"-
long scrap 2 × 3. Screw it to the
miter gauge, as shown in **11–69**.
Remove the saw blade and insert
a dado blade. Set the blade at 0°
(parallel with the miter gauge
grooves), and raise it ¾" above
the surface of the table.

3 To cut the dadoes in the top
beam, position the miter
gauge in the slot to the right of
the blade and set the miter gauge
10° right of center, as shown in
11–69. Before cutting, mark reference lines for the location of the
four dadoes on the *bottom* side of
the 2 × 6, then mark Xs on waste

stock with a pencil to make sure
you cut on the correct side of the
lines. With the 2 × 6 positioned
against the beveled fence, cut
dado no. 1, and then flip the 2 × 6
end for end and cut dado no. 2.
Now, move the miter gauge to the
left side of the blade and set the
miter gauge to 10° left of
center, and cut dadoes nos. 3 and
4 as you did 1 and 2.

4 Remove the scrap 2 × 3 from
the miter gauge, and replace
the dado blade with a regular ⅛"
blade. Set the blade at 15° left of
center, and with the miter gauge
10° right of center and on the left-
hand side of the blade, cut one
end of each leg. (If your saw
blade tilts right from center, you
will need to change the miter
gauge setting to 10° left of center

11–69.

**COMPOUND ANGLE
DODO CUTTING**

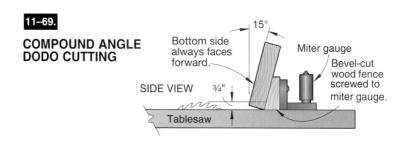

15°

Bottom side
always faces
forward.

Miter gauge

Bevel-cut
wood fence
screwed to
miter gauge.

SIDE VIEW   ¾"

Tablesaw

and make the first cut with the miter gauge on the right-hand side of the blade.) Keep the settings the same, and move the miter gauge to the opposite side of the blade; cut the remaining ends to a finished length of 25".

5 Attach the legs to the top, using 1⅝" drywall screws. (We used three screws for each leg.) Do not use glue on the assembly, because you may need to replace a part if it gets cut accidentally.

6 To make the four gussets, return the blade to 0° and set the miter gauge at 15°. Crosscut the gussets to the dimensions shown in **11–68**, and attach them with 1⅝" drywall screws. If you want a smoother finish, lightly belt-sand all the joinery flush.

11–70.

CUTTING THE TOP BEAM DADOES

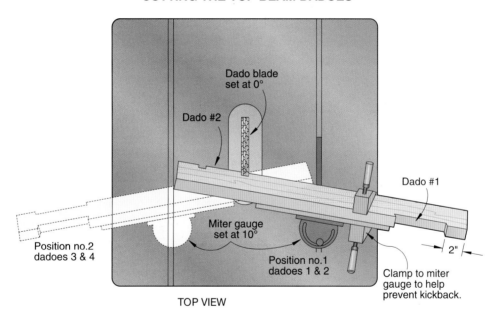

TOP VIEW

METRIC EQUIVALENTS CHART
Inches to Millimeters and Centimeters

MM=MILLIMETERS   CM=CENTIMETERS

INCHES	MM	CM	INCHES	CM	INCHES	CM
⅛	3	0.3	9	22.9	30	76.2
¼	6	0.6	10	25.4	31	78.7
⅜	10	1.0	11	27.9	32	81.3
½	13	1.3	12	30.5	33	83.8
⅝	16	1.6	13	33.0	34	86.4
¾	19	1.9	14	35.6	35	88.9
⅞	22	2.2	15	38.1	36	91.4
1	25	2.5	16	40.6	37	94.0
1¼	32	3.2	17	43.2	38	96.5
1½	38	3.8	18	45.7	39	99.1
1¾	44	4.4	19	48.3	40	101.6
2	51	5.1	20	50.8	41	104.1
2½	64	6.4	21	53.3	42	106.7
3	76	7.6	22	55.9	43	109.2
3½	89	8.9	23	58.4	44	111.8
4	102	10.2	24	61.0	45	114.3
4½	114	11.4	25	63.5	46	116.8
5	127	12.7	26	66.0	47	119.4
6	152	15.2	27	68.6	48	121.9
7	178	17.8	28	71.1	49	124.5
8	203	20.3	29	73.7	50	127

BOARD FEET GUIDELINES

Dealers typically price softwoods by the running (lineal) foot and hardwoods by the board feet (a volume measurement). A board foot includes thickness, width, and length measurements that equal 144 cubic inches. See "How to Determine Board Feet" below to see some sample calculations.

The thickness of lumber, especially hardwoods, is referred to in quarters of an inch, such as 4/4 ("four/quarters" or 1"), 5/4 (1.25"), 6/4 (1.5"), 8/4 (2"), and so on. However, these hardwood thicknesses are designated and the board footage calculated before surfacing. Although you'll pay for the full designated thickness, what you'll actually get in lumber surfaced two sides (S2S) is shown in "Effects of Planing" below. Also, in a lumber store the board footage is rounded up or down to the nearest one-half board foot, except for more costly exotic or imported wood. Exotic wood is calculated to the inch, and will be rounded to the nearest hundredth of an inch.

HOW TO DETERMINE BOARD FEET

Whether you measure a board's length in inches or feet, calculating board footage for that piece of lumber is simple math:

$$\frac{T \times W \times L}{144} = \text{board feet}$$

$$\frac{T \times W \times L}{144} = \text{board feet}$$

1" x 6" x 96" = 576
576 ÷ 144 = 4 board feet

1" x 6" x 8' = 48
48 ÷ 12 = 4 board feet

EFFECTS OF PLANING

When you buy lumber, you pay for its rough thickness before surface-planing on two sides (S2S). Here's how planning affects the thickness of purchased boards. You can save money by buying full-thickness rough stock from mills that offer it, and then plane it yourself.

HARDWOOD THICKNESS AFTER S2S	LUMBER ROUGH THICKNESS	SOFTWOOD THICKNESS AFTER S2S
13/16"	4/4 = 1"	3/4"
11/16"	5/4 = 1¼"	15/32"
15/16"	6/4 = 1½"	113/32"
1¾"	8/4 = 2"	113/16"
2¼"	10/4 = 2½"	2⅜"
2¾"	12/4 = 3"	2¾"

INDEX

E

CREDITS

Special thanks to the following people or companies for their contributions:

Clyde Allison, for project designs used in Chapters 9 and 10

King Au/Studio Au, for photographs used in Chapters 2, 3, and 7

Richard Baker, for project design used in Chapter 10

Marty Baldwin, for photographs used in Chapters 3, 4, 5, 6, 7, 8, 9, 10, and 11

Baldwin Photography, for photographs used in Chapters 3, 5, 6, 7, 9, and 11

Bill Baton, for photographs used in Chapter 1

Joseph Boehm, for Chapter 1 interior design

Jim Boelling, for project designs used in Chapters 7 and 10

Kevin Boyle, for project design used in Chapter 8

David Brennan, for photographs used in Chapter 1

Tim Cahill, for illustration used in Chapter 3

Bob Calmer, for photographs used in Chapters 8, 10, and 11

Cameron Campbell, for project designs used in Chapter 6

Dave Campbell, for text used in Chapter 3 and project design used in Chapter 10

Craig Carpenter, for photographs used in Chapter 4

Donna Chiarelli Studios, for photographs used in Chapter 3

James R. Downing, for laying out and designing workshops seen in Chapter 1; for photographs used in Chapter 2; for project designs and illustrations used in Chapter 3; and for project designs used in Chapters 6, 7, 8, 9, 10, and 11

Jamie Downing, for illustrations used in Chapters 3, 9, 10, and 11

Kim Downing, for illustrations used in Chapters 1, 3, 4, 6, 7, 8, 9, 10 and 11; and text used in Chapter 3

Owen Duvall, for text used in Chapter 8

Mike Gililland, for text used in Chapter 1

Doug Guyer, for text used in Chapter 7

Jim Harrold, for text used in Chapter 11

Chuck Hedlund, for project designs used in Chapters 3, 6, 7, 8, 9, and 10; and text used in Chapter 6

Burdette Heikens, for project design used in Chapter 8

Kevin Heilman, for project design used in Chapter 10

Mike Henry, for illustration used in Chapter 10

Kirk Hesse, for text used in Chapter 3

Hetherington Photography, for photographs used in Chapters 1, 3, 7, 8, 9, 10, and 11

R. B. Himes, for project design used in Chapter 8

William M. Hopkins, for photographs used in Chapters 4, 6, 7, 8, 9, 10, and 11

Marvin Hoppenworth, for project design used in Chapter 11

Iowa State University Extension Service's Department of Human Development and Family Studies, for text used in Chapter 1

Brian Jensen, for illustrations used in Chapters 3, 5, 8, and 11

Lorna Johnson, for illustrations used in Chapters 3, 4, 5, 6, 7, 8, 9, 10, and 11

Jennifer Jordan, for photographs used in Chapter 1

Marlen Kemmet, for producing information used in Chapters 6, 7, 8, 9, 10, and 11; project design used in Chapter 7

Bill Krier, for text used in Chapters 4 and 6

Bill LaHay, for text used in Chapters 8 and 11

Roxanne LeMoine, for illustrations used in Chapters 3, 4, 6, 7, 8, 9, 10, and 11

Bill Lovelace, for project design used in Chapter 8

Jeff Mertz, for project designs used in Chapters 6, 7, 8, and 10

Mike Mittermeier, for illustrations used in Chapters 8

Donovan Nagel, for project design used in Chapter 6

Jack Neff, for text used in Chapter 5

Frank Polimene, for project design used in Chapter 8

Jim Pollock, for text used in Chapters 3 and 7

Radiant Electric Heat, Inc., for photographs used in Chapter 4

Radiant Floor Co., for photographs used in Chapter 4

Reznor, Inc., for photographs used in Chapter 4

Erv Roberts, for photographs used in Chapter 3; project designs used in Chapters 3, 6, and 10; and text used in Chapter 11

Robert Russell, for project design used in Chapter 10

S & S Photography, for photographs used in Chapter 1

John F. Schultz, for photographs used in Chapter 5

Robert J. Settich, for text used in Chapters 6 and 10

Douglas E. Smith, for photographs used in Chapter 10

Merwin Snyder, for project design used in Chapter 10

Peter J. Stephano, for text used in Chapters 1 and 5

Jim Stevenson, for illustrations used in Chapter 4

David Stone, for text used in Chapters 2, 3, and 4

Jan Hale Svec, for project designs and text used in Chapters 6, 7, 8, and 11

Dean Tanner, for photograph used in Chapter 11

Richard Tollesfson, for project designs used in Chapter 10

Steve Uzzell, for photographs used in Chapters 1 and 11

Chad Veach, for project designs used in Chapter 6

Bill Zaun, for illustrations used in Chapters 3 and 11

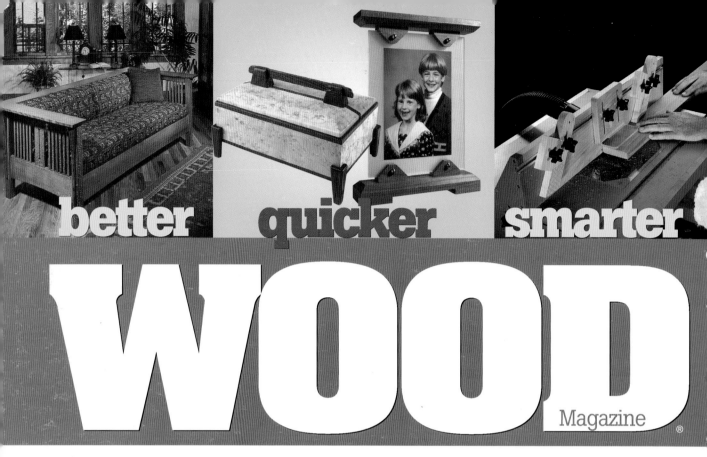